Behavioural Methods to Study Cognitive Capacities of Animals

Behavioural Methods to Study Cognitive Capacities of Animals

Editors

Lucia Regolin
Maria Loconsole

MDPI

Basel • Beijing • Wuhan • Barcelona • Belgrade • Novi Sad • Cluj • Manchester

Editors
Lucia Regolin
Department of General Psychology, University of Padova
Padova
Italy

Maria Loconsole
Department of General Psychology, University of Padova
Padova
Italy

Editorial Office
MDPI
St. Alban-Anlage 66
4052 Basel, Switzerland

This is a reprint of articles from the Topical Collection published online in the open access journal *Animals* (ISSN 2076-2615) (available at: https://www.mdpi.com/journal/animals/topical_collections/behavior_cognitive_capacities_animals).

For citation purposes, cite each article independently as indicated on the article page online and as indicated below:

Lastname, A.A.; Lastname, B.B. Article Title. *Journal Name* **Year**, *Volume Number*, Page Range.

ISBN 978-3-7258-0799-4 (Hbk)
ISBN 978-3-7258-0800-7 (PDF)
doi.org/10.3390/books978-3-7258-0800-7

Contents

About the Editors

Lucia Regolin

Lucia Regolin graduated with a degree in Experimental Psychology from the University of Padova, where she later obtained her Ph.D. in Psychology. Her Ph.D thesis titled "Cognitive, neuropsychological and neuropharmacological aspects of detour behaviour in the domestic chick" was also conducted in the laboratories of Prof. Richard J. Andrew (Sussex University, Brighton, U.K.) and Prof. Steven P. R. Rose (The Open University, Milton Keynes, U.K.). She was a Researcher at the University of Padova until she became full Professor in 2017. At present, she teaches Animal Psychology at the international B.S. level in the Psychological Science program and Animal Cognition at the M.S. level in the Applied Cognitive Psychology program. She has been the Visiting Chair Professor of Wesleyan College at Guangzhou University (Guangzhou, China) and a guest lecturer at ShanghaiTech University, Shanghai (China). Since 1994, she has published approximately 100 peer-reviewed articles and book chapters in international journals and presented over 200 papers at international conferences. She investigated cognitive abilities in several animal species, mostly in the domestic chicken, *Gallus gallus*. She hopes her research will contribute to increasing the general public's awareness of the complexity of non-human species' minds and improve the consideration and the treatment of farm animals, such as the domestic chicken.

Maria Loconsole

Maria Loconsole is a Research Fellow at the Department of General Psychology at the University of Padova, Italy. Dr. Loconsole obtained an M.S. degree in Neuropsychology and Neuropsychological Rehabilitation in 2017 and a Ph.D. in Psychological Sciences in 2022 at the University of Padova, with a final thesis titled "Multimodal information processing in the domestic chick (*Gallus gallus*): role of precocial cognitive abilities and predispositions". During her Ph.D., she also worked in the Comparative Cognition Laboratory at Cambridge University (UK) as a Visiting Researcher. In 2021, she was a Teaching Fellow of Comparative Psychology and an Associate Researcher at the Queen Mary University of London. Her research aims to investigate predisposed cognitive abilities in non-human animals. Although her main research is focused on the domestic chicken (*Gallus gallus*), allowing her to study the early onset of different psychological mechanisms and function, she has broadened her studies to different animal models including dogs (*Canis lupus familiaris*) and corvids (*Corvus corone; Garrulus glandarius*). She has a profound interest in animal welfare, and she hopes that her work can help highlight the importance of animals' perspectives (by describing their perceptual and cognitive worlds) to properly design interventions for improving their wellbeing and living conditions.

animals

MDPI

Editorial

Behavioural Methods to Study Cognitive Capacities of Animals

Lucia Regolin * and Maria Loconsole

Department of General Psychology, University of Padua, 35131 Padua, Italy; maria.loconsole@unipd.it
* Correspondence: lucia.regolin@unipd.it

Over the past 20 years, the scientific community has witnessed a growing interest in the comparative study of mental capabilities. Animal cognition has become an independent field of interdisciplinary investigation, featuring specialized methodologies and paradigms tailored to diverse animal species. Recently, new approaches have emerged, allowing us access to large numbers of data and offering the possibility of highly reliable analysis with enhanced explicatory power. However, this has often been achieved with the disadvantage of an over-simplification of the behavioural response observed. On the one hand, there are the automated and observer-independent techniques, which allow the collection of vast sets of data but may incur the risk of denaturing the meaning and richness of natural behaviours. On the other hand, there is the study of the neurobiological correlates of mental processes, which requires reference to standardized paradigms, posing constraints on the complexity and ecological validity of behaviour. In contrast to these approaches, more traditional behavioural methods are firmly grounded in species ethology, which is considered crucial in grasping the distinctive qualitative features of behavioural repertoires.

This Special Issue aims to offer a collection of contributions employing behavioural methods to investigate the mental abilities of animals or any other behaving organisms. The Special Issue comprises a series of reviews presenting possible solutions to the key issue of "asking a question" to non-human animals.

Evidence is reported from groundbreaking research that has contributed novel techniques and approaches to the study of animal minds. Irene Pepperberg offers a review of the model–rival technique, an innovative paradigm that was developed by the author to overcome the limitations of traditional approaches based on conditioning and that, ultimately, allowed the unveiling of complex cognitive abilities in parrots [1]. Thomas Zentall contributes pertinent examples from the current literature, in particular regarding pigeons, that effectively describe the parallelism between human and non-human cognitive abilities, reflecting on the possible existence of shared (i.e., not uniquely human) mechanisms that affect the behaviours of different species [2].

Pivotal studies that have made it possible to better describe and comprehend the cognitive and perceptual worlds of animals will be also discussed in detail. Lesley Rogers presents the state of the art of the research on brain lateralization, illustrating the advantages of having a lateralized brain at the levels of both the single individual and the population [3]. Maria Santacà and colleagues tackle the topic of investigating visual illusion as a means of accessing animals' perceptual worlds, discussing the results achieved so far in this field and critically examining the different paradigms employed with this aim [4]. Andrea Dissegna and colleagues contribute evidence supporting the idea that habituation, previously considered as a sheer single-event learning process, relies, in fact, on some sophisticated learning mechanisms [5].

The Special Issue also features research articles that contribute novel knowledge to the field by describing instances of complex (or so-called "higher") cognitive abilities in non-human subjects. Sigmundson and colleagues test Japanese macaques with a cooperation task in an experimental setting that closely resembles this species' natural environment and thus allows for the study of macaques' cognitive abilities while maintaining the

Citation: Regolin, L.; Loconsole, M. Behavioural Methods to Study Cognitive Capacities of Animals. *Animals* **2023**, *13*, 3445. https://doi.org/10.3390/ani13223445

Received: 11 October 2023
Accepted: 30 October 2023
Published: 8 November 2023

social factors that help to shape their behaviours in nature [6]. Alexandra Horowitz and colleagues provide experimental evidence of dogs being able to navigate through a changing environment and discuss how this may represent a way of understanding their sense of themselves [7]. Lisa Horn and colleagues test two corvid species, the azure-winged magpie and the carrion crow, in a prosocial task and use their results as the basis for a discussion on the importance of naturalistic methods and comparisons between different species and experimental paradigms [8]. Greta Baratti and colleagues test spatial and orientation abilities in zebrafish and discuss the relevance of this animal model in comparison with mammalian species [9].

Instances of complex cognitive abilities are also discussed, with a particular focus on the importance of accounting for individual variability when evaluating animals' cognitive performance. Loïc Pougnault and colleagues describe experimental tests for evaluating spontaneous attention in songbirds that could also detect individual variations and attentional characteristics [10]. Marie Pelé and colleagues analyse individual differences in orangutans' drawing styles, linking them to differences in behavioural styles, states of mind, and motivations [11].

Lastly, new challenges within the study of behaviour, and critical reflections on, and controversies around, the exploitation of current animal models, are presented. Valentina Simonetti and colleagues discuss an example of a non-canonical model for the study of behaviour, showing that plant movement could match some basic features of animal movement in that it is proven to be adaptive, flexible, anticipatory, and goal-directed [12]. Maria Padrell and colleagues discuss the ethical problems posed by invasive research on primates, comparing the present legislation in different countries and reviewing the current evidence of their sophisticated cognitive abilities [13]. Nereida Bueno-Guerra presents a collective metadisciplinary discussion on the current issues within the discipline, including definitions of the key concepts, model species, the use of novel technology and data manipulation, networking, and the impact of sociocultural and ecological factors [14].

Altogether, these contributions offer a multifaceted picture of the current approaches and the challenges that researchers face (and how to overcome some of them) when dealing with the complex topic of animal behaviour and cognition. However, there is still much knowledge on animal minds that is still to be gathered, and the development of new behavioural and ethical methods will allow new insights into the cognitive worlds of animals.

Conflicts of Interest: The authors declare no conflict of interest.

References

1. Pepperberg, I.M. A Review of the Model/Rival (M/R) Technique for Training Interspecies Communication and Its Use in Behavioral Research. *Animals* **2021**, *11*, 2479. [CrossRef] [PubMed]
2. Zentall, T.R. Comparative Cognition Research Demonstrates the Similarity between Humans and Other Animals. *Animals* **2023**, *13*, 1165. [CrossRef] [PubMed]
3. Rogers, L.J. Brain Lateralization and Cognitive Capacity. *Animals* **2021**, *11*, 1996. [CrossRef] [PubMed]
4. Santacà, M.; Agrillo, C.; Miletto Petrazzini, M.E. The Challenge of Illusory Perception of Animals: The Impact of Methodological Variability in Cross-Species Investigation. *Animals* **2021**, *11*, 1618. [CrossRef] [PubMed]
5. Dissegna, A.; Turatto, M.; Chiandetti, C. Context-Specific Habituation: A Review. *Animals* **2021**, *11*, 1767. [CrossRef] [PubMed]
6. Sigmundson, R.; Stribos, M.S.; Hammer, R.; Herzele, J.; Pflüger, L.S.; Massen, J.J.M. Exploring the Cognitive Capacities of Japanese Macaques in a Cooperation Game. *Animals* **2021**, *11*, 1497. [CrossRef] [PubMed]
7. Horowitz, A.; West, E.; Ball, M.; Bagwell, B. Can Dogs Limbo? Dogs' Perception of Affordances for Negotiating an Opening. *Animals* **2021**, *11*, 620. [CrossRef]
8. Horn, L.; Zewald, J.S.; Bugnyar, T.; Massen, J.J.M. Carrion Crows and Azure-Winged Magpies Show No Prosocial Tendencies When Tested in a Token Transfer Paradigm. *Animals* **2021**, *11*, 1526. [CrossRef]
9. Baratti, G.; Rizzo, A.; Miletto Petrazzini, M.E.; Sovrano, V.A. Learning by Doing: The Use of Distance, Corners and Length in Rewarded Geometric Tasks by Zebrafish (Danio Rerio). *Animals* **2021**, *11*, 2001. [CrossRef] [PubMed]
10. Pougnault, L.; Cousillas, H.; Heyraud, C.; Huber, L.; Hausberger, M.; Henry, L. Experimental Tests for Measuring Individual Attentional Characteristics in Songbirds. *Animals* **2021**, *11*, 2233. [CrossRef] [PubMed]
11. Pelé, M.; Thomas, G.; Liénard, A.; Eguchi, N.; Shimada, M.; Sueur, C. I Wanna Draw Like You: Inter- and Intra-Individual Differences in Orang-Utan Drawings. *Animals* **2021**, *11*, 3202. [CrossRef] [PubMed]

12. Simonetti, V.; Bulgheroni, M.; Guerra, S.; Peressotti, A.; Peressotti, F.; Baccinelli, W.; Ceccarini, F.; Bonato, B.; Wang, Q.; Castiello, U. Can Plants Move Like Animals? A Three-Dimensional Stereovision Analysis of Movement in Plants. *Animals* **2021**, *11*, 1854. [CrossRef] [PubMed]
13. Padrell, M.; Llorente, M.; Amici, F. Invasive Research on Non-Human Primates—Time to Turn the Page. *Animals* **2021**, *11*, 2999. [CrossRef] [PubMed]
14. Bueno-Guerra, N. Where Is Ethology Heading? An Invitation for Collective Metadisciplinary Discussion. *Animals* **2021**, *11*, 2520. [CrossRef] [PubMed]

animals

MDPI

Article

Can Dogs Limbo? Dogs' Perception of Affordances for Negotiating an Opening

Alexandra Horowitz *, Eloise West, Molly Ball and Blakeley Bagwell

Dog Cognition Lab., Department of Psychology, Barnard College, New York, NY 10027, USA; ewest@psych.ubc.ca (E.W.); miball@me.com (M.B.); bmb2192@barnard.edu (B.B.)
* Correspondence: ahorowit@barnard.edu

Simple Summary: Recent behavioral research with domestic dogs has focused largely on their social cognition: how they interact with and interpret both other dogs and humans. Less well studied are the various aspects of their perceptual experience which might provide knowledge of how they understand the non-social world and themselves. In two studies, we look at how dogs navigate their environment. We first set up a situation to test whether dogs understand when they are too big to go through an opening; we also look at how they adjust their bodies to increasingly smaller (shorter) openings. We then also look at how dogs navigate an opening when their body width is effectively increased by their holding a stick in their mouth. We find that dogs show more hesitation approaching openings that are too small than ones through which they comfortably fit. Dogs of all sizes also change their behavior in a uniform way to negotiate short openings. When holding a stick, dogs did not initially change their behavior but are able to negotiate through an opening with experience. Researching how dogs navigate through a changing environment may be a fruitful way to begin to understand their sense of themselves.

Citation: Horowitz, A.; West, E.; Ball, M.; Bagwell, B. Can Dogs Limbo? Dogs' Perception of Affordances for Negotiating an Opening. *Animals* **2021**, *11*, 620. https://doi.org/10.3390/ani11030620

Academic Editor: Lucia Regolin

Received: 1 February 2021
Accepted: 24 February 2021
Published: 26 February 2021

Abstract: Very little research has focused on canines' understanding of their own size, and their ability to apply this understanding to their surroundings. The current study tests domestic dogs' judgment of their body size in relation to a changing environment in two novel experimental situations: when encountering an opening of decreasing height (Study 1) and when negotiating an opening when carrying a stick in their mouth (Study 2). We hypothesized that if dogs understand their own body size, they will accurately judge when an opening is too small for their body to fit through, showing longer latencies to approach the smaller openings and adjusting their body appropriately to get through—although this judgment may not extend to when their body size is effectively increased. In line with these hypotheses, we found that the latency for subjects to reach an aperture they could easily fit through was significantly shorter than to one which was almost too small to fit through. We also found that the order of subjects' adjustments to negotiate an aperture was invariant across individuals, indicating that dogs' perception of affordances to fit through an aperture is action-scaled. Preliminary results suggest that dogs' approach behavior is different when a horizontal appendage is introduced, but that dogs were able to alter their behavior with experience. These results are consistent with the hypothesis that dogs understand their own body size and the affordances of their changing environment.

Keywords: domestic dogs; affordances; behavior; body size; size sense; size perception; sense of self; animal cognition; self-representation

1. Introduction

Research since the late 1990s has substantially broadened our understanding of the cognitive abilities of the domestic dog—in particular, how dogs perceive the world and interact with others. The greatest majority of research focuses on social cognitive abilities, such as gaze- and point-following [1,2] and use of others' attention [3,4]; other research looks at the possible result of such social cognitive skill, such as perspective-taking [5,6] and representation of self [7]. Less well studied are the contributions of foundational perceptual abilities to cognition.

Recent work has begun to investigate the cognitive consequences of olfactory perception [8,9]; very few research programs have examined perception of affordances and body-size awareness [10–12]. Animals' information about their body size, especially relative to environmental changes and constraints—body sense—is critical to navigating new environments as well as social success. An animal's size guides their behavior and, for many species, can be a key determinant in both their short- and long-term survival [13,14]. Relative body size also relates to reproductive output [15,16] and fighting behavior [17]. A size sense is also integral to navigating changing environments: it has been proposed that the capacity for self-concept evolved to enable large animals to navigate complex environments [18]. While most animals' environments fluctuate with naturalistic changes, such as a fallen tree blocking a habitual route, the environments of species who live among humans are additionally driven by human behavior. For an owned dog living in a human household, various aspects of their environment may be altered daily: a chair pushed away from a table or a door closed partway. In these scenarios, an animal's sense of their own size would facilitate negotiation of their changing surroundings [19,20].

Previous research with dogs has investigated their perception of size primarily of conspecifics, not themselves. In a looking-time paradigm, Taylor et al. [21] found that dogs could match growls of different pitches with corresponding images of dogs of different sizes. Other work has found that dogs accurately matched food-guarding growls with an image of the source of the growl, but looked toward the image of a dog larger than the source when hearing playful growls [22]. Some research addresses a dog's understanding of their body size with respect to the size of other dogs. Dogs self-handicap with smaller dogs in play [23] and small dogs perform higher raised-leg displays relative to their body size than large dogs do [24].

Having an understanding of one's size relative to features in the environment also enables determination of what actions are possible in a given scenario, or the "affordances" of the environment [20,25]. Human studies show that perceptual input—primarily visual—is used to guide decisions about what actions can be performed [20], sensitive to changes of one's own body and of one's environment [26]. Subjects attempt to reach or move through openings larger, but not smaller, than their appendage or body size, and make appropriate adjustments to move through narrow openings [27,28]. Moreover, subjects wearing size-increasing prosthetics (such as of the hand or stomach) made appropriate adjustments, with experience [27,29]. In research with canine subjects, Wagman et al. [12] evaluated dog behavior when attempting to acquire a treat placed at increasingly higher points along a wall; they found that the height at which subjects transition from reaching for a treat with their head only to rearing was different across individuals but with the same ratio of shoulder height to treat height [30]. Two projects looked at dogs' behavior when approaching differently sized openings. Maeda and Fujita [11] found that when simultaneously given a larger or smaller doorway opening, subjects preferred to go through the larger opening. Lenkei et al. [10] found longer latencies for subjects to approach too narrow or short openings than openings through which they could fit; moreover, subjects did not attempt to pass through openings that were too small [10]. These studies provide preliminary evidence that dogs can perceive affordances of a given environment and integrate information about the external world and their own body to organize their behavior.

The present research extends this literature into dogs' knowledge of their body size, and further explores their understanding of the affordances of the environment. In the first study, we assess whether dogs understand their size relative to an opening which they are asked to pass through by examining behavioral modifications and approach time as the opening becomes increasingly shorter. As in Lenkei et al. [10], we assess subjects' judgment of their ability to fit through the openings by measuring their latency to approach openings of decreasing height. We predict that if dogs have a knowledge of their own body size, then they will accurately judge which openings are passable, showing longer latencies to approach smaller openings. In addition, we analyze dogs' physical adjustments as they attempt to pass through increasingly smaller openings, and examine how their behavior relates to their height at withers [12], eye-height [31], or elbow joint-height (which may differ by breed [32]). As the opening is adjusted incrementally along one dimension, we can determine the ratio at which dogs first adjust their behavior to negotiate the opening, relative to subjects' height at withers. We predict that the order of physical adjustments will be invariant across dogs, and thus consistent across different body sizes, as will the ratio of body-height to the height of the opening at which dogs make their first bodily adjustment.

Given the importance of tactile information, acquired via experience, to subject behavior [29], in a second study, we examine subject behavior when their body size is effectively increased. To naturalistically and temporarily increase their width, we offer dogs sticks of various lengths to hold in their mouths while passing through a fixed-width opening. We measure subject modifications in approach time and body position with variously sized sticks, and offer subjects the chance to choose a stick that will easily fit through the aperture. We predict subjects will inaccurately judge affordances as stick length increases—leading to an inability to fit through the opening—but will be able to update their perception of possible actions with experience, and alter their approach so to fit.

2. Materials and Methods

2.1. Participants

Dogs and owners were recruited through the Barnard Dog Cognition Lab database. Forty-four domestic dogs (23M, 21F) and their owners began the study. Subject dogs were required to be at least 8 months old, healthy, fully vaccinated, and comfortable with new people and in new environments. The mean subject age was 5.5 years (range: 1 year 4 months to 11 years 4 months); all but 2 dogs were spayed or neutered. Twenty-five dogs were described by their owners as purebred and 19 were mixed breeds (Table 1).

Table 1. All subjects in Study 1; shaded subjects were also in Study 2.

Name	Age(Yrs)	Breed	Sex	Status	Bin	Height (cm) Withers	Elbow	Eye	Final Aperture Height
Donald	5.5	Mixed	M	neutered	S	31	12	33	29
Aspen	9.2	Siberian Husky	F	spayed	L	60	28	66	18
Maki	4.5	Chihuahua	M	neutered	XS	22	9	22	8
Charley	7.3	Mini Poodle	F	spayed	M	40.5	20.5	52.5	n/a
Ice Cream Louie	2.7	Mixed	M	neutered	M	48	23	50	16
Morticia	5.9	Greyhound	F	spayed	L	67.5	38	76	61.5
Molly	6.5	Mixed	F	spayed	S	36	20	45	15
Roland	7.6	Yorkshire Terrier	F	spayed	XS	22	8	28	12
Sammy	3.0	Labrador retriever	M	neutered	L	56	28	68	20
Layla	8.1	Mixed	F	spayed	M	46	23	52	16
Enzo	1.3	Mixed	M	neutered	S	–	–	–	10
Jackson	11.1	Mixed	M	neutered	M	53	25	60	23
Wigs	3.0	Springer Spaniel	M	neutered	M	53	27	64	16
Wally	5.8	Bearded Collie	M	neutered	L	58	28	63	28
Maloney	2.6	Mixed	M	neutered	M	53	28	62	15
Jesse	6.1	Shi Tzu	M	neutered	S	31	16	37	28
Nemo	4.8	Mixed	M	neutered	S	26	10.5	31	8
Penny	3.6	Labrador retriever	F	spayed	L	57	33	66	21

Table 1. *Cont.*

Name	Age(Yrs)	Breed	Sex	Status	Bin	Withers	Height (cm) Elbow	Eye	Final Aperture Height
Artemis	1.7	Border Collie	F	intact	M	45	24	54	15
Bobby	5.6	Labrador retriever	M	neutered	L	61	32	77	21
Lulu	9.9	Yorkshire Terrier	F	spayed	S	26	14	24	8
River	8.8	German Shepherd	F	spayed	L	68	33	72	44
Jameson	7.2	Mixed	M	neutered	M	45	26	54	15
Milo	3.9	Mixed	M	neutered	L	62	30	68	26
Gotham	3.7	Keeshond	F	spayed	M	41	19	52	16
Walter	5.2	Golden retriever	M	neutered	L	65	37	82	23
Amy	7.2	Bull Terrier	F	spayed	M	48	25.5	54	18
Pepper	4.3	Mixed	F	spayed	L	57	32	60	21
Oliver	2.4	Mixed	M	neutered	M	47	19	54	17
Francie	11.3	Chihuahua	F	spayed	XS	25	11	23	13
Buttons	2.9	Mixed	M	neutered	M	40	19	51	10
Bear	5.3	Mixed	M	neutered	L	67	35	70	31
Izzy	6.0	Mixed	M	neutered	S	33	15	39	21
Lia	8.7	Yorkshire Terrier	F	spayed	XS	23	12	29	11
Arrow	3.3	Airedale Terrier	M	neutered	XL	70	37	80	40
Indy	4.4	Labrador retriever	F	spayed	L	54	22	63	18
Gracie	1.8	Yorkshire Terrier	F	spayed	XS	24	8	29	18
Wyatt	2.3	Mixed	M	neutered	M	44	17	48	14
Mango	8.5	Mixed	F	spayed	S	31	16	36	13
Wellington	4.9	Springer Spaniel	M	neutered	M	50	25	64	15
Camuggi	7.3	Miniature poodle	F	spayed	S	34	17	37	13
Ollie	6.3	Labrador retriever	M	intact	L	60	31	68	42
Mara	5.9	Mixed	F	spayed	L	68	35	80	26
Luz	2.8	Mixed	F	spayed	L	65	33	73	35

2.2. Design

Prior to participating, owners completed a questionnaire about their dog's breed, age, health, size, and training history, as well as behavioral characteristics like temperament, ability to follow commands, food motivation, and stick-carrying behavior. Based on owners' measurement of their dogs' height at withers, subjects were assigned to a category from XS to XL (see Table 1) in order to determine aperture heights in Study 1; this height was confirmed when the subject arrived at the lab. Dogs described by their owners as both habitually "carrying sticks," as well as engaging in two or more other stick-related behaviors ("retrieves stick after it's thrown"; "keeps sticks at home"; "has a favorite stick"; "presents a stick to you to throw or tug on"; "chews stick") were also invited to participate in Study 2. Owners brought their dogs to the Dog Cognition Lab on Barnard College's New York City campus, completed a consent form, and were informed about their role in the studies, outlined below. Owners with a dog in Study 2 additionally brought a favored dog toy from home. Participants were scheduled to arrive in sequence, so that only one dog was present in the room for each trial. Testing took place from January 2020 to March 2020.

2.3. Apparatus and Stimuli

The testing room at the Barnard Dog Cognition Lab is 3.53 × 3.35 m, with a single door and no windows. Three cameras (Lorex 1080p, Lorex, Markham, ON, Canada) are situated to record trials: two capturing room views, from the northeast and southwest corners of the room; a third low on the northern wall, pointing toward the experimental theater (Figure 1).

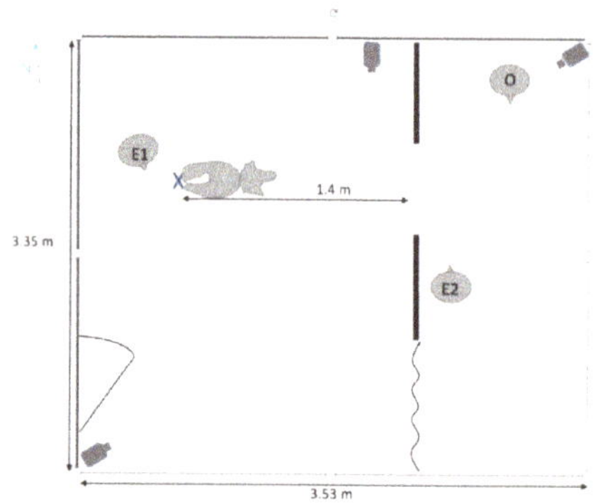

Figure 1. Layout of Dog Cognition Lab with position of subject, apparatus, experimenters (E1 and E2), owner (O), cameras, and magnetic north (N) indicated.

The experimental apparatus consisted of two upright wooden members serving as walls of 1.8 m width and 1.2 m height around an opening (hereinafter "aperture") of width 61 cm (Figure 2). The edges of the walls around the aperture were fitted with tracks that allowed a Plexiglas panel to be raised or lowered, enabling experimenters to adjust the height of the aperture to specific increments relative to the dog's height at withers. On one side of the apparatus, a curtain was pulled across a second opening serving as an indirect, alternative route around the apparatus. The alternate route was available so that subjects did not become overly frustrated if they could not reach their owners. The owner sat in a chair behind the apparatus, while the dog was held on a short leash 1.4 m in front of the apparatus.

Figure 2. Experimental apparatus from the subject's point of view, with aperture and Plexiglas (marked with blue tape) within it.

A transparent Plexiglas was used to enable the dogs to see their owners seated on the other side of the apparatus; blue painter's tape was applied to the Plexiglas panel in order to make it more visible (no subjects ran into the barrier).

For Study 2, sticks of three lengths—small enough to fit in (46 cm), just the size of (61 cm), or too large to fit in (76 cm) the aperture—were collected from Riverside Park in New York City.

To control for unintended odors, the floor was cleaned with a solution of 70% isopropyl alcohol between subjects. The temperature of the room was recorded at the beginning of trials, and varied from 70.1 to 79 degrees Fahrenheit (mean = 74); humidity levels ranged from 11 to 34% (mean = 24%).

2.4. Experimental Procedure

An experimenter met the owner and dog at a street-level entrance to the College and walked them upstairs to the testing space. Upon arrival in the lab, the owner was asked to remove their dog's leash and the dog was allowed to explore the room independently and become acclimated to the testing space while the experimenters explained the owner's role in the studies. When the dog appeared calm, the owner was given a short leash to attach to their dog's collar. While the owner held the dog's leash, experimenters measured the height of each dog at their withers, eyes, and front elbows.

2.4.1. Study 1: Size Sense

Warm-up trial. The owner was seated behind the experimental apparatus; the first experimenter (E1) sat on the opposite side of the apparatus from the owner, holding the dog by the leash. A second experimenter (E2) stood behind the apparatus to move the Plexiglas to adjust the aperture size (height of the opening) in each trial. In order to be as neutral as possible, experimenters avoided staring at, talking to, or petting the subjects.

The Plexiglas was removed so that the aperture had no upper limit. The owner was given a treat for their dog, and E2 announced the start of the trial by asking the owner to call their dog by saying "[Name], come here!" or as they usually did. When their dog passed through the aperture, the owner gave them the treat. This procedure was repeated until a trial was successfully completed (the dog passed through the aperture). The experiment was aborted if the dog did not pass through the aperture after five attempts (only 1 dog did not pass through the aperture in five attempts).

Test trials. The experimental set-up was identical to the warm-up trial, except that the Plexiglas panel was slid downward in the tracks to create the upper edge of the aperture at specific increments relative to the subjects' height at withers (HW; or an averaged size, should the dog not cooperate with being measured (n = 1)). Dogs' bin category (from XS to XL) was used to determine the specific aperture sizes on the first trial and second trial, and the increments by which the Plexiglas panel were lowered, to make consistently smaller apertures, for subsequent trials (Table 2).

Table 2. Dimensions used to determine the size of the aperture on the first and subsequent trials.

Dog Size	First Trial Aperture Size (Height of Opening)	Second Trial Size (Height of Opening)	Increments by Which Plexiglas Lowered for Subsequent Trials
XS: 14–25.4 cm	1.5 × HW or 30 cm	HW + 2 cm	2 cm
S: 25.5–39.4 cm	1.5 × HW or 48 cm	HW + 3 cm	3 cm
M: 39.5–53.4 cm	1.5 × HW or 70 cm	HW + 5 cm	5 cm
L: 53.5–68 cm	1.5 × HW or 90.5 cm	HW + 6 cm	6 cm
XL: >68.1 cm	1.5 × HW or 108 cm	HW + 7 cm	7 cm

HW—height at withers.

When E2 announced the start of the trial, the owner, seated behind the apparatus, called their dog by saying "[Name], come here!" or as they usually did. The owner gave their dog a treat after the dog reached their side of the apparatus, regardless of whether they passed through the aperture or arrived via an alternative method. After completing each trial, E1 retrieved the dog and brought them back to the starting position. E2 then lowered the Plexiglas panel by the predetermined increment to make an aperture of the designated height. This procedure was repeated until the dog refused to pass through the aperture or pursued an alternate route and walked behind the curtain. Following refusal or pursuit of an alternate route, the trial was repeated once with the same sized aperture. If the dog went through the aperture on this trial, the experiment resumed and the Plexiglas panel was lowered once more. If the dog did not go through the aperture on the repeated trial, the experiment was concluded.

2.4.2. Study 2: Stick-Carrying

Following Study 1, those dogs (n = 13) described by their owners as stick-carrying dogs (see Design, Section 2.2, above) participated in Study 2. They were allowed to rest or play off leash as E2 explained the next study's procedure.

Warm-up trial. The starting positions for the subject, owner, and apparatus were the same as those in Study 1, except that the aperture was completely open (i.e., had no height ceiling) throughout the trials. As in Study 1, the aperture width was constant (at 61 cm) through the trials. E1 gave the dog a toy that the owner had brought from home. When the dog was holding the toy, E2 announced the start of the trial. The owner called the dog through the aperture using their typical request (such as "[Name], come here!"). If the dog passed entirely through the aperture while still holding the toy, the owner gave the dog a treat. E1 then brought the dog to the starting position. This procedure was repeated until the dog completed two warm-up trials successfully. If the dog did not pass through the aperture while holding the toy across five attempts, the experiment was concluded.

Test trials. With the aperture width held constant, the dogs were offered sticks of three different sizes (i.e., when held horizontally, three different widths). E1 offered the dog the 46 cm stick. When the dog was holding the stick in their mouth, E2 announced the start of the trial. The owner called their dog and rewarded them if and when they passed all the way through the aperture while still holding the stick. E1 then returned the dog to the starting position. The same procedure was followed with the 61 cm stick and then with the 76 cm stick. At each stage, the trial was repeated until the dog successfully passed through the aperture, refused to pass through the aperture, dropped and abandoned the stick, or pursued an alternate route. In the latter three cases, the trial was repeated once with the same sized stick. If the dog successfully completed the repeated trial, the sequence was resumed. If they again refused to pass, dropped the stick, or pursued an alternate route, the experiment was concluded.

Following the final trial, E1 returned the dog to the starting position and placed all three sticks on the ground in front of the dog. E1 then asked, "What's this?" If the dog picked up a stick, E1 noted the stick size. The owner called the dog through the aperture and gave their dog a treat if and when the dog passed through the aperture completely.

2.5. *Behavioral Coding*

Video cameras captured all subject behavior in both studies for later frame-by-frame playback (30 fps) and coding by one of the authors (BB). On each trial, we calculated the subject's latency to reach the aperture—the time from E1's release of the dog to the dog's arrival at the opening—as well as the order of behavioral modifications made by the subject in negotiating the aperture: head duck, head or body turn, front elbow bend, rear elbow bend, and alternate routes taken (Figure 3). We also coded each subject's pass-through successes and pass-through attempts on each trial, and noted the point to which subjects get through the aperture on aborted and unsuccessful attempts. Additionally, in Study 2, we noted the number of head and body rotations; number of times the subject knocked the side of the aperture with the stick; and stick choice on final trial, if any. For Study 1, subjects were divided into two height groups post hoc, per previous height binning [12]: short (height at withers less than 54 cm) and tall (height at withers greater than or equal to 54 cm). We applied a linear regression analysis for latencies to reach the aperture, chi-squared goodness-of-fit tests for order of modifications, and independent-sample t-tests to compare subjects' smallest aperture by bin and to calculate ratios of dog height to height of aperture at first adjustment. Reliability of coding was gauged against a second, independent coder of the videos. Inter-observer agreement was high for latency to reach the aperture (n =

118 trials; Spearman's rho = 0.51, $p < 0.001$) and for the first bodily adjustment on ($n = 11$ subjects, Cohen's Kappa = 0.76).

Figure 3. Example of head duck (**a**), front-elbow bend (**b**), and back-elbow bend (**c**).

3. Results

3.1. Study 1

Forty-three dogs participated in 424 trials ($n = 1$ dog refused to participate in any of the trials). Subjects were categorized into five size bins on the basis of height at withers: XS ($n = 5$), S ($n = 9$), M ($n = 13$), L ($n = 16$), XL ($n = 1$) (see Table 2 for height ranges per category, and Table 1 for individual participant characteristics). For analysis, the single XL dog was included in the L bin category.

Subjects attempted to pass through the aperture an average of 9.65 times (SD = 2.14), and successfully passed through the aperture an average of 7.57 times (SD = 1.98). A linear regression of subjects' age and the number of successful trials completed showed no significant correlation between age and performance ($r^2 = 0.01$, $p > 0.05$). The subjects modified their behavior on an average of 6.09 trials (SD = 1.76), or an average of 61.8% of all the trials they completed (SD = 15.6%). Subject latency to reach the aperture depended on its height: latency was significantly longer on subjects' final successful trial (M = 5.78 s,

SD = 6.02) than on the first trial (M = 1.15 s, SD = 0.73) t(42) = −4.95, $p < 0.001$) (Figure 4). There was no significant difference in latency between the first trial and the first trial with a behavioral modification (t(42) = −1.40, $p = 0.16$).

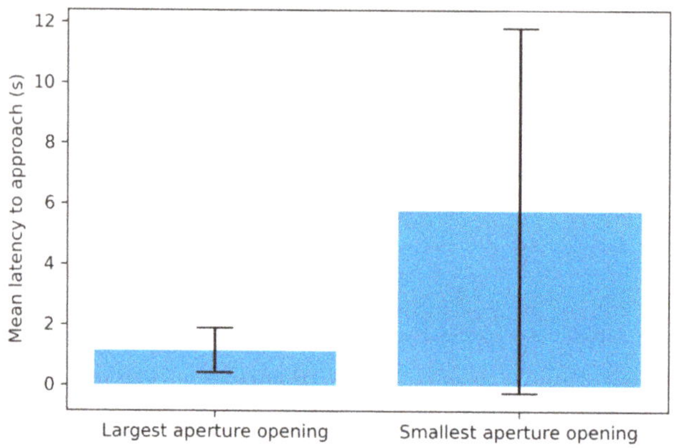

Figure 4. Mean latency for subjects to reach the aperture at its largest opening (first trial) and at its smallest opening (final trial with successful passage) ($p < 0.001$). Subjects took significantly longer to reach the aperture at the final trial than at the first trial. Error bars represent standard deviation.

Across subjects, the final aperture height negotiated varied according to the subjects' size. A linear regression of subjects' height at withers and the aperture height on the final successful trial—the trial on which the subject passed through the aperture—showed a positive correlation ($r^2 = 0.52$, $p < 0.001$; Figure 5). Eye height ($r^2 = 0.44$, $p < 0.001$) and elbow height ($r^2 = 0.50$, $p < 0.001$) were also positively correlated, less strongly, with final aperture height.

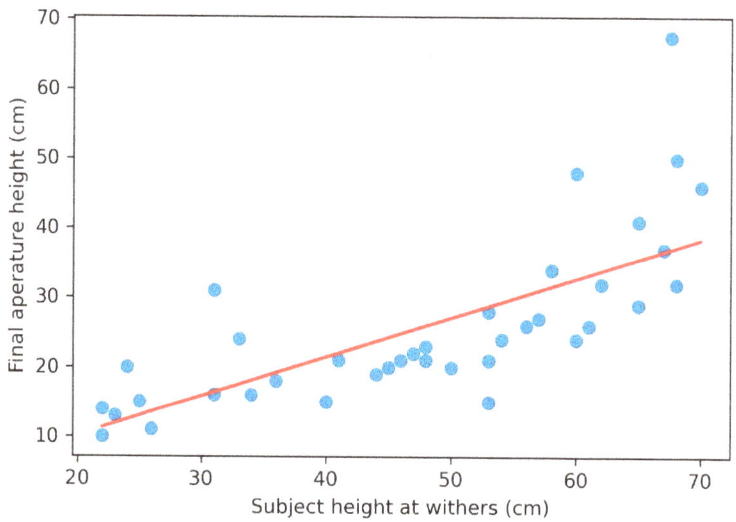

Figure 5. The relationship between subjects' final aperture height and height at withers (HW). A significant positive correlation was found between the final aperture height subjects successfully passed through and subjects' height at withers ($p < 0.001$).

The final aperture height successfully negotiated was significantly higher for dogs in the tall group ($M = 62.22$ cm, SD = 4.90) than for dogs in the short group ($M = 37.58$ cm,

SD = 10.54), $t(41) = -8.57$, $p < 0.001$; Figure 6). Considered by their bin category, the final aperture height successfully negotiated was significantly higher for dogs in the large group (M = 35.66 cm, SD = 11.70) than for dogs in the medium group (M = 20.5 cm, SD = 3.15), $t(41) = -4.40$, $p < 0.001$), the small group (M = 19.0 cm, SD = 7.42), $t(41) = -3.70$, $p = 0.001$), and the extra-small group (M = 14.4 cm, SD = 3.26), $t(41) = -3.82$, $p = 0.001$). Furthermore, the final aperture height was significantly higher for dogs in the medium group than dogs in the extra-small group ($t(41) = -3.41$, $p = 0.003$).

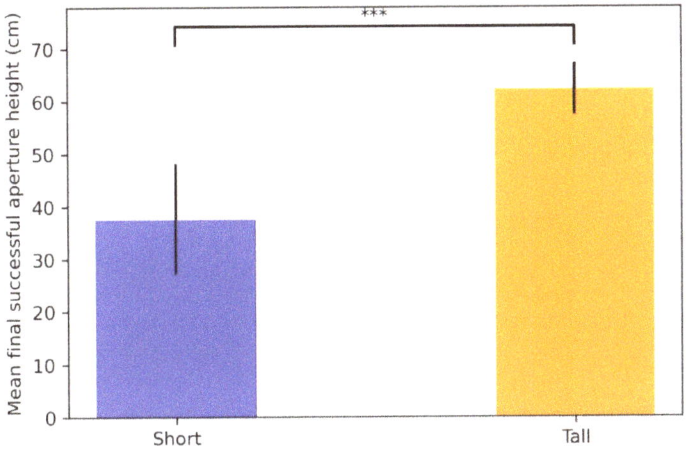

Figure 6. Mean height of aperture at final successful trial for short and tall dogs (*** $p \leq 0.001$).

To determine if the height at which dogs made their first adjustment was the same across dogs of different size, we calculated the mean aperture height for the first adjustment, divided by dog's HW, for both short and tall dogs. Comparison of these ratios found no significant difference for dog sizes (Mshort = 0.91; Mtall = 0.91; $t(41) = -0.07$, $p > 0.05$).

To examine whether dogs used a common strategy to negotiate the aperture, Chi-squared goodness of fit tests were computed on frequencies of the use of specific behavioral adjustments to fit through the aperture—as well as the frequencies of overall adjustment sequences across trials. Subjects' first behavioral adjustment was significantly more likely to be a head duck than any of the other behaviors ($X^2(3, n = 43) = 97.05$, $p < 0.001$). The second adjustment was often either a front-elbow bend or a back-elbow bend ($X^2(3, n = 43) = 34.02$, $p < 0.001$), with no significant difference between the two ($X^2(1, n = 39) = 0.02$, $p = 0.63$). Similarly, the third behavior was most often either a front elbow bend or back elbow bend ($X^2(3, n = 43) = 28.56$, $p < 0.001$), with no significant difference in frequency between the two ($X^2(1, n = 37) = 0.68$, $p = 0.41$). The fourth behavioral adjustment was significantly more likely to be the body turn ($X^2(3, n = 43) = 39.78$, $p < 0.001$) than any other behavior.

Moreover, adjustment sequence "head duck–front elbow–back elbow–body turn" and "head duck–back elbow–front elbow–body turn" were significantly more likely to occur than any other order of behavioral adjustments ($X^2(3, n = 41) = 23.10$, $p < 0.001$) (Figure 7). There was no significant difference between the frequency of these two sequences. An independent Chi-square test revealed no significant relationship between the adjustment sequence chosen and subject body size ($X^2(16, n = 41) = 7.94$, $p = 0.54$).

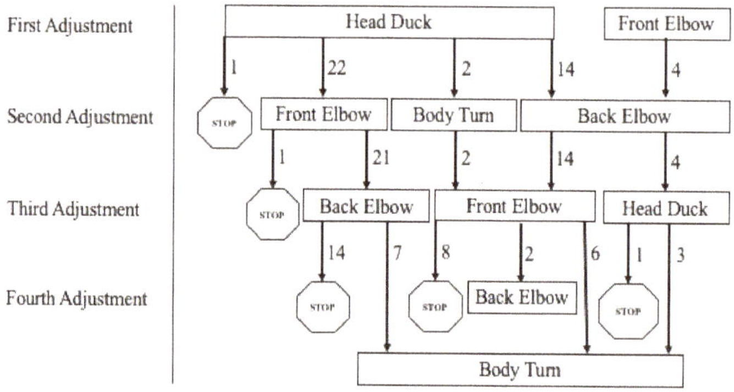

Figure 7. Transition diagram of behavioral adjustments used by all subjects to negotiate the aperture. Order of adjustments goes from top downward, with arrows indicating the number of subjects who followed each sequence to either another adjustment or to a stop (no further adjustment).

A linear regression of subjects' height at withers on the aperture height when subjects made specific behavioral adjustments revealed a strong positive correlation with height at head duck, with height at withers accounting for 75% of variance in the height of aperture ($r^2 = 0.75$, $p < 0.001$) (Figure 8). A significant correlation is also seen between height at withers and aperture height at which subjects first bent their elbows (either front or back) ($r^2 = 0.87$, $p < 0.001$) (Figure 9).

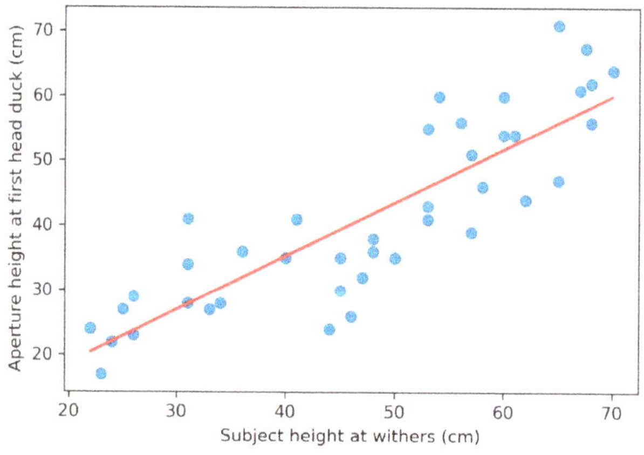

Figure 8. The relationship between the aperture height at subjects' first head adjustment and subjects' height at withers (HW). A significant positive correlation was found between the aperture height the first time subjects ducked their head to pass through and subjects' height at withers ($p < 0.001$).

Three dogs used the alternate route on one trial: two on their final height, after failing to go through the aperture, and one who used it but later returned to navigating the opening on the next trial.

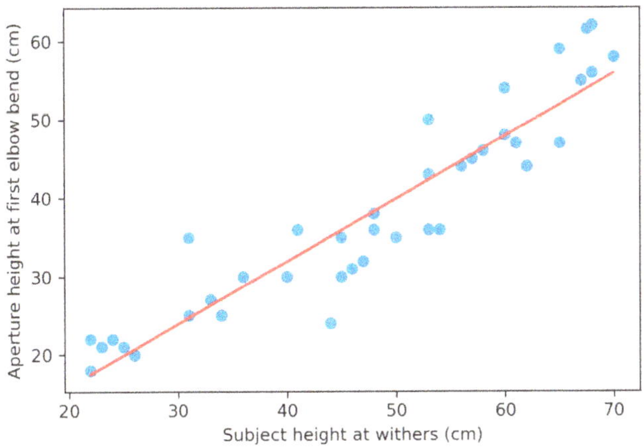

Figure 9. The relationship between the aperture height at subjects' first elbow adjustment and subjects' height at withers (HW). A significant positive correlation was found between the aperture height the first time subjects bent their elbows to pass through and subjects' height at withers ($p < 0.001$).

3.2. *Study 2*

Thirteen dogs were selected to participate in Study 2; three dogs did not successfully complete the training trials, and four dogs did not successfully complete the first experimental trial. Thus, data were analyzed from $n = 6$ dogs. Four of the six dogs fully completed the procedure: one did not make it through the aperture successfully on all trials.

Due to the high attrition rate, inferential statistical analysis was not conducted; all analyses are descriptive. All six subjects successfully passed through the aperture with the small and large stick; all but one successfully passed through with the medium stick. Subjects attempted to pass through the aperture an average of 1.5 times with the small stick, 1.17 times with the medium stick, and 1.67 times with the large stick.

Latency to reach the aperture with the small stick was 21.79 s (SD = 27.58), excluding one subject outlier who had an approach time of 1020 s. Latency to approach with the medium stick was 1.54 s (SD = 0.76); with the large stick, 4.06 s (SD = 5.21).

Behavioral adjustments on these trials included head turns (small stick: 1; medium stick: 1; large stick, 2) and dropping the stick (mean for small stick: 1; medium stick: 0.83; large stick, 1.16). Subjects more often hit the side of the aperture with the large stick (mean = 1.33) than with the small or medium sized sticks (mean = 0.5). When all six subjects were given a final choice of stick on the final trial, three chose the large stick, one chose the small, one chose the medium, and one made no choice.

4. Discussion

These studies were designed to address subject dogs' behavior when asked to go through an opening (aperture) of decreasing height, until they no longer fit through the opening. The results speak to dogs' perception of their body size with respect to their environment. This suggests that dogs not only perceive others' body sizes, as previously found [21,22,24], but also their own. The dogs in this study used knowledge of their body size when judging whether they would fit through an opening: the latency for subjects to reach an aperture they could easily fit through was significantly shorter than to one which was almost too small to fit through. This finding is consistent with Lenkei et al. [10], who also found a longer latency to approach a too-small opening. Both our results suggest that dogs distinguish between easily passable and difficult to pass openings before acting.

As expected, the final height of the aperture differed significantly between short and tall dogs and across size bins (XS, S, M, L) and correlated with several measurements: the dog's height at withers, elbows, and eyes. This indicates that what is considered "too small" to fit through is dependent on the dog's sense of their own size. Moreover, as all measurements of the dog's body appear to influence their perception of affordances, it is possible that the dog uses a crude representation of their body to determine what actions are possible. Notably, at the same time, subjects' increase in latency across trials cannot be explained by, for instance, a decrease in motivation as the aperture diminished in height, as most subjects continued to attempt to squeeze through the opening even when it was impossible for them to pass through it.

Interestingly, in the present study, 12 subjects passed through an aperture shorter than one-third of their height at withers; this is smaller in height than what was considered "too small" in Lenkei et al. [10]. Many more subjects unsuccessfully attempted to pass through an opening that was in fact too small for them to fit through even with bodily modifications. This difference between subject behavior in the present study and in Lenkei et al. [10] might be explained as resulting from the latter's adjustment of both the height and width of the aperture (in their Experiment 2), while the present study only adjusted the height. An aperture which is not wide enough to pass through may be more obviously "too small", as subjects cannot adjust their behavior to significantly reduce their width. By contrast, dogs can adjust their height by, for example, ducking their head or bending their elbows. In addition, decreasing the aperture size incrementally may have enabled more precision in determination of what size was too small for the subjects.

Additionally, the subjects' different behavior may be related to the cost or benefit associated with negotiating the experimental apparatus in the two studies. Studies with human infants have found that subjects decide whether to use a small doorway according to the related risk: many infants err and get stuck when squeezing into too small openings, but they do not attempt a too small aperture when such an error would result in falling [33]. This indicates that infants can perceive the affordances allotted by an aperture, but also use information about the cost of using the aperture when planning their actions. Dogs may be similarly evaluating the associated risk during their decision-making process: there was very little risk in passing through the similar apparatuses used in Lenkei et al. [10] and in this study. However, in addition, in the present study, there was an added benefit: their owners held a treat on the opposite side of the apparatus and were reachable only by successfully negotiating the apparatus. While Lenkei et al. [10] suggested that dogs' motivation to get through their opening was high even without the owners (Experiment 1) or without a treat (Experiment 2 and 3), the present study points to the added motivation present when both owner and treat are reached by completing the task.

To further understand the dog's decision-making process, we sought to determine not only when a dog decides whether they can or cannot fit through an aperture, but also *how* dogs fit through an aperture. Thus, we extended our analyses to the order of behavioral adjustments both across and within trials. Through examination of the sequence of bodily modifications made by subjects as they negotiated ever-smaller openings, we can suggest that subjects' perception of affordances to fit through an aperture is action-scaled. Our results revealed that subjects' first adjustment came at the same ratio of dogs' height to aperture height across sizes. Moreover, we found a preferred sequence to adjustments: subjects most often first ducked their heads, then bent front or back elbows, then turned their body. This sequence was relatively invariant across individuals, regardless of body size, providing evidence that dogs know which actions are necessary in the face of environmental constraints. Across subjects, there was a positive correlation between the size of the aperture opening and the size of the subjects' withers, eye-height, and elbow height. Dog behavior reflects an appreciation of the task—navigating the aperture—relative to the capabilities and constraints of their own bodies. The timing of specific bodily adjustments correlated with subject size: dogs' height at withers predicts when subjects would adjust their bodies by ducking their head or bending their elbow to fit through ever-diminishing apertures. Similarly, Wagman et al. [12] found that dogs' perception of the affordances for reaching was action-scaled, with the ratio of the stimulus height to body height at which subjects moved from reaching to rearing the same for tall dogs and small dogs. The present work adds to the small body of literature examining non-human animals' navigation of novel environments, which has found, for instance, that snakes change behavioral strategies depending on the affordances of their environment [34], and frogs prefer jumping through horizontal openings than same-size vertical openings, in line with their physical profile [35].

Our results in Study 2 must be considered preliminary because of the small sample size. Indeed, subject data provide no clear message. Subject latencies to reach the aperture were longer with a stick that fit than with a stick that did not fit, against expectations. Subjects made multiple attempts to pass through the aperture with all stick sizes. In line with our prediction that subjects may inaccurately judge affordances when their bodies were effectively vertically extended, the sticks often hit the side of the aperture—more with the large-sized stick than the small- or medium-sized sticks. Certainly, subject behavior was not consistent with the idea that they had foreknowledge of the way the stick had changed their size. Similarly, it has been reported that dogs often bump into objects when navigating a familiar environment with an unfamiliar neck appendage like the Elizabethan collar, a flexible cone usually of plastic that is secured around a dog's neck after surgeries in order to prevent the dog's oral manipulation of a surgical site. This is suggestive that dogs' perception of their size does not extend to appendages adding to their height or width.

Instead, modifications to their body (by turning their head) were performed only after knocking the apparatus: possible evidence that subjects updated their perception of affordances with experience. Future research examining these questions may want to include training for dogs to reliably hold or carry sticks in order to assure a robust sample size.

These results indicate that, much like humans [26,28], dogs integrate complex information about their body size with knowledge of how adjustment behaviors can alter their size when determining which actions are possible. Future research may explore the role of specific experiences in developing this "size sense". Humans require experience with their artificially altered body size to update their perception of affordances [29], and preliminary evidence from Experiment 2 suggests that dogs are, at least initially, similarly inaccurate at determining affordances when their width is effectively increased.

This work suggests that dogs' representation of their body size extends beyond the olfactory modality [8] and past actions [7]. Prior research on self-representation, using the mirror-mark test, has found mixed results with non-humans, some of which may be due to problems resulting from adapting primate-centric cognitive paradigms to non-primates [8]. Insofar as a sense of one's size may be described as a representation of oneself [10,18], testing an animal's sense of their size in a changing environment may be a novel methodological way to investigate the notoriously elusive sense of self.

5. Conclusions

Subject dogs demonstrate an understanding of their size via their differential latencies to approach openings of various sizes. This representation may not update when their size is temporarily increased by holding an object wider than their body width. Dogs' perceptions of the affordances of this novel environment were action-scaled: subjects' body size correlated with their bodily adjustments to the shortening aperture. Examining subject behavior in a novel environment may be a fruitful way to investigate self-representation in non-human animals.

Author Contributions: Conceptualization, A.H. and E.W.; methodology, A.H. and E.W.; software, M.B.; validation, A.H.; formal analysis, M.B.; investigation, E.W., M.B., B.B.; resources, A.H., E.W., M.B., and B.B.; data curation, M.B.; writing—original draft preparation, A.H., E.W., and M.B.; writing—review and editing, A.H., E.W., M.B., and B.B.; visualization, M.B.; supervision, A.H.; project administration, A.H.; funding acquisition, A.H. All authors have read and agreed to the published version of the manuscript.

Funding: This research was funded by the April Benson research grant at Barnard College.

Institutional Review Board Statement: This research was approved by the Institutional Animal Care and Use Committee (IACUC) at Columbia University and the Institutional Review Board (IRB) at Barnard College. Participation was voluntary. Subject owners signed consent forms to participate, and were allowed to withdraw from participation at any time.

Data Availability Statement: The data referenced in this manuscript will be made available by the authors, without undue reservation, to any qualified researcher upon request.

Acknowledgments: Thank you to all dog and owner participants, whose efforts are critical to the research. Thank you to the other members of the Dog Cognition Lab involved in running the experiment, including Sophie Baum, Leah Kim, Gabriella Privitelli, and Katherine Ziska, and Courtney Garrity for managing the trials. Special thanks to Gabriella Privitelli and Greg Winkler, for constructing the apparatus, and Scarlett Boulting, for creating the certificates for participation. Thank you to April Benson for support of this research.

Conflicts of Interest: The authors declare no conflict of interest. The funders had no role in the design of the study; in the collection, analyses, or interpretation of data; in the writing of the manuscript, or in the decision to publish the results.

References

1. Agnetta, B.; Hare, B.; Tomasello, M. Cues to food location that domestic dogs (Canis familiaris) of different ages do and do not use. *Anim. Cogn.* **2000**, *3*, 107–112. [CrossRef]
2. Soproni, K.; Miklósi, Á.; Topál, J.; Csányi, V. Dogs' (*Canis familaris*) responsiveness to human pointing gestures. *J. Comp. Psychol.* **2002**, *116*, 27. [CrossRef] [PubMed]
3. Call, J.; Bräuer, J.; Kaminski, J.; Tomasello, M. Domestic dogs (*Canis familiaris*) are sensitive to the attentional state of humans. *J. Comp. Psychol.* **2003**, *117*, 257–263. [CrossRef]
4. Schwab, C.; Huber, L. Obey or Not Obey? Dogs (*Canis familiaris*) Behave Differently in Response to Attentional States of Their Owners. *J. Comp. Psychol.* **2006**, *120*, 169–175. [CrossRef] [PubMed]
5. Bräuer, J.; Call, J.; Tomasello, M. Visual perspective taking in dogs (*Canis familiaris*) in the presence of barriers. *Appl. Anim. Behav. Sci.* **2004**, *88*, 299–317. [CrossRef]
6. Catala, A.; Mang, B.; Wallis, L.; Huber, L. Dogs demonstrate perspective taking based on geometrical gaze following in a Guesser–Knower task. *Anim. Cogn.* **2017**, *20*, 581–589. [CrossRef]
7. Fugazza, C.; Pongrácz, P.; Pogány, Á.; Lenkei, R.; Miklósi, Á. Mental representation and episodic-like memory of own actions in dogs. *Sci. Rep.* **2020**, *10*, 1–8. [CrossRef]
8. Horowitz, A. Smelling themselves: Dogs investigate their own odours longer when modified in an "olfactory mirror" test. *Behav. Process.* **2017**, *143*, 17–24. [CrossRef]
9. Brauer, J.; Belger, J. A ball is not a Kong: Odor representation and search behavior in domestic dogs (Canis familiaris) of different education. *J. Comp. Psychol.* **2018**, *132*, 189–199. [CrossRef]
10. Lenkei, R.; Faragó, T.; Kovács, D.; Zsilák, B.; Pongrácz, P. That dog won't fit: Body size awareness in dogs. *Anim. Cogn.* **2020**, *23*, 337–350. [CrossRef]
11. Maeda, T.; Fujita, K. Do dogs (Canis familiaris) recognize their own body size? In Proceedings of the 2nd Canine Science Forum, Vienna, Austria, 25–28 July 2010.
12. Wagman, J.B.; Langley, M.D.; Farmer-Dougan, V. Doggone affordances: Canine perception of affordances for reaching. *Psychon. Bull. Rev.* **2017**, *24*, 1097–1103. [CrossRef] [PubMed]
13. Austad, S.N. Comparative Biology of Aging. *J. Gerontol. Ser. A Boil. Sci. Med Sci.* **2009**, *64*, 199–201. [CrossRef]
14. Schmidt-Nielsen, K. Scaling in biology: The consequences of size. *J. Exp. Zool.* **1975**, *194*, 287–307. [CrossRef] [PubMed]
15. Janis, C.; Carrano, M. Scaling of reproductive turnover in archosaurs and mammals: Why are large terrestrial mammals so rare? *Ann. Zool. Fenn.* **1992**, *28*, 201–216.
16. Werner, J.; Griebeler, E.M. Reproductive Biology and Its Impact on Body Size: Comparative Analysis of Mammalian, Avian and Dinosaurian Reproduction. *PLOS ONE* **2011**, *6*, e28442. [CrossRef]
17. Parker, G. Assessment strategy and the evolution of fighting behaviour. *J. Theor. Biol.* **1974**, *47*, 223–243. [CrossRef]
18. Povinelli, D.J.; Cant, J.G. Arboreal clambering and the evolution of self-conception. *Q. Rev. Biol.* **1995**, *70*, 393–421. [CrossRef] [PubMed]
19. Adolph, K.E. The growing body in action: What infant locomotion tells us about perceptually guided action. In *Embodiment, Ego-Space, and Action*; Klatzy, R., Behrmann, M., MacWhinney, B., Eds.; Erlbaum: Mahwah, NJ, USA, 2008; pp. 275–321.
20. Gibson, J.J. Visually Controlled Locomotion And Visual Orientation In Animals*. *Br. J. Psychol.* **1958**, *49*, 182–194. [CrossRef]
21. Taylor, A.M.; Reby, D.; McComb, K. Cross Modal Perception of Body Size in Domestic Dogs (*Canis familiaris*). *PLOS ONE* **2011**, *6*, e17069. [CrossRef] [PubMed]
22. Bálint, A.; Faragó, T.; Dóka, A.; Miklósi, Á.; Pongrácz, P. 'Beware, I am big and non-dangerous!'—Playfully growling dogs are perceived larger than their actual size by their canine audience. *Appl. Anim. Behav. Sci.* **2013**, *148*, 128–137. [CrossRef]

23. Horowitz, A. Attention to attention in domestic dog (*Canis familiaris*) dyadic play. *Anim. Cogn.* **2008**, *12*, 107–118. [CrossRef] [PubMed]
24. McGuire, B.; Olsen, B.; Bemis, K.E.; Orantes, D. Urine marking in male domestic dogs: Honest or dishonest? *J. Zool.* **2018**, *306*, 163–170. [CrossRef]
25. Bornstein, M.H.; Gibson, J.J. The Ecological Approach to Visual Perception. *J. Aesthet. Art Crit.* **1980**, *39*, 203. [CrossRef]
26. Adolph, K.E.; Berger, S.E. Motor development. In *Handbook of Child Psychology: Vol. 2. Cognition, Perception, and Language*, 6th ed.; Kuhn, D., Siegler, R.S., Eds.; John Wiley & Sons: New York, NY, USA, 2006; pp. 161–213.
27. Ishak, S.; Adolph, K.E.; Lin, G.C. Perceiving affordances for fitting through apertures. *J. Exp. Psychol. Hum. Percept. Perform.* **2008**, *34*, 1501–1514. [CrossRef] [PubMed]
28. Franchak, J.M.; van der Zalm, D.J.; Adolph, K.E. Learning by doing: Action performance facilitates affordance perception. *Vis. Res.* **2010**, *50*, 2758–2765. [CrossRef] [PubMed]
29. Franchak, J.M.; Adolph, K.E. Gut estimates: Pregnant women adapt to changing possibilities for squeezing through doorways. *Atten. Percept. Psychophys.* **2014**, *76*, 460–472. [CrossRef] [PubMed]
30. Wagman, J.B.; Langley, M.D.; Farmer-Dougan, V. Carrying their own weight: Dogs perceive changing affordances for reaching. *Q. J. Exp. Psychol.* **2018**, *71*, 1040–1044. [CrossRef]
31. Warren, W.H.; Whang, S. Visual guidance of walking through apertures: Body-scaled information for affordances. *J. Exp. Psychol. Hum. Percept. Perform.* **1987**, *13*, 371–383. [CrossRef] [PubMed]
32. Sutter, N.B.; Mosher, D.S.; Gray, M.M.; Ostrander, E.A. Morphometrics within dog breeds are highly reproducible and dispute Rensch's rule. *Mamm. Genome* **2008**, *19*, 713–723. [CrossRef]
33. Franchak, J.M.; Adolph, K.E. What infants know and what they do: Perceiving possibilities for walking through openings. *Dev. Psychol.* **2012**, *48*, 1254–1261. [CrossRef]
34. Jayne, B.C.; Lehmkuhl, A.M.; Riley, M.A. Hit or miss: Branch structure affects perch choice, behaviour, distance and accuracy of brown tree snakes bridging gaps. *Anim. Behav.* **2014**, *88*, 233–241. [CrossRef]
35. Ingle, D. Spontaneous shape discrimination by frogs during unconditioned escape behavior. *Physiol. Psychol.* **1973**, *1*, 71–73. [CrossRef]

animals

MDPI

Review

The Challenge of Illusory Perception of Animals: The Impact of Methodological Variability in Cross-Species Investigation

Maria Santacà [1,*], Christian Agrillo [1,2] and Maria Elena Miletto Petrazzini [3]

1 Department of General Psychology, University of Padova, 35131 Padova, Italy; christian.agrillo@unipd.it
2 Padua Neuroscience Center, University of Padova, 35131 Padova, Italy
3 Department of Biomedical Sciences, University of Padova, 35131 Padova, Italy; mariaelena.milettopetrazzini@gmail.com
* Correspondence: santacamaria@gmail.com; Tel.: +39-049-827-7424

Simple Summary: Research in neurobiology and ethology has given us a glimpse into the different perceptual worlds of animals. More recently, visual illusions have been used in behavioural research to compare the perception between different animal species. The studies conducted so far have provided contradictory results, raising the possibility that different methodological approaches might influence illusory perception. Here, we review the literature on this topic, considering both field and laboratory studies. In addition, we compare the two approaches used in laboratories, namely spontaneous choice tests and training procedures, highlighting both their relevance and their potential weaknesses. Adopting both procedures has the potential to combine their advantages. Although this twofold approach has seldomly been adopted, we expect it will become more widely used in the near future in order to shed light on the heterogeneous pattern observed in the literature of visual illusions.

Citation: Santacà, M.; Agrillo, C.; Miletto Petrazzini, M.E. The Challenge of Illusory Perception of Animals: The Impact of Methodological Variability in Cross-Species Investigation. *Animals* **2021**, *11*, 1618. https://doi.org/10.3390/ani11061618

Academic Editor: Alan Tilbrook

Received: 25 March 2021
Accepted: 27 May 2021
Published: 30 May 2021

Abstract: Although we live on the same planet, there are countless different ways of seeing the surroundings that reflect the different individual experiences and selective pressures. In recent decades, visual illusions have been used in behavioural research to compare the perception between different vertebrate species. The studies conducted so far have provided contradictory results, suggesting that the underlying perceptual mechanisms may differ across species. Besides the differentiation of the perceptual mechanisms, another explanation could be taken into account. Indeed, the different studies often used different methodologies that could have potentially introduced confounding factors. In fact, the possibility exists that the illusory perception is influenced by the different methodologies and the test design. Almost every study of this research field has been conducted in laboratories adopting two different methodological approaches: a spontaneous choice test or a training procedure. In the spontaneous choice test, a subject is presented with biologically relevant stimuli in an illusory context, whereas, in the training procedure, a subject has to undergo an extensive training during which neutral stimuli are associated with a biologically relevant reward. Here, we review the literature on this topic, highlighting both the relevance and the potential weaknesses of the different methodological approaches.

Keywords: visual illusions; comparative perception; motion illusions; distortion illusions; subjective contours

1. Introduction

Understanding how the different animal species see the world around them has long interested researchers from ancient times to today. For many animals, vision is the primary link to the world that allows them to seek out food, communicate, avoid predators, or find a mate to reproduce. Since their appearance, animals have colonized nearly every ecological niche on Earth and, thus, have evolved different visual systems to assimilate the surrounding information. For example, some species can only see in shades of grey, whereas other species have colour vision even in a near-total darkness condition. Other species can see different parts of the light spectrum that are invisible to humans, such as ultraviolet and infrared light [1]. As a consequence, the animal visual worlds are highly different from the world that humans take for granted. However, perceiving the world is not a simple and passive acquisition of the images of the surrounding environment, but it is based on past experience and stored information [2]. In fact, a three-dimensional image has to be translated into a bi-dimensional retinal representation that, subsequently, is interpreted by cognitive and neural processes. The perception of a visual stimulus can, therefore, differ considerably from its physical counterpart. In the psychology literature, some of these "errors" of perception are referred to as visual illusions. Contrary to this indirect approach, it has been proposed that the nature of visual perception is not limited to simple geometric or physical properties. In fact, according to the Gibson's ecological approach, perception is a direct contact with the environment and, thus, not mediated by mental images or other mental representations [3–5]. Gibson also stated that the content of perception must be already relevant for action, and it is primarily comprised of opportunities for behaviour. According to this direct approach, vision must be considered in terms of the whole visual system and activity over time, and it is insufficient to look at its retinal image to understand what an animal perceives.

The research regarding the perception of visual illusions in animals began in the 1920s [6], and since then, visual illusions have become an important tool to compare visual perception in animals. This research, in fact, allows researchers to assess whether animals interpret and alter visual inputs as humans do or if they detect visual inputs with little or no variability. In addition, visual illusions can be also used to comprehend the psychological and cognitive processes underlying visual perception and to shed light on the impact of environmental and evolutionary pressures on visual perception and processing [7]. In fact, according to the traditional theories of perception, the evolutionary assumption underlying these studies is that, if the susceptibility to the same illusory pattern is shared between two species, it is possible to infer that the two species share similar perceptual mechanisms to perceive the world around them. Interspecific comparisons have led to a high heterogeneity in the results. Some studies reported a similar perception to the same illusory pattern in distantly related species. For instance, different monkeys (capuchin monkeys, *Cebus apella* [8]; rhesus monkeys, *Macaca mulatta* [9]), birds (ringneck dove, *Streptopelia risorii* [10]; African grey parrot, *Psittacus erithacus* [11]) and fish (guppies, *Poecilia reticulata* [12]; redtail splitfin fish, *Xenotoca eiseni* [13]) demonstrated an ability to perceive the Müller-Lyer illusion as humans do. Figure 1a shows the classical version of this illusion in which two parallel lines are presented; one line has inward-pointing arrows on the ends, whereas the other has outward-pointing arrows. Humans and the abovementioned species were shown to underestimate the length of the target line ending with the outward-pointing arrows and to overestimate the length of the target line ending with the inward-pointing arrows. In addition, rhesus monkeys [14], felines (lions, *Panthera leo* [15]; cats, *Felis catus* [16]) and fish (guppies and zebrafish, *Danio rerio* [17]) seem to be susceptible to motion illusions, perceiving the Rotating Snake illusion (Figure 1b). Lastly, chimpanzees (*Pan troglodytes* [18]) and other monkeys (for a review, see [19]), cats [20], birds (barn owl, *Tyto alba* [21]), fish (bamboo sharks, *Chiloscyllium griseum* [22]; goldfish, *Carassius auratus* [23]; redtail splitfin [24]) were shown to perceive illusory contours of the Kanizsa illusory picture (Figure 1c).

Other studies, however, highlighted a lack of perception of an illusory pattern in different vertebrates. Guinea baboons (*Papio papio* [25]) and starlings (*Sturnus vulgaris* [26]) do not see the Ebbinghaus illusion, whereas dogs (*Canis lupus familiaris* [27,28]) and red-footed tortoises (*Chelonoidis carbonaria* [29]) do not see the Delboeuf illusion. Both illusions occur when the size of a target item is misperceived depending on its surrounding context. Figure 1d,e show the classical versions of both illusions: two identical target circles encompassed by a larger and smaller context, namely rings in the Delboeuf illusion and circles in the Ebbinghaus illusion. In both cases, humans typically underestimate the size of the target circle encompassed by the larger context and tend to overestimate the size of the target circle encompassed by the smaller context. In addition, dogs [30] and bamboo sharks [22] do not perceive the Müller-Lyer illusion.

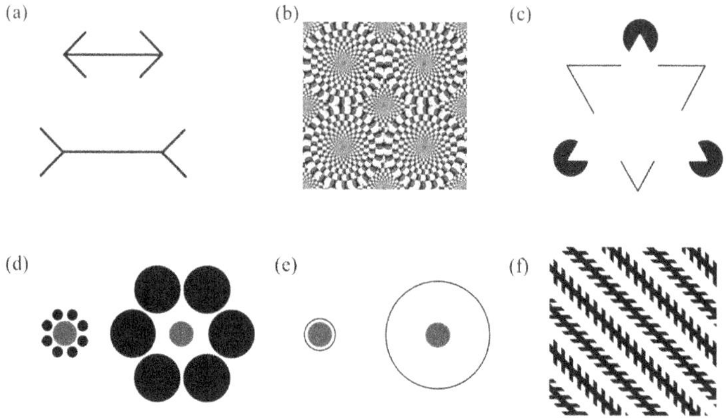

Figure 1. Müller-Lyer illusion (**a**), Rotating Snake illusion (**b**), Kanizsa's triangle (**c**), Ebbinghaus illusion (**d**), Delboeuf illusion (**e**) and Zöllner illusion (**f**).

Lastly, other studies reported the perception of visual illusions in the opposite direction compared to humans, also defined as a reversed perception. Pigeons (*Columba livia* [31]) and bamboo sharks [32] experienced a reverse Ebbinghaus illusion, whereas different species of teleost fish perceive a reverse Delboeuf illusion [33,34]. In addition, both pigeons [35] and bantam chickens (*Gallus gallus* [36]) demonstrated a reverse Zöllner illusion that consists in a set of parallel lines that appear non-parallel due to series of short crosshatches superimposed on the lines (Figure 1f).

In summary, some studies found a similar perception of the same visual illusion among vertebrates, suggesting similar perceptual mechanisms, whereas others highlighted interspecific differences. The origin of the observed variability in the perception of visual illusions remains unclear. The most accredited hypothesis is that this observed variability might be ascribed to different contextual factors, such as the adopted methodology, the different stimuli, the age and the sex of the subjects (e.g., [7,37]). In fact, there is evidence that different methods of investigating the Ebbinghaus illusion can lead to different results in the same species, as in bantam chickens. Rosa Salva and colleagues [38] demonstrated that four-days chicks perceived the illusion as humans do, whereas Nakamura and colleagues [39] concluded the opposite testing 6-month-old chickens. In addition, in some cases (e.g., the odd task used with olive baboons or *Papio anubis* in the case of the Zöllner illusion [40]), the methodology adopted provides a yes/no response in terms of susceptibility to an illusory phenomenon, but the information about the direction of the illusion (human-like or reverse) is not available.

In the literature on illusory perception, almost every study, with a couple of exceptions, has been conducted in laboratories adopting two different methodological approaches, namely spontaneous choices tests and training procedures. Due to the primary relevance of the abovementioned methodological questions, here, we critically examine and compare behavioural studies that investigated the perception of visual illusions in non-human animals. In this review, after a brief description and presentation of various studies using the different methodologies, we compare these different methods, highlighting the pros and cons of each.

2. Field Studies

When breeding, feeding and moving within their environment, wild animals interact with their physical surroundings and the biological world. In almost every field study, researchers simply observed animals in their natural environment without any manipulation. In few cases, researchers could adopt a minimally invasive approach deciding to move or manipulate some physical characteristics of the environment to observe any consequent change in an animal's behaviour. Field studies have been conducted also to understand whether animals perceive or use visual illusions. In fact, visual illusions are naturally present in real-life contexts and might be advantageous and positively selected in animal species able to generate them by manipulating the environment or their coloration and movement.. For example, illusions of size or brightness may be present on the body and influence different social context such as choosing a mate or intrasex conflicts. In fact, in both sexual and conflict contexts, animals typically compare locally available conspecifics and make choices based on size and colour that may signal quality [41]. Such illusions may change the perceived quality of an individual and, thus, influence animals' behaviour. Instead, motion illusions or illusions that obscure the body shape might play an important part in animal camouflage to reduce the risk of being predated or to increase the possibility to predate [42]. Despite their high ecological value, there is only a little and mainly indirect evidence of illusory phenomena in real-life contexts with only one exception.

One indirect evidence regards male fiddler crabs (*Uca mjoebergi*) that actively manipulate their social environment, namely the presence of neighbours, that resembles the Ebbinghaus illusion, to potentially increase their relative attractiveness [43]. In addition, male guppies may actively utilize the same illusory pattern to select the most appropriate context relative to their coloration to increase their reproductive success [44]. The only direct evidence is the study in the great bowerbird (*Ptilinorhynchus nuchalis*), in which the authors demonstrated that males construct bower courts with forced visual perspective from the audience view to manipulate their size perception and, thus, to increase their reproductive success [45]. To create this perspective illusion, the males of this species arrange grey and white objects in a positive size–distance gradient forced perspective. The authors demonstrated that, when the forced perspective is reversed, males actively restore it within a couple of days [45]. In addition, males vary consistently in the quality of the perspective illusion and that the latter is positively correlated with mating success [46,47].

3. Laboratory Studies

The environmental conditions cannot be neatly controlled and documented; in fact, the field studies are characterized by a lack of control and the difficulty of precisely distinguishing the several environmental factors. This is the main reason why the majority of the studies regarding the perception of visual illusions have been conducted in laboratories. In this controlled environment, researchers adopted two different methodological approaches that differ in several respects which we will later discuss.

3.1. Spontaneous Choice Tests

In spontaneous choice tests, subjects are typically presented with biologically relevant stimuli that, in most cases, consist of food items (Table 1). However, the experimental setup considerably differs depending upon the type of visual illusion.

Table 1. Summary of the existing works on visual illusions adopting a spontaneous choice paradigm.

Visual Illusion	Reference	Sample	Stimuli	Susceptibility
Delboeuf illusion	[48]	3 *Pan troglodytes* (chimpanzees)	Food	Yes
	[49]	9 *Lemur catta* (ring-tailed lemurs)	Food	n/a
	[50]	18 *Felis catus* (cats)	Food	Yes
	[28]	13 *Canis lupus familiarias* (dogs)	Food	No
	[51]	12 *Pogona vitticeps* (bearded dragons)	Food	Yes
	[29]	8 *Chelonoidis carbonaria* (red-footed tortoises)	Food	No
	[33]	12 *Poecilia reticulata* (guppies)	Food	Reversed

Table 1. *Cont.*

Visual Illusion	Reference	Sample	Stimuli	Susceptibility
	[34]	12 *Betta splendens* (Siamese fighting fish), 12 *Danio rerio* (zebrafish), 12 *Pterophyllum scalare* (angelfish), 12 *Trichopodus trichopterus* (three-spot gourami) 12 *Xenotoca eiseni* (redtail splitfin)	Food	Reversed
Müller-Lyer illusion	[52]	12 *Pogona vitticeps* (bearded dragons)	Food	Yes
	[53]	9 *Equus caballus* (horses)	Food	Yes
Rotating Snake illusion	[16]	11 *Felis catus* (cats)	Printed stimuli	Yes
	[15]	3 *Panthera leo* (lions)	Printed stimuli	Yes

Regarding all the illusions in which the visual context induces a distortion in size, the procedure consists in the presentation of two arrays containing two identical food portions in two different contexts that resemble a specific illusory pattern. The first food portion selected by the subject is recorded as the dependent variable. This approach exploited animals' natural tendency to maximise the food intake. Different studies proved the spontaneous preference for larger food portions in several animal species such as chimpanzees [54], cats [55], African grey parrots [56], Italian wall lizards (*Podarcis sicula* [57]), Hermann's tortoises (*Testudo hermanni* [58]) and guppies [59]. Moreover, this approach relies on the evidence that humans' perception of food size is influenced by the context in which it is presented, in particular, in a Delboeuf illusory context (Figure 2).

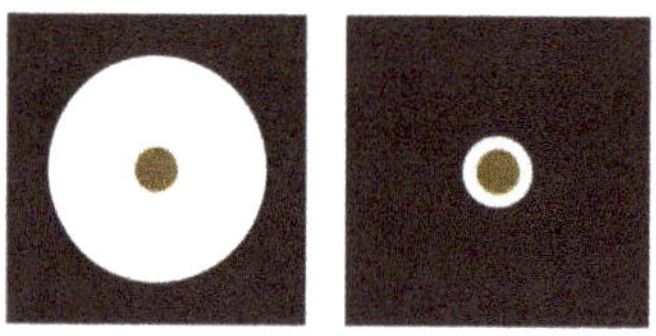

Figure 2. An example of biologically relevant stimuli (i.e., food) that could be used in spontaneous choice studies investigating the Delboeuf illusion in animals.

Different studies demonstrated that humans overestimate a food portion size and, hence, under-serve when it is presented on a smaller dish (e.g., [60–62]). The pioneering study adopting this procedure in non-human animals was conducted with chimpanzees by Parrish and Beran [48] to assess the existence of the Delboeuf illusion in this species. When presented with two identical food portions on two different-sized plates, chimpanzees significantly selected the food portion inserted in the smaller plate, thus perceiving the Delboeuf illusion as humans do. Chimpanzees were also presented with control trials consisting of two different-sized food portions on two identical plates to assess their tendency to maximise the food intake in the experimental context. The assumption of this approach is that, if a species spontaneously selects the larger quantity in order to maximize food intake, it is expected to choose the portion that appears larger in illusory trials. Other studies adopted the same methodological approach of Parrish and Beran [48] to investigate the perception of the Delboeuf illusion in cats [50], dogs [28], reptiles [29,51] and fish [33,34]. The same approach has been used to study other visual illusions, such as the Müller-Lyer illusion in reptiles [52] and horses (*Equus caballus* [53]). A crucial aspect of this methodological approach consists in the performances in control trials in which two different-sized food portions are presented. As abovementioned, this type of procedure exploits animals' natural tendency to maximise the food intake. Nonetheless, animals could also not choose the larger of two rewards. This could occur when both food portions are large enough for the species under investigation and, thus, animals do not need to maximize their intake (e.g., [28,49]). Alternatively, animals could be trying to maximize food intake, but the physical difference between the food portions could be too subtle to be detected due to their visual acuity (e.g., [28,49]). In this scenario, since animals are not maximising the food intake in the experimental context, no conclusion can be drawn regarding their performances in illusory trials.

There is also a simpler procedure to investigate the perception of visual illusions adopting a spontaneous approach; however, this can be applied only to motion illusions, such as the Rotating Snake illusion (Figure 1b). The assumption is that a subject perceives the motion illusion, and it is expected to approach, more than chance, the illusory stimulus to pursue movement. In the only two studies of this type, the researchers investigated the perception of the Rotating Snake illusion in felines, namely cats [16] and lions [15], adopting a preferential looking experiment. In both studies, researchers placed different visual stimuli in their environment, respectively, a "cat-café" in Fukuoka-city and an Italian zoological garden. The stimuli consisted of the Rotating Snake illusion and a control stimulus that did not evoke any motion perception, even if the overall configuration was identical to that of the illusory pattern. Researchers recorded the interacting time or the number of interactions with the illusory stimuli and found that both cats and lions were specifically attracted by the Rotating Snake illusion to pursue movement. In addition, Regaiolli and colleagues [15] also found that this illusion has a positive effect on the animals' welfare, consisting of a reduction in self-directed behaviours and an increase in attentive behaviours.

However, confounding factors can potentially affect spontaneous choice tests. In fact, animals may have a bias for choosing the illusory stimulus more than chance, even if they do not perceive the illusion. This emerged in a study investigating the perception of the Delboeuf illusion in ring-tailed lemurs (*Lemur catta* [49]). As in the previous studies, lemurs were presented with both control and illusory trials. Despite they exhibited poor performance in control trials compared to other mammals previously observed, one subject consistently selected the food portion in the larger plate in the illusory trials. This might lead researchers to believe that this lemur was highly sensitive to trials in which the plates were differing in size. In fact, this condition was less frequent compared to control trials, and it may have led to a bias for choosing one array (in this case, the food portion on the larger plate), which was not observed when identical plates were presented. The same hypothesis of a spontaneous bias for the surrounding context, itself, also emerged in the study with guppies that demonstrated to strongly perceive a reverse Delboeuf illusion [33]. However, in this investigation, the researchers conducted a control experiment in which the test trials consisted in presenting the large array and the small array without any food stimuli. If guppies exhibited a spontaneous preference for the larger array in the presence of different-sized arrays, they were expected to select this one more than chance. The results of this control experiment excluded the possibility that the guppies' performance in illusory trials was due to any sort of spontaneous bias for the context in which the food portion were presented [33]. Another confounding factor that can potentially affect a spontaneous choice test consists in the foraging habits of a species. For example, in the reptiles' studies, tortoises exhibited a lower performance compared to bearded dragons in size estimation with vertically arranged items [63]. An intriguing possibility is that tortoises pay less attention to vertical stimuli in the natural environment due to their ecological niche, since they live on the ground layer. Bearded dragons, instead, are considered to be semi-arboreal, and they quite readily climb and bask at height. Thus, the configuration or the position of the stimuli could influence the test result [37,63].

In summary, studies using the spontaneous procedure reported the perception of visual illusions in a wide range of species. However, as the natural behaviour of the animals is observed, it exists the possibility that animals' choices are not based on the target stimulus, i.e., food portion, but on the context in which the stimuli are presented, potentially resulting in a "false" illusory perception. For better control this confounding factors, other studies preferred to use training procedures with inanimate objects as stimuli.

3.2. Training Procedures

In a training procedure, a subject undergoes an extensive training during which neutral stimuli are associated with a biologically relevant reward. In fact, a subject is trained to select a target stimulus in order to obtain a reward (i.e., food or social companions). This procedure necessarily requires several training phases before the subject is presented with a visual illusion. In the first phase, a subject usually familiarizes with the apparatus and the procedure undergoing a shaping procedure in order to learn the association of the reward with the correct response. Subsequently, the subject is presented with a discrimination task until it reaches a learning criterion—normally a significant percentage of correct choices in two consecutive sessions or an overall significant performance considering all training trials (e.g., [12,27]).

In this type of procedure, the stimuli can be presented on a monitor, as it has been successfully done with several species such as primate and non-primate mammals (e.g., [27,30,64,65]), birds (e.g., [35,36]) and fish (e.g., [17,22,32]) (Table 2). Alternatively, the stimuli can be printed on specific presentation cards; this setup has mainly been adopted with fish (e.g., [12,13,24]) but also with birds (e.g., [38]).

Table 2. Summary of the existing works on visual illusions adopting a training procedure.

Visual Illusion	Reference	Sample	Stimuli	Task type	Reward	Susceptibility
Delboeuf illusion	[65]	7 *Macaca mulatta* (rhesus monkeys)	Presented on a monitor	Two-choice discrimination task/ absolute classification task	Food	No (two-choice task). Yes (absolute classification task)
	[27]	8 *Canis lupus familiaris* (dogs)	Presented on a monitor	Two-choice discrimination task	Food	No
Ebbinghaus illusion	[27]	8 *Canis lupus familiaris* (dogs)	Presented on a monitor	Two-choice discrimination task	Food	Reverse
	[39]	3 *Gallus gallus* (bantams)	Presented on a monitor	Absolute classification task	Food	Reverse
	[31]	6 *Columba livia* (pigeons)	Presented on a monitor	Absolute classification task	Food	Reverse
	[38]	24 *Gallus gallus* (chicks)	Printed stimuli	Two-choice discrimination task	Food	Yes
	[26]	5 *Sturnus vulgaris* (starlings)	Presented on a monitor	Absolute classification task	Food	No
	[32]	4 *Chiloscyllium griseum* (bamboo sharks), 5 *Chromis chromis* (damselfish)	Presented on a monitor	Two-choice discrimination task	Food	Reverse (sharks). Yes (damselfish)
	[13]	8 *Xenotoca eiseni* (redtail splitfin)	Printed stimuli	Two-choice discrimination task	Social	Yes
Müller-Lyer illusion	[30]	7 *Canis lupus familiaris* (dogs)	Presented on a monitor	Two-choice discrimination task	Food	No
	[66]	4 *Columba livia* (pigeons)	Presented on a monitor	Absolute classification task	Food	Yes
	[67]	6 *Xenotoca eiseni* (redtail splitfin)	Printed stimuli	Two-choice discrimination task	Social	Yes
	[12]	12 *Poecilia reticulata* (guppies)	Printed stimuli	Two-choice discrimination task	Food	Yes
	[22]	9 *Chiloscyllium griseum* (bamboo sharks)	Presented on a monitor	Two-choice discrimination task	Food	No
Kanizsa figures	[18]	2 *Pan troglodytes* (chimpanzees)	Presented on a monitor	Two-choice discrimination task	Food	Yes
	[24]	7 *Xenotoca eiseni* (redtail splitfin)	Printed stimuli	Two-choice discrimination task	Social	Yes
	[22]	9 *Chiloscyllium griseum* (bamboo sharks)	Presented on a monitor	Two-choice discrimination task	Food	Yes

Table 2. *Cont.*

Visual Illusion	Reference	Sample	Stimuli	Task type	Reward	Susceptibility
Zöllner illusion	[64]	6 *Macaca mulatta* (rhesus monkeys)	Presented on a monitor	Two-choice discrimination task	Food	Yes
	[40]	2 *Papio anubis* (olive baboons)	Printed stimuli	Oddity task	Food	Yes
	[35]	6 *Columba livia* (homing pigeons)	Presented on a monitor	Two-choice discrimination task	Food	Reversed
	[36]	3 *Gallus gallus* (bantam chickens)	Presented on a monitor	Two-choice discrimination task	Food	Reversed
Rotating Snake illusion	[17]	12 *Danio rerio* (zebrafish), 12 *Poecilia reticulata* (guppies)	Presented on a monitor	Two-choice discrimination task	Food	Yes

The majority of the studies adopted a training procedure consisting of a two-choice discrimination task (Table 2). In particular, a subject is simultaneously presented with two stimuli that differ in regard to their size, brightness or other physical characteristics. Parrish and colleagues [65] trained rhesus monkeys and capuchin monkeys to investigate whether they perceive the Delboeuf illusion. In the first experiment, monkeys were presented with a relative discrimination task requiring them to choose the larger of two target dots that were sometimes encircled by concentric rings. All monkeys successfully learned the discrimination rule; instead, in illusory trials, they exhibited heterogenous results. In fact, two monkeys demonstrated to perceive the illusion as humans do; twelve monkeys seemed not to perceive the illusion, whereas six appeared to perceive the illusion in a reversed manner [65]. As the abovementioned spontaneous choice examples, the researchers were concerned that the contrasting results were driven by a bias for the dot encircled by a large context considering the contexts and dots as one stimulus. For this reason, they conducted a second experiment with an absolute classification task in which subjects had to classify a target dot of variable size as either "large" or "small" in comparison with a never-presented central target size. In this task, capuchin and rhesus monkeys showed sensitivity to the Delboeuf illusion as reported for humans, demonstrating that the perception of this illusion is influenced by the procedure adopted. Despite the absolute classification procedure proved to overcome the aforementioned concerns, it has been less adopted than the relative two-choice discrimination procedure. To date, it had been mainly used with birds, namely pigeons and bantams (e.g., [31,39,66]). Regarding this issue, a recent study from Qadri and Cook [26] with starlings suggests that the choice for the array that is supposed to reflect an illusory perception could be a consequence of the training procedure adopted. In fact, the researchers discovered that the classical training procedure used to investigate this illusion in pigeons leads the starlings to integrate the irrelevant context into their decision process, precluding the study of illusory perception [26]. Thus, the absolute discrimination task also cannot completely resolve or eliminate the abovementioned concerns.

Training procedures have been used to study the perception of visual illusions in non-human animals with a food reward or a social reward (Table 2). The food reward has been used with several different species, from primates to fish. Just to give some examples, Fagot and Tomonaga [18] trained chimpanzees, delivering a piece of apple after a correct choice, whereas Fuss and colleagues [22] trained sharks, rewarding them with dropped food, namely a bait, for a correct choice during the training phase. To date, a social reward was only used in the studies investigating the perception of the Kanizsa illusory picture, the Ebbinghaus illusion and the Müller-Lyer illusion in redtail splitfin fish [13,24,67]. In these experiments, fish were removed from their social group and placed in an unfamiliar square environment. They could rejoin companions only by passing through one of two identical tunnels, that were associated with geometric figures, at opposite corners. Only the door associated with the reinforced stimulus allowed the subject to rejoin its companions.

4. A Comparison of the Different Methodological Approaches

Field research is research conducted in a natural environment or in the real world or simply observing, interpreting and explaining what already exists or manipulating some physical characteristics of the environment to observe any consequent change in an animal's behaviour. Field study subjects may not be aware that they are being tested. Thus, field studies have the advantage of representing real-life circumstances, because they consist of a range of situations and environments that are encountered in the natural habitat, such as in the studies with the great bowerbird [45–47]. For these reasons, this type of study has the highest ecological validity. Laboratory research, on the other hand, is performed in an environment specifically designed for research. This type of research is sometimes characterized as a tightly regulated investigation in which the researcher manipulates the factor under investigation to see whether the manipulation causes a shift in the subjects. Laboratory studies have the advantage of providing better control over irrelevant variables that may otherwise influence the findings, as well as clearer clues to the observed behaviour (e.g., [12,30]). However, laboratory research, like any other form of research, may have drawbacks. On one hand, the experimental setup can be meticulously monitored and recorded, but, on the other, it could reflect a simulated environment that influences how subjects act and, as a result, alters their performance. Nonetheless, since researchers can usually exert a greater control in a laboratory setting than in a naturalistic setting, the perception of visual illusions in non-human animals has been mainly investigated in laboratories.

Laboratory studies adopt two different methodological approaches: a spontaneous choice test or a training procedure. The two approaches are different in many aspects and also reveal different aspects of visual processing [68]. Animals are expected to behave naturally in spontaneous choice experiments, and their performance is thought to reflect the cognitive and perceptual processes they would activate in nature. However, this procedure has some weakness. In fact, since the stimulus to discriminate corresponds with the reward, the performance of the subjects in this approach is heavily influenced by their motivation. For example, in a food discrimination task, animals may get a certain amount of food regardless of the correctness of their choices; thus, their motivation can decrease trial after trial, leading to a null result, as in the investigation of the Delboeuf illusion in lemurs [49]. In training procedures, on the other hand, animals can only obtain a food reward if they make the right choice. Since only those subjects that successfully learn the discrimination rule (e.g., choose the larger stimulus) are presented with illusory trials, their performances in this type of trials are thought to reflect their real illusory perception. In spontaneous choice tests, animals can also differ in their individual preferences for the stimuli's distinctive features. This has been found both in interspecific and intraspecific studies. For example, despite the willingness to adopt the same methodological approach, tortoises and bearded dragons had to be tested with different preferred food, respectively, mango jelly and vegetable extract (kale, cucumber and mint) jelly, when tested for the perception of the Delboeuf illusion [29,52]. In the investigation of the same illusion in cats, prior to the experiment, the authors presented two different types of food stimuli, namely canned tuna and dry cat food, to assess any individual difference in the food preference [50]. This allowed the researchers to test each cat with its preferred food [50]. Other sensory modalities, such as olfactory cues, may also affect the animals' performance in spontaneous choice experiments. In fact, there are high interspecific differences in the relative importance of vision and other senses, such as olfaction, in solving different cognitive tasks. In the field of illusory perception, this could have a crucial impact on the performance in illusory trials. As a matter of fact, the illusions in which the visual context induces a distortion in size are resembled by presenting the same food portion but in two different contexts, as reported above. In such illusory trials, animals that mainly rely on olfaction to select the larger food portion may not be fooled by the visual illusion, leading to a null result [28]. On the contrary, in training procedures, the problem of individual preferences or using other sensory modalities is generally not a concern, since the stimuli are bi-dimensional (i.e., static objects presented on a monitor or printed stimuli). Finally, since the same neutral stimuli may be used with distantly related species, training procedures allow a fine interspecific comparison, as in the studies with the Rotating snake illusion [14–17]. In contrast, the most appropriate stimuli for the species under investigation are needed in spontaneous choice experiments, as mentioned above. More practical considerations, such as sample size, distinguish the two methods. A larger sample size is normally evaluated in spontaneous choice experiments to evaluate the group performance and, thus, overcome any individual preference for the stimuli. Finally, because motivation may decrease during the trials, a between-subjects design with each subject tested in a relatively short period of time should be preferred in spontaneous choice tests (e.g., [28,33,48]). In training procedures, on the other hand, each subject may require a significant amount of time to meet the learning criterion; thus, for this second methodological approach, a within-subjects design is preferred, as in the dogs study regarding their perception of the Delboeuf and Ebbinghaus illusion [27].

The two approaches are clearly complementary. Only a few studies adopted both procedures to have an integrated result regarding the illusory perception of a visual illusions. This is the case of the Delboeuf illusion in guppies [33] and in dogs [27,28]. In a first experiment, Lucon-Xiccato and colleagues [33] trained guppies in a two-choice task to discriminate between two different-sized printed stimuli. Upon achieving the learning criterion, guppies passed to the test phase in which they could face different types of trials, such as congruent (i.e., a large context included the larger target circle and a small context included the smaller target circle) and incongruent (i.e., a large context that included the smaller target circle and a small context that included the large target circle) size discrimination trials. If guppies were susceptible to the Delboeuf illusion, they were expected to be facilitated in size judgments in incongruent trials compared to the congruent trials. The guppies' performances in these trials revealed that guppies showed a facilitation effect due to the Delboeuf illusion in the reversed direction compared to humans [33]. However, in the illusory trials of the test phase, the authors found no significant preference, both in group analysis considering the entire sample of subjects and in a tentative analysis at the individual level. To further investigate and understand their perception of the Delboeuf illusion, the researchers conducted a second experiment, adopting a spontaneous food choice test in which guppies demonstrated a preference for the larger food item in control trials with a higher accuracy compared to the first experiment (66% vs. 57% [33]). In the illusory trials of this second experiment, guppies showed a marked preference for the food item presented in the large context, revealing a Delboeuf illusory effect greater than observed in the training experiment. This could be related to the higher salience of the stimuli in the spontaneous choice experiment; in addition, guppies could use different strategies to distinguish between printed stimuli and food items, and the difference in illusory effect could be due to distinct strategies [33]. Considering the dogs studies, Miletto Petrazzini and colleagues [28] found that, when dogs could choose between two identical food portions but presented on a Delboeuf illusion context in a spontaneous preference paradigm, they did not seem to be susceptible to the Delboeuf illusion. In line with this conclusion, Byosiere and colleagues [27] used an operant conditioning procedure to train dogs to make a fine discrimination between a larger and a smaller black target circle. The dogs in this experiment also performed randomly at the group level in the presence of the Delboeuf pattern. In both cases, it is intriguing and suggestive that two experiments that used two different approaches (untrained behaviour with biologically relevant stimuli vs. trained behaviour with two-dimensional figures on the screen) came up with the same result. To date, no other illusory perception has been investigated with both procedures, leaving uncertain the obtained results.

5. Conclusions

The findings obtained until now prompted a debate as to whether animal species share similar perceptual mechanisms underlying visual perception and if these are homologous to our perceptual mechanisms. This issue becomes even more relevant regarding the possibility of developing animal models to investigate visual perception disorders, to find out general principles necessary to build successful artificial visual systems and, more generally, to study critical research topics in perception that, for practical and ethical reasons, cannot be easily deepened in our species, such as the ontogeny of perceptual mechanisms and the exact neural circuits underlying visual perception. The origins of the heterogeneous perception of visual illusion could be ascribed to the different contextual factors. In this review, it was demonstrated that different methods can lead to different results in the same species. We also identified different aspects of the same methodological approach that should be carefully taken into account when comparing the results of different studies. Lastly, we highlighted both the relevance and the potential weaknesses of the different methodological approaches. Regarding the laboratory studies, we suggest that adopting the two different procedures, namely spontaneous choice tests or training procedures, has the potential to combine their advantages we have discussed in this review. On this basis, despite the fact that spontaneous choice tests or training procedures have been poorly adopted in the same investigation, we expect this twofold approach will become more widely used in the near future in order to shed light on the heterogeneous pattern observed in the literature of visual illusions.

Author Contributions: Conceptualization, C.A. and M.E.M.P.; writing—original draft preparation, M.S.; writing—review and editing, M.S., C.A. and M.E.M.P. All authors have read and agreed to the published version of the manuscript.

Funding: This study was supported by the University of Padua under the STARS Grants programme (MetaZeb) to M.E.M.P. The present work was carried out within the scope of the research programme "Dipartimenti di Eccellenza" (art.1, commi 314–337 legge 232/2016), which was supported by a grant from MIUR ("Ministero dell'istruzione, dell'università e della ricerca") to the Department of General Psychology, University of Padova.

Institutional Review Board Statement: Not applicable.

Informed Consent Statement: Not applicable.

Data Availability Statement: Not applicable.

Acknowledgments: We would like to thank three anonymous Reviewers whose suggestions greatly enriched the manuscript.

Conflicts of Interest: The authors declare no conflict of interest.

References

1. Lazareva, O.F.; Shimizu, T.; Wasserman, E.A. *How Animals See the World: Comparative Behavior, Biology, and Evolution of Vision*; Oxford University Press: Oxford, UK, 2012.
2. Gregory, R.L. *The Intelligent Eye*; Weidenfeld and Nicolson: London, UK, 1970.
3. Gibson, J.J. *The Perception of the Visual World*; Houghton Mifflin: Boston, MA, USA, 1950.
4. Gibson, J.J. *The Ecological Approach to Visual Perception*; Houghton Mifflin Harcourt: Boston, MA, USA, 1979.
5. Gibson, J.J. A theory of direct visual perception. In *Vision and Mind: Selected Readings in the Philosophy of Perception*; Noë, A., Thompson, E., Eds.; MIT Press: Cambridge, MA, USA, 2002; pp. 77–90.
6. Révész, G. Experiments on animal space perception. *Brit. J. Psychol.* **1924**, *14*, 387–414.
7. Feng, L.C.; Chouinard, P.A.; Howell, T.J.; Bennett, P.C. Why do animals differ in their susceptibility to geometrical illusions? *Psychon. Bull. Rev.* **2017**, *24*, 262–276. [CrossRef]
8. Suganuma, E.; Pessoa, V.F.; Monge-Fuentes, V.; Castro, B.M.; Tavares, M.C.H. Perception of the Müller–Lyer illusion in capuchin monkeys (*Cebus apella*). *Behav. Brain Res.* **2007**, *182*, 67–72. [CrossRef] [PubMed]
9. Tudusciuc, O.; Nieder, A. Comparison of length judgments and the Müller-Lyer illusion in monkeys and humans. *Exp. Brain Res.* **2010**, *207*, 221–231. [CrossRef] [PubMed]
10. Warden, C.J.; Baar, J. The Müller-Lyer illusion in the ring dove, *Turtur risorius*. *J. Comp. Psychol.* **1929**, *9*, 275. [CrossRef]
11. Pepperberg, I.M.; Vicinay, J.; Cavanagh, P. Processing of the Müller-Lyer illusion by a Grey parrot (*Psittacus erithacus*). *Perception* **2008**, *37*, 765–781. [CrossRef]
12. Santacà, M.; Agrillo, C. Perception of the Müller–Lyer illusion in guppies. *Curr. Zool.* **2020**, *66*, 205–213. [CrossRef]
13. Sovrano, V.A.; Albertazzi, L.; Rosa Salva, O. The Ebbinghaus illusion in a fish (*Xenotoca eiseni*). *Anim. Cogn.* **2015**, *18*, 533–542. [CrossRef] [PubMed]

14. Agrillo, C.; Gori, S.; Beran, M.J. Do rhesus monkeys (*Macaca mulatta*) perceive illusory motion? *Anim. Cogn.* **2015**, *18*, 895–910. [CrossRef] [PubMed]
15. Regaiolli, B.; Rizzo, A.; Ottolini, G.; Miletto Petrazzini, M.E.; Spiezio, C.; Agrillo, C. Motion illusions as environmental enrichment for zoo animals: A preliminary investigation on lions (*Panthera leo*). *Front. Psychol.* **2019**, *10*, 2220. [CrossRef] [PubMed]
16. Bååth, R.; Seno, T.; Kitaoka, A. Cats and illusory motion. *Psychology* **2014**, *5*, 1131–1134. [CrossRef]
17. Gori, S.; Agrillo, C.; Dadda, M.; Bisazza, A. Do fish perceive illusory motion? *Sci. Rep. UK* **2014**, *4*, 1–6. [CrossRef]
18. Fagot, J.; Tomonaga, M. Effects of element separation on perceptual grouping by humans (*Homo sapiens*) and chimpanzees (*Pan troglodytes*): Perception of Kanizsa illusory figures. *Anim. Cogn.* **2001**, *4*, 171–177. [CrossRef] [PubMed]
19. Nieder, A. Seeing more than meets the eye: Processing of illusory contours in animals. *J. Comp. Physiol. A* **2002**, *188*, 249–260. [CrossRef]
20. Bravo, M.; Blake, R.; Morrison, S. Cats see subjective contours. *Vision Res.* **1988**, *28*, 861–865. [CrossRef]
21. Nieder, A.; Wagner, H. Perception and neuronal coding of subjective contours in the owl. *Na.t Neurosci.* **1999**, *2*, 660–663. [CrossRef]
22. Fuss, T.; Bleckmann, H.; Schluessel, V. The brain creates illusions not just for us: Sharks (*Chiloscyllium griseum*) can "see the magic" as well. *Front. Neur. Circ.* **2014**, *8*, 24. [CrossRef]
23. Wyzisk, K.; Neumeyer, C. Perception of illusory surfaces and contours in goldfish. *Vis. Neurosci.* **2007**, *24*, 291. [CrossRef] [PubMed]
24. Sovrano, V.A.; Bisazza, A. Perception of subjective contours in fish. *Perception* **2009**, *38*, 579–590. [CrossRef]
25. Parron, C.; Fagot, J. Comparison of grouping abilities in humans (*Homo sapiens*) and baboons (*Papio papio*) with the Ebbinghaus illusion. *J. Comp. Psychol.* **2007**, *121*, 405. [CrossRef]
26. Qadri, M.A.; Cook, R.G. Perception of Ebbinghaus–Titchener stimuli in starlings (*Sturnus vulgaris*). *Anim. Cogn.* **2019**, *22*, 973–989. [CrossRef] [PubMed]
27. Byosiere, S.E.; Feng, L.C.; Woodhead, J.K.; Rutter, N.J.; Chouinard, P.A.; Howell, T.J.; Bennett, P.C. Visual perception in domestic dogs: Susceptibility to the Ebbinghaus–Titchener and Delboeuf illusions. *Anim. Cogn.* **2017**, *20*, 435–448. [CrossRef] [PubMed]
28. Miletto Petrazzini, M.E.; Bisazza, A.; Agrillo, C. Do domestic dogs (*Canis lupus familiaris*) perceive the Delboeuf illusion? *Anim. Cogn.* **2017**, *20*, 427–434. [CrossRef]
29. Santacà, M.; Miletto Petrazzini, M.E.; Wilkinson, A.; Agrillo, C. Red-footed tortoises (*Chelonoidis carbonaria*) do not perceive the Delboeuf illusion. *Can. J. Exp. Psychol.* **2020**, *74*, 201. [CrossRef] [PubMed]
30. Keep, B.; Zulch, H.E.; Wilkinson, A. Truth is in the eye of the beholder: Perception of the Müller-Lyer illusion in dogs. *Learn. Behav.* **2018**, *46*, 501–512. [CrossRef]
31. Nakamura, N.; Watanabe, S.; Fujita, K. Pigeons perceive the Ebbinghaus-Titchener circles as an assimilation illusion. *J. Exp. Psychol. Anim. Behav. Process.* **2008**, *34*, 375. [CrossRef] [PubMed]
32. Fuss, T.; Schluessel, V. The Ebbinghaus illusion in the gray bamboo shark (*Chiloscyllium griseum*) in comparison to the teleost damselfish (*Chromis chromis*). *Zoology* **2017**, *123*, 16–29. [CrossRef] [PubMed]
33. Lucon-Xiccato, T.; Santacà, M.; Miletto Petrazzini, M.E.; Agrillo, C.; Dadda, M. Guppies, *Poecilia reticulata*, perceive a reversed Delboeuf illusion. *Anim. Cogn.* **2019**, *22*, 291–303. [CrossRef]
34. Santacà, M.; Lucon-Xiccato, T.; Agrillo, C. The Delboeuf illusion's bias in food choice of teleost fishes: An interspecific study. *Anim. Behav.* **2020**, *164*, 105–112. [CrossRef]
35. Watanabe, S.; Nakamura, N.; Fujita, K. Pigeons perceive a reversed Zöllner illusion. *Cognition* **2011**, *119*, 137–141. [CrossRef]
36. Watanabe, S.; Nakamura, N.; Fujita, K. Bantams (*Gallus gallus domesticus*) also perceive a reversed Zöllner illusion. *Anim. Cogn.* **2013**, *16*, 109–115. [CrossRef]
37. Byosiere, S.E.; Feng, L.C.; Wuister, J.; Chouinard, P.A.; Howell, T.J.; Bennett, P.C. Do dogs demonstrate susceptibility to a vertically presented Ponzo illusion. *Anim. Behav. Cogn.* **2018**, *5*, 254–267. [CrossRef]
38. Rosa Salva, O.; Rugani, R.; Cavazzana, A.; Regolin, L.; Vallortigara, G. Perception of the Ebbinghaus illusion in four-day-old domestic chicks (*Gallus gallus*). *Anim. Cogn.* **2013**, *16*, 895–906. [CrossRef]
39. Nakamura, N.; Watanabe, S.; Fujita, K. A reversed Ebbinghaus–Titchener illusion in bantams (*Gallus gallus domesticus*). *Anim. Cogn.* **2014**, *17*, 471–481. [CrossRef] [PubMed]
40. Benhar, E.; Samuel, D. Visual illusions in the baboon (*Papio anubis*). *Anim. Learn. Behav.* **1982**, *10*, 115–118. [CrossRef]
41. Andersson, M. *Sexual Selection*; Princeton University Press: Princeton, NJ, USA, 1994.
42. Gomez, D.; Théry, M. Simultaneous crypsis and conspicuousness in color patterns: Comparative analysis of a neotropical rainforest bird community. *Am. Nat.* **2007**, *169*, S42–S61. [CrossRef]
43. Kelley, L.A.; Kelley, J.L. Animal visual illusion and confusion: The importance of a perceptual perspective. *Behav. Ecol.* **2014**, *25*, 450–463. [CrossRef]
44. Gasparini, C.; Serena, G.; Pilastro, A. Do unattractive friends make you look better? Context-dependent male mating preferences in the guppy. *Proc. R. Soc. B-Biol. Sci.* **2013**, *280*, 20123072. [CrossRef]
45. Endler, J.A.; Endler, L.C.; Doerr, N.R. Great bowerbirds create theaters with forced perspective when seen by their audience. *Curr. Biol.* **2010**, *20*, 1679–1684. [CrossRef]
46. Kelley, L.A.; Endler, J.A. Illusions promote mating success in great bowerbirds. *Science* **2012**, *335*, 335–338. [CrossRef]

47. Kelley, L.A.; Endler, J.A. Male great bowerbirds create forced perspective illusions with consistently different individual quality. *Prac. Natl. Acad. Sci. USA* **2012**, *109*, 20980–20985. [CrossRef]
48. Parrish, A.E.; Beran, M.J. When less is more: Like humans, chimpanzees (*Pan troglodytes*) misperceive food amounts based on plate size. *Anim. Cogn.* **2014**, *17*, 427–434. [CrossRef]
49. Santacà, M.; Regaiolli, B.; Miletto Petrazzini, M.E.; Spiezio, C.; Agrillo, C. Preliminary study to investigate the Delboeuf illusion in ring-tailed lemurs (*Lemur catta*): Methodological challenges. *Anim. Behav. Cogn.* **2017**, *4*, 365–377. [CrossRef]
50. Szenczi, P.; Velázquez-López, Z.I.; Urrutia, A.; Hudson, R.; Bánszegi, O. Perception of the Delboeuf illusion by the adult domestic cat (*Felis silvestris catus*) in comparison with other mammals. *J. Comp. Psychol.* **2019**, *133*, 223. [CrossRef] [PubMed]
51. Santacà, M.; Miletto Petrazzini, M.E.; Agrillo, C.; Wilkinson, A. Can reptiles perceive visual illusions? Delboeuf illusion in red-footed tortoise (*Chelonoidis carbonaria*) and bearded dragon (*Pogona vitticeps*). *J. Comp. Psychol.* **2019**, *133*, 419. [CrossRef] [PubMed]
52. Santacà, M.; Miletto Petrazzini, M.E.; Agrillo, C.; Wilkinson, A. Exploring the Müller-Lyer illusion in a nonavian reptile (*Pogona vitticeps*). *J. Comp. Psychol.* **2020**, *134*, 391. [CrossRef] [PubMed]
53. Cappellato, A.; Miletto Petrazzini, M.E.; Bisazza, A.; Dadda, M.; Agrillo, C. Susceptibility to Size Visual Illusions in a Non-Primate Mammal (*Equus caballus*). *Animals* **2020**, *10*, 1673. [CrossRef]
54. Beran, M.J.; Evans, T.A.; Harris, E.H. Perception of food amounts by chimpanzees based on the number, size, contour length and visibility of items. *Anim. Behav.* **2008**, *75*, 1793–1802. [CrossRef]
55. Bánszegi, O.; Urrutia, A.; Szenczi, P.; Hudson, R. More or less: Spontaneous quantity discrimination in the domestic cat. *Anim. Cogn.* **2016**, *19*, 879–888. [CrossRef]
56. Al Aïn, S.; Giret, N.; Grand, M.; Kreutzer, M.; Bovet, D. The discrimination of discrete and continuous amounts in African grey parrots (*Psittacus erithacus*). *Anim. Cogn.* **2009**, *12*, 145–154. [CrossRef]
57. Miletto Petrazzini, M.E.; Fraccaroli, I.; Gariboldi, F.; Agrillo, C.; Bisazza, A.; Bertolucci, C.; Foà, A. Quantitative abilities in a reptile (*Podarcis sicula*). *Biol. Lett. UK* **2017**, *13*, 20160899. [CrossRef] [PubMed]
58. Gazzola, A.; Vallortigara, G.; Pellitteri-Rosa, D. Continuous and discrete quantity discrimination in tortoises. *B. Biol. Lett. UK* **2018**, *14*, 20180649. [CrossRef] [PubMed]
59. Lucon-Xiccato, T.; Miletto Petrazzini, M.E.; Agrillo, C.; Bisazza, A. Guppies discriminate between two quantities of food items but prioritize item size over total amount. *Anim. Behav.* **2015**, *107*, 183–191. [CrossRef]
60. Davis, B.; Payne, C.R.; Bui, M. Making small food units seem regular: How larger figure size reduces calories to be consumed. *J. Assoc. Consum. Res.* **2016**, *1*, 115–124. [CrossRef]
61. Wansink, B.; Cheney, M.M. Super bowls: Serving bowl size and food consumption. *Jama-J. Am. Med. Assoc.* **2005**, *293*, 1723–1728. [CrossRef]
62. Wansink, B.; Van Ittersum, K. Portion size me: Downsizing our consumption norms. *J. Acad. Nutr. Diet.* **2007**, *107*, 1103–1106. [CrossRef] [PubMed]
63. Santacà, M.; Miletto Petrazzini, M.E.; Wilkinson, A.; Agrillo, C. Anisotropy of perceived space in non-primates? The horizontal-vertical illusion in bearded dragons (*Pogona vitticeps*) and red-footed tortoises (*Chelonoidis carbonaria*). *Behav. Process.* **2020**, *176*, 104117. [CrossRef]
64. Agrillo, C.; Parrish, A.E.; Beran, M.J. Do rhesus monkeys (*Macaca mulatta*) perceive the Zöllner illusion? *Psychon. B. Rev.* **2014**, *21*, 986–994. [CrossRef] [PubMed]
65. Parrish, A.E.; Brosnan, S.F.; Beran, M.J. Do you see what I see? A comparative investigation of the Delboeuf illusion in humans (*Homo sapiens*), rhesus monkeys (*Macaca mulatta*), and capuchin monkeys (*Cebus apella*). *J. Exp. Psychol. Anim. Learn. Cogn.* **2015**, *41*, 395. [CrossRef]
66. Nakamura, N.; Fujita, K.; Ushitani, T.; Miyata, H. Perception of the standard and the reversed Müller-Lyer figures in pigeons (*Columba livia*) and humans (*Homo sapiens*). *J. Comp. Psychol.* **2006**, *120*, 252. [CrossRef]
67. Sovrano, V.A.; Da Pos, O.; Albertazzi, L. The Müller-Lyer illusion in the teleost fish *Xenotoca eiseni*. *Anim. Cogn.* **2016**, *19*, 123–132. [CrossRef] [PubMed]
68. Agrillo, C.; Bisazza, A. Spontaneous versus trained numerical abilities. A comparison between the two main tools to study numerical competence in non-human animals. *J. Neurosci. Meth.* **2014**, *234*, 82–91. [CrossRef] [PubMed]

animals

MDPI

Article

Can Plants Move Like Animals? A Three-Dimensional Stereovision Analysis of Movement in Plants

Valentina Simonetti [1,*], Maria Bulgheroni [1], Silvia Guerra [2], Alessandro Peressotti [3], Francesca Peressotti [4], Walter Baccinelli [1], Francesco Ceccarini [2], Bianca Bonato [2], Qiuran Wang [2] and Umberto Castiello [2,*]

1 Ab.Acus srl, 20155 Milan, Italy; mariabulgheroni@ab-acus.com (M.B.); walterbaccinelli@ab-acus.eu (W.B.)
2 Dipartimento di Psicologia Generale, Università di Padova, 35131 Padova, Italy; silvia.guerra.5@phd.unipd.it (S.G.); ceccarini1980@gmail.com (F.C.); bianca.bonato.1@phd.unipd.it (B.B.); qiuran.wang@phd.unipd.it (Q.W.)
3 Dipartimento di Scienze Agroalimentari, Ambientali e Animali, Università degli Studi di Udine, 33100 Udine, Italy; alessandro.peressotti@uniud.it
4 Dipartimento di Psicologia dello Sviluppo e della Socializzazione, Università degli Studi di Padova, 35131 Padova, Italy; francesca.peressotti@unipd.it
* Correspondence: valentinasimonetti@ab-acus.eu (V.S.); umberto.castiello@unipd.it (U.C.)

Simple Summary: Intrigued by the ability of climbing peas to detect and grasp structures such as garden reeds, we adapted a method classically used to investigate the grasping movement of animals to the study of grasping movements in plants. We used time-lapse photography to document the behavior of pea plants, grown in the vicinity of a support pole. Using this footage, we analyzed the kinematics of tendrils growth and found that their approach and grasp exhibited movement signatures comparable to those characterizing the reach-to-grasp movement of animals. Through our method it may be possible to demonstrate that plants may be more sentient than we give them credit for: namely, they may possess the ability to act intentionally.

Abstract: In this article we adapt a methodology customarily used to investigate movement in animals to study the movement of plants. The targeted movement is circumnutation, a helical organ movement widespread among plants. It is variable due to a different magnitude of the trajectory (amplitude) exhibited by the organ tip, duration of one cycle (period), circular, elliptical, pendulum-like or irregular shape and the clockwise and counterclockwise direction of rotation. The acquisition setup consists of two cameras used to obtain a stereoscopic vision for each plant. Cameras switch to infrared recording mode for low light level conditions, allowing continuous motion acquisition during the night. A dedicated software enables semi-automatic tracking of key points of the plant and reconstructs the 3D trajectory of each point along the whole movement. Three-dimensional trajectories for different points undergo a specific processing to compute those features suitable to describe circumnutation (e.g., maximum speed, circumnutation center, circumnutation length, etc.). By applying our method to the approach-to-grasp movement exhibited by climbing plants (*Pisum sativum* L.) it appears clear that the plants scale movement kinematics according to the features of the support in ways that are adaptive, flexible, anticipatory and goal-directed, reminiscent of how animals would act.

Keywords: kinematics; circumnutation; plant behavior; plant movement

Citation: Simonetti, V.; Bulgheroni, M.; Guerra, S.; Peressotti, A.; Peressotti, F.; Baccinelli, W.; Ceccarini, F.; Bonato, B.; Wang, Q.; Castiello, U. Can Plants Move Like Animals? A Three-Dimensional Stereovision Analysis of Movement in Plants. *Animals* **2021**, *11*, 1854. https://doi.org/10.3390/ani11071854

Academic Editor: Clive J. C. Phillips

Received: 15 April 2021
Accepted: 17 June 2021
Published: 22 June 2021

1. Introduction

Although terrestrial plants are unable to move from one place to another, they are very much in tune with their environment and are very capable of a variety of movements. Plants may not move as far or as quickly as animals, but some of their movements achieve many of the same functional ends as those of animals [1,2]. Here we focus on circumnutation, a helical movement of elongating plant organs that has been investigated for decades [3]. Changes in circumnutation patterns can provide useful insights for auto-ecological and physiological studies (e.g., [4]), and contribute to explaining the mechanisms that drive gravitropism [5–7], circadian rhythms [8–10] and growth rate [11]. Despite its importance as a plant physiological signature, the standard approach for circumnutation analysis still needs to be consolidated. Some studies make use of time-lapse images and adopt video processing to extract movement-related features [5,6,8,10,12–15]. For instance, Stolarz and colleagues [16] described a structured approach for 2D plant circumnutation analysis and developed a software (i.e., Circumnutation Tracker) for the extraction of key parameters with a standard set up that includes time-lapse video acquisition of the plant from a top view of the circumnutation movement, manual harvesting of the coordinates and parameters calculation. Their time-lapse video acquisition works on standard video formats, but it does not take into account possible lens distortion of the cameras that may introduce discrepancies between the real movement and recorded movement. A top view positioning of the camera has been adopted in a variety of studies [5,10,14,17], but it determines constraints on camera position and the type of points that can be investigated (points with circumnutating movement along with the camera view). A two-dimensional top view also limits the trajectory analysis of the camera's field of view, precluding the recording of movements in other directions. In this respect, some studies tried to improve this set up by adding side view contextual acquisition [5,14,17]. As manual harvesting of coordinates can become a heavy process with the increase of video duration and points to track, Stolarz and colleagues [16] considered the automatization of this process as a next step for a future development of their system.

Johnsonn et al. [13] described a possible approach for 3D movement reconstruction with a system that records stereovision pictures by rotating two mirrors with a control mechanism. Their system is still under development and its use is strongly linked to the hardware, which has been built to meet the specific needs of their experimental setup, making its adaptation to other experimental conditions difficult.

More recently, Raja et al. [15] have proposed a minimalistic 1D approach that relies on nonlinear methods of behavioral analysis to uncover the dynamics of plant nutation. This approach focuses on the time dependencies characterizing the processes that give rise to circumnutation patterns and may help for the identification of alternative parameters than those derived from classic kinematics.

Continuing on this analysis, Bastien and colleagues ([18]; see also Porat et al. [19], Gallentine et al. [20]) developed computational 3D models to emulate growth dynamics in rod-like organs, including the effect of external stimuli in the environment. They focused on the limits of experimental measurements of nutation that use the projection of the apical part of the organ in the plane orthogonal to the gravitational field. They show that a complete kinematic description should consider geometrical and local effects in the 3D space and should not restrict the analysis to the position of the apical tip. They also stress the importance of future development of measurements considering proper 3D kinematics.

Here, we present an easy-to-scale system solution that: (1) leverages a pair of commercial cameras for 3D plant movement reconstruction; (2) compensates for possible lens distortion artifacts allowing a more robust computation of coordinates; and (3) implements semi-automatic trajectories that harvest and extract the circumnutation parameters. This approach is adaptable to different plants and experimental setups, which permits the investigation of motion in plants and allows for a comparison with the kinematical ways used to investigate movements in animal species.

2. Materials and Methods

The solution proposed in this paper is based on the acquisition and analysis of the plant's movement through a stereovision system that generates time-lapses of the plant and reconstructs its 3D movement. The system workflow (Figure 1) envisages the acquisition of temporally equally spaced images of the plant by two fixed calibrated cameras. A semi-automatic point tracking process is applied to reconstruct the 3D trajectory of custom landmarks over the entire acquisition. Finally, the descriptive features of the 3D movement of the plant are computed from the analysis of the trajectories. The following sections illustrate the process details. The system has been extensively used for a specific experimental setup designed to investigate circumnutation pattern of climbing peas (*Pisum sativum* L. var. saccharatum cv Carouby de Maussane; hereafter *P. sativum*)

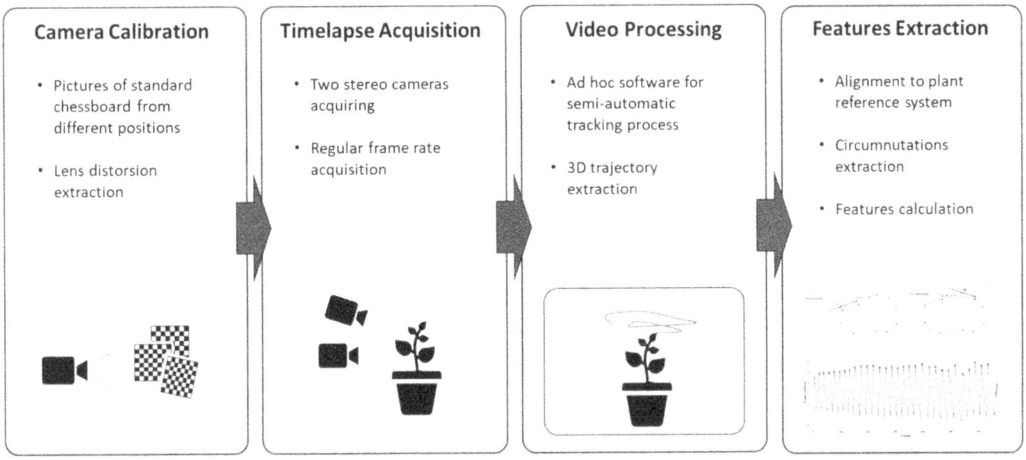

Figure 1. Sequence of the steps for 3D reconstruction and processing of the movement of the plant.

2.1. Instrumentation

A pair of RGB-infrared cameras has to be set for each plant under investigation. In our experimental setup we installed two cameras (i.e., IP 2.1 Mpx outdoor varifocal IR 1080 P) inside each thermo-light-controlled growth chamber. Both cameras were placed on the same side of the chamber to obtain a complete view of the organ to track, thus enabling a stereovision image recording. The cameras were set to acquire standard RGB images in high light conditions and automatically switch to infrared recording mode in low light condition. Each camera was wired with Ethernet cables to a wireless router (D-link Dsr-250n, D-Link Corporation Ltd., Taiwan) connected via Wi-Fi to a PC used as a data storage unit. A dedicated software, CamRecorder (Ab.Acus s.r.l. Milan, Italy), was developed to synchronize acquisition and customize frame rate (0.0056 Hz for the experimental protocols tested) for each pair of cameras.

2.2. Cameras Calibration and Stereovision

We developed a Matlab procedure (Camera Calibrator app) to reduce the cameras' optical lens distortions and avoid geometrical position errors [21–24]. This procedure is run for each of the two lenses by ingesting 20 pictures of a paper-printed chessboard (squares' side 18 mm, 10 columns, 7 rows) from multiple angles and distances, in natural, non-direct light conditions. The reprojection error is evaluated and images with high mean residual error (>1 pixel) are discarded and substituted with new pictures. This process is repeated until a set of 20 images with a low reprojection error is obtained. The lens distortion parameters computed for each camera are used to compensate the effects on the acquired images.

In addition to the aforementioned process for the computation of cameras' intrinsic parameters, the extrinsic parameters for each pair of cameras have to be computed as to obtain the cameras' relative positions. From the 2D positions of the images acquired by the two cameras, the 3D position for each point is reconstructed. The cameras' relative positions were calculated with the same Matlab procedure from simultaneous pictures of a chessboard target with known landmarks. The target was placed in the middle of the growing chamber before plant growing onset.

2.3. Video Processing

The dedicated software SPROUTS (Ab.Acus s.r.l. Milan, Italy) was developed by using Python 3.7 to enable a user-friendly tracking for the considered key landmarks. The software is designed to work for any kind of growing setup that provides image streams from two cameras. Through a simple interface users can: (1) perform a semi-automatic tracking of custom number of key points, (2) compute 3D trajectories of each tracked point in real word dimensions and (3) save them as .cvs files containing the coordinates of the point in each frame. Initially, the user is asked to identify the points to be tracked on the first image, then a tracking algorithm estimates the position of the points on the following frames. The user is allowed to check for position estimation errors eventually introduced by the automatic tracking procedure: the user can re-mark the point to be tracked on the first available image, supervise the tracking process and eventually adjust the trajectory of the point being tracked. At the end of the tracking process, the user can review the position of the tracked points in all the images and eventually correct the positioning errors. Such semi-automatic tracking is implemented using the Lucas Kanade computer vision method [25] based on optical flow, using a size of the search window of 10 by 10 pixels, a maximal pyramid level number (iterative lowering of image resolution) equal to 20 and the termination criteria of 30 maximum iterations. Three-dimensional trajectories are obtained from 2D trajectories acquired for both the left and right-side cameras using the Matlab triangulating function [26]. A sample for the 2D trajectory extracted for one camera is shown in the Supplementary video 1, whereas an example of the reconstructed 3D trajectory is shown in Supplementary video 2.

2.4. Features Extraction

Finally, the 2D and 3D positions and kinematic features are calculated from coordinates of landmarks with a dedicated procedure. The extraction module was developed in Matlab 2020a and was designed to process the 3D trajectory files obtained from the previously introduced Video Processing steps. The 3D point position reconstruction algorithms provide the x, y and z coordinates of each point of the trajectory in a reference system with its origin on the lower left corner of the calibration chessboard. Since the positioning of the chessboard cannot be consistent on all the experiments, to enable the extraction computation of the correct features, the points' reference system is roto-translated to the plant's reference system.

The new reference system is built such that the system origin coincides with one of the points along the vertical axis of the plant not showing relevant movement during the acquisition. The y-axis of the new 3D system corresponds to the plant's vertical axis, while the x and z axes lay on the plane perpendicular to the y-axis containing the system origin. To correctly identify the plant's vertical axis and origin, the user is required to position two points along the plant's axis during the video processing. The vertical axis is the line passing through these two points. In the case of pea plants (*P. sativum*), the y-axis was set as the line passing through the first and second internode, with the system origin being the first internode (Figure 2).

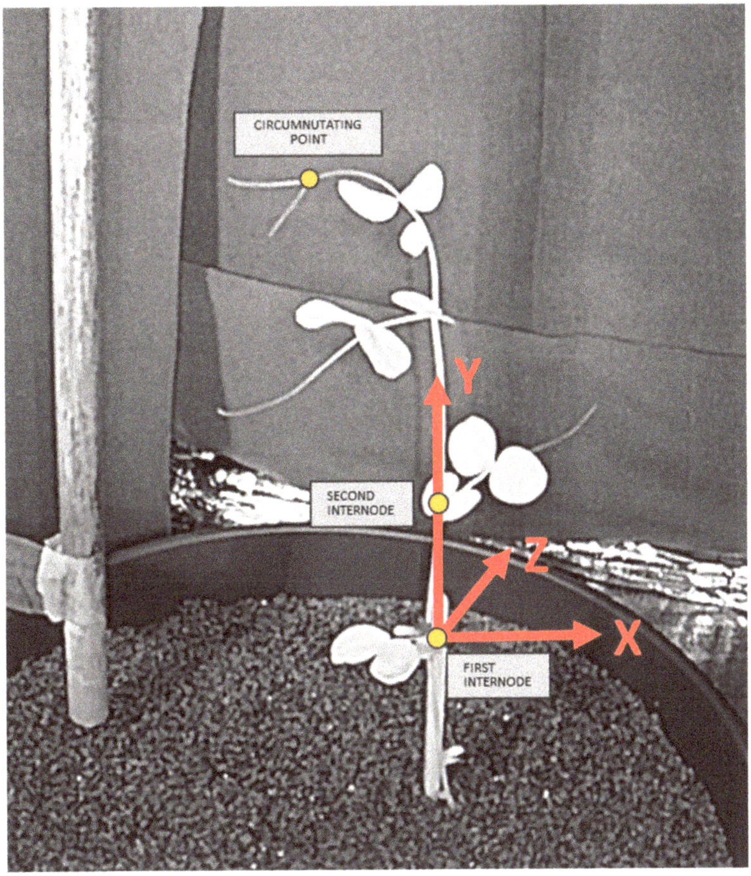

Figure 2. Plant reference system. The y-axis corresponds to the plant vertical axis and the x-axis and z-axis describe the perpendicular planes. The picture shows a pea plant (*P. sativum*) where the vertical axis has been identified as the line passing through the first and the second internode, with the first internode being the origin.

Once the reference system is aligned to the plant's system, the software detects circumnutating movements. Taking the X–Z components of the 3D movement, the software computes the angle (α) between the horizontal axis (x-axis) and the movement vector of each frame, considered as the vector between the tracked point at frame t (P2) and the same point at frame t-1 (P1) as shown in Figure 3a. Figure 3b shows the result of the whole trajectory, with a five-sample moving average. A single circumnutation movement is considered as the movement occurring between two local maxima of the α angle. Figure 4 shows the result of the extraction of all the circumnutations from the trajectory obtained from a *P. sativum* in 35 h of movement.

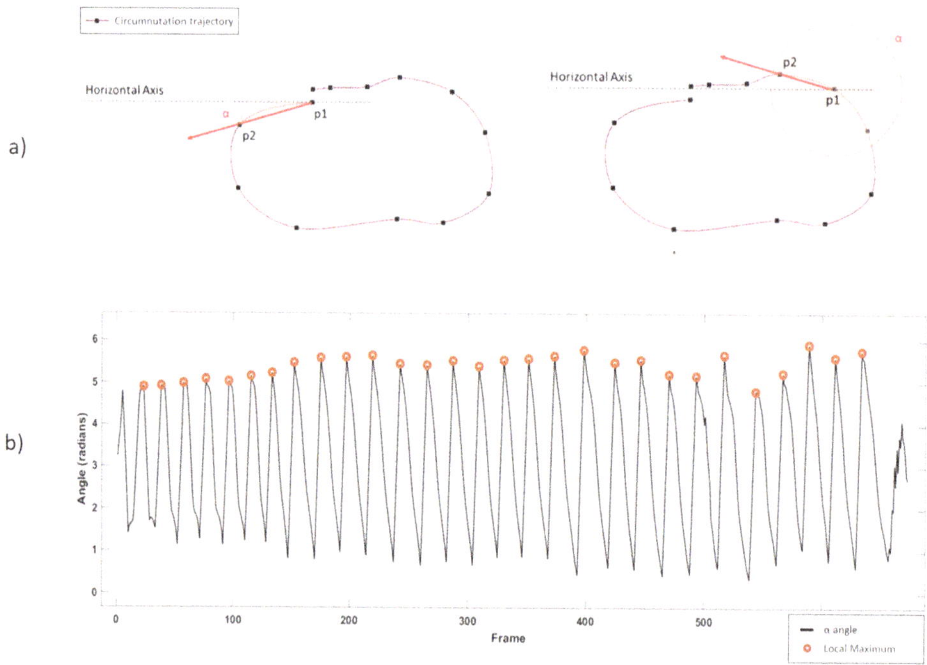

Figure 3. (**a**) α angle computed as the angle between the horizontal axis and the vector of movement between consecutive frames. The movement vector is identified as the vector between the point position in a specific frame (p1) and the point position in the next frame (p2). Movement vector is shown in red for two sample cases: between first and second frame (left) and between ninth and tenth frame (right). (**b**) α angle along the whole movement after a five-sample moving average application. Red dots represent local maxima, two consecutive local maxima encompass a single circumnutation.

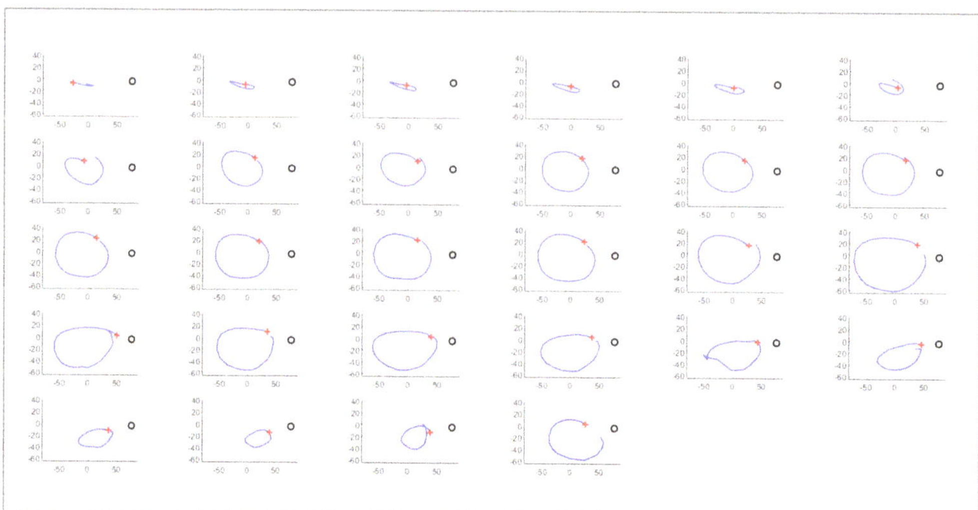

Figure 4. All circumnutations extracted from a complete trajectory of the circumnutating point shown in Figure 2 on the X–Z plane after 35 h of movement. In each box, the trajectory of the tracked point is shown for an identified circumnutation. A red cross represents the starting point of the circumnutation. A black circle represents the support. The reported measurement units of all axes are in millimeters.

After the circumnutations segmentation, a set of features aimed at describing the movement of plants is computed. Such features can be classified in four different functional categories: circumnutation related features, whole movement related features, point related features and experimental specific features. Along with the features provided by the literature [8,11,12,16], we enlarged the panorama of measures available with extra features representative of the movement of the plant. Below are the details on each computed feature.

2.5. Circumnutation Related Features

Circumnutation related features provide a quantitative description of each circumnutation extracted from the movement segmentation. Along with the metrics related to circumnutation's kinematics, the algorithm also computes indices related to circumnutation's shape and orientation. Such features are:

1. Duration: seconds needed to complete the whole circumnutation movement. It is computed as the time between two local maxima of the α angle shown in Figure 3a.
2. Mean/Max/Min Circumnutation Speed: values for mean value, maximum value and minimum value for the speed traveled by the circumnutating point along each circumnutation.
3. Circumnutation Path Length: the overall 3D path travelled by the circumnutating point. Computed as the sum of all the Euclidean distances between subsequent point positions during the single circumnutation:

$$Circumnutation\ Path\ Length = \sum_{t=1}^{T-1} \sqrt{(x(t) - x(t+1))^2 + (y(t) - y(t+1))^2 + (z(t) - z(t+1))^2} \quad (1)$$

where:

T: total number of points in the circumnutation; $x(t), y(t), z(t)$: point 3D coordinates at time t; $x(t+1), y(t+1), z(t+1)$: point 3D coordinates at time $t + 1$

4. Circumnutation Center: geometric center of gravity in the X–Z plane computed as the mean of each coordinate for all the points constituting the circumnutation.
5. Circumnutation Center Distance from Plant origin: euclidean distance between the Circumnutation Center and the plant origin in the X–Z plane.
6. Circumnutation Center Speed: speed of the Circumnutation Center computed as the distance traveled by the Circumnutation Center in two consecutive frames on the X–Z plane divided by the time between frames.
7. Circumnutation Centroid: considering the points of the trajectory described by the circumnutation on the X–Z plane, the algorithm identifies the region of interest as the closed line obtained by linking such points. The pixels in the polygon area are set to 1, while the others are set to 0. The Circumnutation Centroid is calculated as the geometric center of gravity in the X–Z plane computed as the mean of each coordinate for all the points with value 1 contained in the circumnutation area.
8. Circumnutation Centroid Distance from Plant origin: the Euclidean distance between the Circumnutation Centroid and the plant origin in the X–Z plane.
9. *Circumnutation Centroid Speed:* speed of the Circumnutation Centroid computed as the distance traveled by the Circumnutation Centroid in two consecutive frames on the X–Z plane divided by the time between frames.
10. Circumnutation Main Axis: the principal axis of the ellipsoid of the circumnutation, computed as the maximum distance between two points of the circumnutation trajectory in the X–Z plane.
11. Circumnutation Area: the sum of pixels with a value equal to 1 obtained from the binarization of the circumnutation trajectory on the X–Z plane as described for the calculation of Circumnutation Centroid.
12. Direction: clockwise or counterclockwise. For each circumnutation, the software computes the sum of all the angles between the movement vector at time t and the movement vector at time t+1. The direction, then, is determined according to

the following logic: if the resulting sum is equal to $2\pi \pm 1.2$, then the direction is counterclockwise, or else if the resulting sum is equal to $-2\pi \pm 1.2$, then the direction is clockwise. For all other cases, no direction is assigned. The pseudocode shown below describes the direction estimation logic.

if:

$$\sum_{movementsC} angleBetween(m(t), m(t+1)) = 2\pi \pm 1.2$$

then: *direction = "counter – clockwise"*
else if:

$$\sum_{movementsC} angleBetween(m(t), m(t+1)) = -(2\pi \pm 1.2)$$

then: *direction = "clockwise"*
else:*direction = "None"*

where:

movementsC: set of all movement vectors in a single circumnutation; $m(t)$: movement vector at time t; $m(t+1)$: movement vector at time t + 1.

2.6. Whole Movement Related Features

Whole movement related features provide descriptions of the plant's complete movement, from circumnutation onset to movement end. Such features are:

1. Max Circumnutation—Circumnutation Path Length: maximum value of the 3D Circumnutation Path Length for all the circumnutations. Example shown in Figure 5.

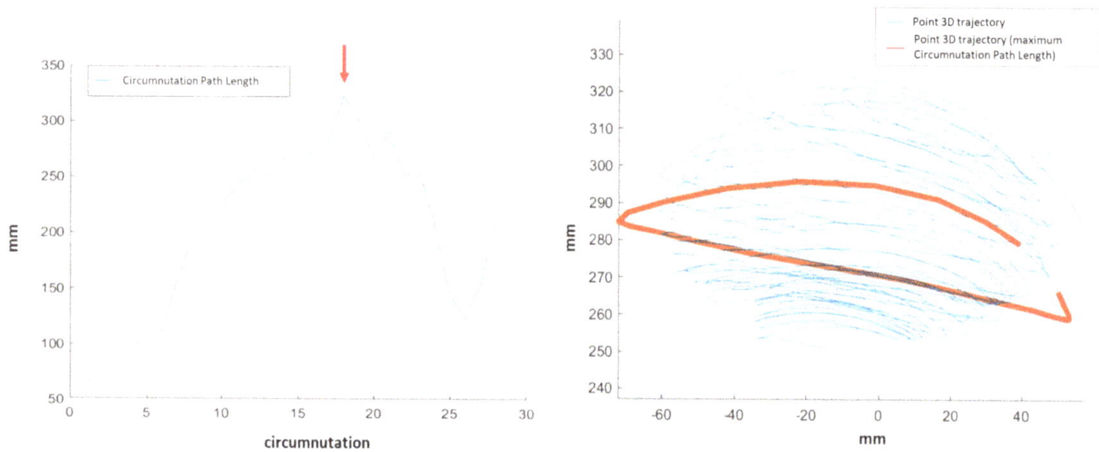

Figure 5. Left: value of circumnutation path length along the circumnutations. Each point represents the 3D path travelled by the circumnutating point during a circumnutation. The red arrow points at the maximum value. Right: circumnutation with maximum value of circumnutation path length displayed in red inside the whole movement trajectory.

2. Max Circumnutation–Circumnutation area: maximum value of Circumnutation Area in the X–Z plane for all the circumnutations.
3. Direction switches: sum of all direction switches in circumnutations along the whole movement. Example shown in Figure 6.

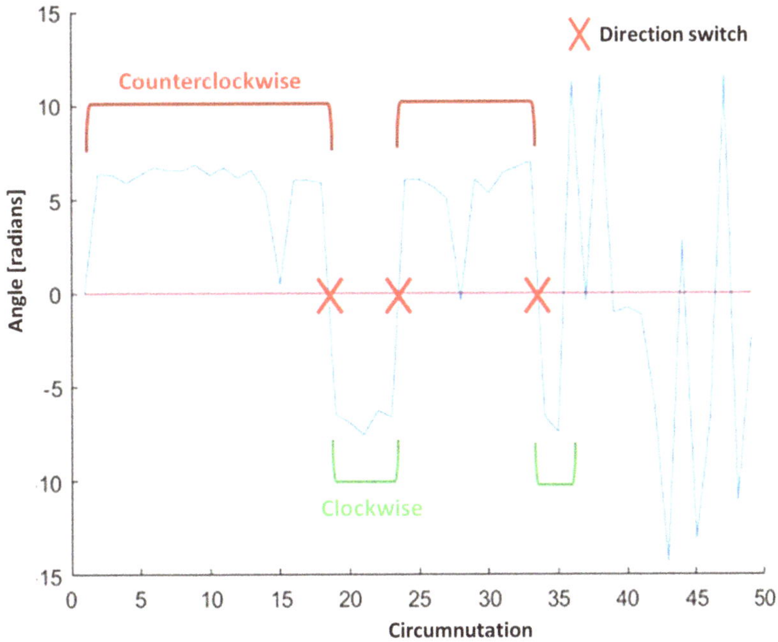

Figure 6. Example of trajectory with three direction switches. Counterclockwise circumnutations are highlighted with red brackets, whereas clockwise circumnutations are highlighted with green brackets. No brackets mean no main direction identified. Red crosses represent changes in the direction of the switches (from counterclockwise to clockwise and vice versa).

2.7. Point Related Features

Point related features are metrics related to circumnutating point kinematics along the movement. Such features are:

1. Mean and Maximum Speed: mean and maximum speed of the point reached along with the whole movement. Speed is computed as the distance traveled by the point between consecutive frames divided by the time between frames.
2. Time of maximum speed: the time at which maximum speed is reached both as absolute time and as a percentage with respect to the whole movement.

2.8. Experimental-Specific Features

The system was used intensively on the climbing pea plant (*P. sativum*) to quantify its circumnutating movement toward a support (e.g., a stimulus). To exploit aspects related to this experimental setup the software was evolved to compute extra experimental specific features.

In details:

1. Center/Centroid distance from stimulus: the Euclidean distance between Circumnutation Center/Centroid and the stimulus in the X–Z plane.
2. Angle between circumnutation main axis and plant stimulus axis: the angle between Circumnutation Main Axis and plant-stimulus axis computed as the line passing through the plant origin and the stimulus on the X–Z axis.
3. Minimum distance from stimulus: the 3D Euclidean distance between the stimulus and its closest point of the circumnutation trajectory.

3. Testing the Methodology

The complete system has been tested by different operators on a population of 49 snow peas (*P. sativum*) in eight individual growth chambers equipped with eight different camera couples without major issues reported.

A possible shortcoming of the system is the manual positioning of the points' positions, both in the first frame and when positioning corrections are required.

Indeed, this manual intervention could result in an operator-dependent effect on the data. In order to verify the independence of measures extracted from the operator, two different users have been asked to perform a complete tracking of a circumnutating point on the same timelapse of a growing chamber and processing the same circumnutating point in order to compare trajectories obtained and the impact of operator intervention on features extracted. The two 3D trajectories obtained have been compared to check that the procedure was stable across users. For each tracked frame, the distance between the point's position obtained by the two operators has been computed. The histogram of differences between the two trajectories is shown in Figure 7. The main differences observed are in the range between 0 and 5 mm with 90% of the points tracked showing less than 5 mm difference.

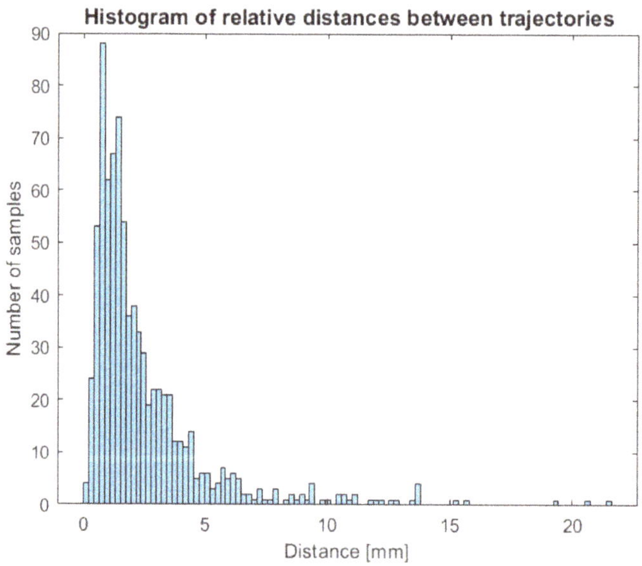

Figure 7. Histogram of relative distances between trajectories obtained from two different users showing that main differences between points are in the range between 0 and 5 mm.

The correlation coefficients were computed between the trajectories obtained by the two users. Values of the correlation coefficients are reported in Table 1.

Table 1. Correlation coefficients obtained for the three axes of the 3D trajectories obtained from different users.

Axis	Correlation Coefficient
x	0.999
y	0.998
z	0.993

The level of agreement between the two operators on circumnutation related features has been evaluated computing Intraclass Correlation Coefficient (ICC) between features extracted from the first user and the second user. The ICC has been computed using the absolute agreement of the two-way random effect model following [27]. Results obtained on 29 circumnutations are shown in Table 2.

Table 2. ICC values obtained comparing circumnutation related features over 29 circumnutations obtained from the same 3D trajectory extracted from two different users.

Feature	ICC
duration_circumnutation	0.995
x_center_coordinates_XZ	0.947
z_center_coordinates_XZ	0.841
x_centroid_coordinates_XZ	0.937
z_centroid_coordinates_XZ	0.784
center_distance_from_origin	0.55
centroid_distance_from_origin	0.72
length_major_axis	0.998
line_integral	0.999
area_circumnutation	0.999
speed_center_XZ	0.955
speed_centroid_XZ	0.981
min_speed_circumnutations	0.967
max_speed_circumnutations	0.989
mean_speed_circumnutations	0.999
center_distance_from_stimulus	0.836
centroid_distance_from_stimulus	0.853
angle_axis_stimulus	0.862
min_dist_stimulus	0.986

In addition to the computation of the ICC, a hypothesis test has been performed with the null hypothesis that ICC = 0, and all features considered scored a p-Value < 0.05.

From the ICC values obtained, all the features, except for center_distance_from_origin and centroid_distance_from_origin, were scored higher than 0.75, showing an excellent correlation. Centroid_distance_from_origin scored a 0.72 ICC value, showing good correlation, and center_distance_from_origin scored a 0.55 ICC value, showing fair correlation. In all cases, the p-Value obtained was <0.05, reporting the significance of values obtained. Overall, the analysis of the manual interventions impact revealed that the system results are stable across different users, indicating the reliability of the data obtained through the proposed methodology.

4. Discussion

Although the processes by which circumnutation occurs are well understood, the controlling mechanism underlying circumnutation is still unknown [8]. Informative content of circumnutation plays a key role in the study of the behavior of plants, hence the importance of having tools to ease and standardize the extraction and processing of circumnutation's key features in 3D space can be crucial. Highly variable characteristics between different plant species require a flexible method to allow wider applicability of a standard approach to different plant structures and moving behaviors. The system proposed in this article addresses this need and makes a step forward to define a standard approach for 3D movement extraction and analysis in plants. An omni-comprehensive approach to the problem has been proposed and implemented from the data acquisition protocol and instrumentation to the data extraction and the features processing software and methodologies. A set of descriptors for the movement of plants has been proposed, starting from the circumnutation concept, with the aim of providing a quantification of the motion behavior of the plants. Some of these features have been already proposed and validated in previous studies, and a step forward was taken in the approach proposed in this paper, by extending their applicability from 2D to 3D space. In addition to this set of features, new descriptors inspired by the circumnutating movement characteristics and the standard kinematic analysis, have been proposed and implemented. Even though the system allows for a 3D reconstruction in time of custom selected points, the descriptive features extracted so far mainly focuses on the kinematic description of the organ tip, that may give an incomplete picture of motion (as also suggested by Bastien et al., [18]). We feel that a precise 3D reconstruction of different key points as identified on the plant may provide a more precise tool for the validation of computational 3D models [18,19]. A future development of our system should focus on the relative movement of selected points, the first approximation of the overall 3D structure or on a 3D surface reconstruction of the organ. Recently, the kinematical approach for the study of movements in plants has been disputed. Raja et al. [15] criticize what they name "kinematics of nutation" approaches, supporting the thesis that kinematical patterns are blind to temporal dependencies versus what they define a "dynamical methodology" for the study of plants dynamic of nutation. Despite the term "dynamics", they do not consider a model of forces that drives the movement, as used in classical mechanics, but mostly focus on the temporal evolution of kinematic variables and the analysis of time series. They compute other kinematics-based descriptive features of behavioural dynamics such as harmonicity, predictability and complexity. Raja's criticism builds on the interpretation of kinematics approaches as methodologies that merely provide movement patterns through the averaging of features along plant's movement, supposing the absence of temporal dependencies. Overcoming the aforementioned interpretation, in this paper we have clarified that kinematic approaches also address the evolution of the movements of plants across time, building upon the computation and extraction of time series. Furthermore, some inherent limitations are present in the approach proposed by Raja et al. [15], where the information reduction through a single-point and single-coordinate analysis do not allow an evaluation of the behavior of different plant's segments and their interrelation for the movement production. Oppositely, we show that the analysis of standard kinematics measures in a multi-point evaluation can reveal specific behavioural patterns [2]. In that study, the focus was on the relative movements of multiple tendrils belonging to the same circumnutating organ in climbing plants, where zoomorphic references (e.g., wrist, digits) were adopted with the intent of ease the understanding of the experimental conditions and to make a direct comparison with a well-understood animal movement [28].

To elaborate, Charles Darwin and his son Francis [3] observed that the tendrils of climbing plants during circumnutation tend to assume the shape of whatever surface they come into contact with; that is, they learn progressively the shape of potential support characteristics [29]. Implicitly this signifies that they perceive the support and plan the movement accordingly. In this view climbing plants might represent actions in terms of their perceivable consequences. This is a strong inference because studying the interaction among organisms and objects is grounded on the requirement of a central nervous system (CNS). What the observations collected by the Darwins suggest is that other options are available, and that they do not require a CNS for adaptive perceptuomotor transformations to be happening. Tendrils can develop in different forms with some of them resembling a kind of digit characterized as changes in aperture as the bending towards a potential support progresses. This kind of reach-to-grasp behaviour resembles that exhibited by human and non-human primates [28] as well as by other animal species such as tetrapods [30–32], which have also evolved significant forelimb prehensile capabilities. In all cases during the course of a reach-to-grasp movement, there is first a progressive opening of the appendage, followed by a gradual closure of the appendage until it matches the to-be-grasped object [33]. Research on reach-to-grasp kinematics has proven insightful in revealing how specific kinematic landmarks modulate with respect to object properties, including, for instance, object thickness [28]. Results which hold across animal species are that the velocity of hand opening during reaching is lower and the maximum aperture of the hand is smaller for thinner than for thicker stimuli. In this respect, 'thickness' offers an ideal opportunity to parallel the ways of grasping in animal and climbing plants given that the success of grasping a support by a climber heavily depends on the support's diameter [34]. Differences in support thickness can determine changes in energy expenditure, which are visible on parameters characterizing the support-finding process [35]. With this in mind, we used our method to ascertain whether plants of *P. sativum* have the ability to perceive a potential support in the environment and modulate the kinematics of movement of the tendrils according to its thickness during the approach phase. The question is whether they are endowed with a purposeful behavior that is anticipatory in nature. We reasoned that if the principles of motion planning at the basis of animals' and plants' approach-to-grasp behavior are based on similar basic mechanisms, then the intrinsic properties of a support such as its thickness might have considerable effects on the kinematics of tendrils' aperture during the approach-to-grasp behavior. The results speak clearly. Not only did the plants acknowledged the presence of the support, but they exhibited a different kinematic patterning depending on stimulus thickness ([2]; see also [36,37]; see Supplementary video 3).

As shown in Figure 8, the average tendril's velocity was significantly greater for the thin than for the large support, and the maximum velocity was significantly higher for the thin than for the thick stimulus. The time at which the tendrils reached the maximum velocity, calculated as a percentage of total movement time, was significantly earlier for the thick than for the thin support. The maximum distance between the tendrils was scaled with respect to the size of the support. It was significantly greater for the thin than for the large support. In addition, the time at which the tendrils reached the maximum aperture calculated as a percentage of the total movement time occurred significantly earlier for the thick than for the thin support. Based on these findings, the plants appear to behave in ways that are adaptive, flexible, anticipatory and goal-directed similar to how animals do. During the movement towards the support the plants scale, the maximum aperture and the velocity of the tendrils' opening with respect to the thickness of the support increased. This evidence suggests that plants are able to process the properties of the support and are endowed with a form of perception underwriting a goal-directed and anticipatory behavior (Guerra et al., [2]). A caveat of these findings, however, is that our results indicate an opposite pattern of that reported in previous animal literature (e.g., Castiello & Dadda, [28]). Remember that the reach-to-grasp in animals demonstrates consistency across studies with regard to results such as a lower maximum peak velocity and an earlier and smaller maximum hand aperture for smaller stimuli relative to larger stimuli [28]. In general, this pattern has been explained in the terms that smaller stimuli require more accuracy and therefore lowering down the velocity allows for dealing with accuracy requirements.

A possible explanation for this discrepancy may reside in the fact that for plants reaching to grasp thick supports is a more energy consuming process than grasping thinner ones. Indeed, the grasping of a thick support implies that plants have to increase the tendrils' length in order to efficiently coil the support [38] and to strengthen the tensional forces to resist gravity [35]. Since these processes are characterized by a high Adenosine triphosphate (ATP) consumption, coiling thicker supports results in more energy expenditure [34]. Coherently, the reduction of movement velocity during the approaching maneuver may allow climbing plants to preserve energy for the coiling phase so as to reduce the risk of errors and assure a firm attachment to the support. In this sense, the accuracy trade-off of plants may be mediated by the consumption of energy.

The reduction of movement velocity may also serve to lengthen the time window within which tendrils establish contact points with the support. Previous literature has shown that climbing plants modify strategically contact points when twining around supports of different diameters [34]. Therefore, the extra time needed to reach a thicker support may allow climbing plants to correct tendrils trajectories and select more accurately contact points in order to twine more firmly the support. Indeed, when plants have no other choice but climbing a thick support, they could have the necessity to slow down the movement so as to accumulate more evidence about its the physical characteristics and implement corrective adjustments to reduce the scatter of tendrils end-position. Overall, these data support the hypothesis that despite plants and animals have two very unique evolutionary adaptations for multicellular life, each depending on unique kingdom-specific sets of cells, tissues and organs, they might have evolved signaling networks and mechanisms based on a common toolset from our unicellular common ancestor.

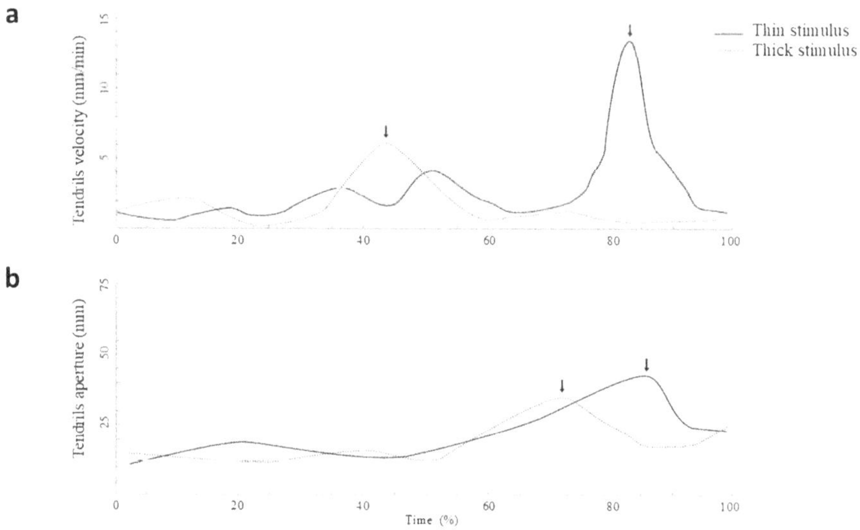

Figure 8. The tendrils' kinematics was scaled with respect to the size of the supports. Velocity (**a**) and tendrils aperture (**b**) profiles for the movements performed toward either the thick or the thin supports. Arrows indicate the occurrence of maximum peak velocity and maximum grip aperture depending on the thickness of the supports. Please note that when the support is thicker, the peak velocity is anticipated, and the maximum aperture of the tendrils is reached earlier for the thicker with respect to the thinner support. Reprinted from Guerra et al. [2].

5. Conclusions

In conclusion, here we describe a kinematical system solution for the investigation of the mechanisms underlying the movements of plants. This described approach has proved sensitive enough to provide key information regarding various aspects which drive circumnutation patterns. Furthermore, the system defines the framework for a systematic investigation on plant's nutation. The very fact that it allows to extract key parameters allowing a comparison between plants and animals in terms of movement planning and control makes it a valuable tool for comparative biology.

Supplementary Materials: The following are available online at https://www.mdpi.com/article/10.3390/ani11071854/s1, Supplementary video 1: A sample for the 2D trajectory extracted from one camera; Supplementary video 2: Example of the reconstructed 3D trajectory; Supplementary video 3: Different kinematic patterning depending on stimulus thickness.

Author Contributions: V.S., M.B. and U.C. conceived the reported methodology. W.B. helped with the implementation of the methodology. B.B., F.C., S.G. and Q.W. ran the experiments. V.S. and U.C. wrote the manuscript. A.P. and F.P. helped in the establishment of the growth conditions of the plants used in the experiments and with data interpretation. All authors critically revised the manuscript and contributed to the design of the experiments in which the methodology has been used.

Funding: This work was supported by a grant from MIUR to the Department of General Psychology, University of Padua (art.1, commi 314–337 legge 232/2016).

Institutional Review Board Statement: Not applicable.

Data Availability Statement: Data describing 3D trajectory obtained by 2 different users on the same plant video are available here: http://doi.org/10.5281/zenodo.5006765.

Acknowledgments: The present work was carried out within the scope of the research program Dipartimenti di Eccellenza.

Conflicts of Interest: The authors declare no conflict of interest.

References

1. Huey, R.B.; Carlson, M.; Crozier, L.G.; Frazier, M.R.; Hamilton, H.; Harley, C.; Hoang, A.; Kingsolver, J.G. Plants Versus Animals: Do They Deal with Stress in Different Ways? *Integr. Comp. Biol.* **2002**, *42*, 415–423. [CrossRef] [PubMed]
2. Guerra, S.; Peressotti, A.; Peressotti, F.; Bulgheroni, M.; Baccinelli, W.; D'Amico, E.; Gómez, A.; Massaccesi, S.; Ceccarini, F.; Castiello, U. Flexible control of movement in plants. *Sci. Rep.* **2019**, *9*, 16570. [CrossRef] [PubMed]
3. Darwin, C.; Darwin, F. *The Power of Movement in Plants*; Cambridge University Press: Cambridge, UK, 1880; ISBN 9780511693670.
4. Parise, A.G.; Reissig, G.N.; Basso, L.F.; Senko, L.G.S.; Oliveira, T.F.d.C.; de Toledo, G.R.A.; Ferreira, A.S.; Souza, G.M. Detection of Different Hosts From a Distance Alters the Behaviour and Bioelectrical Activity of Cuscuta racemosa. *Front. Plant Sci.* **2021**, *12*, 12. [CrossRef] [PubMed]
5. Hatakeda, Y.; Kamada, M.; Goto, N.; Fukaki, H.; Tasaka, M.; Suge, H.; Takahashi, H. Gravitropic response plays an important role in the nutational movements of the shoots of Pharbitis nil and Arabidopsis thaliana. *Physiol. Plant.* **2003**, *118*, 464–473. [CrossRef]
6. Kitazawa, D.; Miyazawa, Y.; Fujii, N.; Nitasaka, E.; Takahashi, H. Characterization of a novel gravitropic mutant of morning glory, weeping2. *Adv. Space Res.* **2008**, *42*, 1050–1059. [CrossRef]
7. Mugnai, S.; Azzarello, E.; Masi, E.; Pandolfi, C. Mancuso, Nutation in plants. In *Rhythms in Plants*; Mancuso, S., Shabala, S., Eds.; Springer: Berlin, Germany, 2007.
8. Charzewska, A.; Zawadzki, T. Circadian Modulation of Circumnutation Length, Period, and Shape in Helianthus annuus. *J. Plant Growth Regul.* **2006**, *25*, 324–331. [CrossRef]
9. Charzewska, A. The rhythms of circumnutation in higher plants. In *Floriculture, Ornamental and Plant Biotechnology*; Teixeira da Silva, J.A., Ed.; Global Science Books: London, UK, 2006; pp. 268–275.
10. Niinuma, K.; Someya, N.; Kimura, M.; Yamaguchi, I.; Hamamoto, H. Circadian Rhythm of Circumnutation in Inflorescence Stems of Arabidopsis. *Plant Cell Physiol.* **2005**, *46*, 1423–1427. [CrossRef] [PubMed]
11. Stolarz, M.; Krol, E.; Dziubinska, H.; Zawadzki, T. Complex relationship between growth and circumnutations in Helianthus annuus stem. *Plant Signal. Behav.* **2008**, *3*, 376–380. [CrossRef] [PubMed]
12. Buda, A.; Zawadzki, T.; Krupa, M.; Stolarz, M.; Okulski, W. Daily and infradian rhythms of circumnutation intensity in Helianthus annuus. *Physiol. Plant.* **2003**, *119*, 582–589. [CrossRef]
13. Johnsson, A.; Solheim, B.G.B.; Iversen, T. Gravity amplifies and microgravity decreases circumnutations in Arabidopsis thaliana stems: Results from a space experiment. *New Phytol.* **2009**, *182*, 621–629. [CrossRef]
14. Kosuge, K.; Iida, S.; Katou, K.; Mimura, T. Circumnutation on the water surface: Female flowers of Vallisneria. *Sci. Rep.* **2013**, *3*, 1–7. [CrossRef] [PubMed]
15. Raja, V.; Silva, P.L.; Holghoomi, R.; Calvo, P. The dynamics of plant nutation. *Sci. Rep.* **2020**, *10*, 1–13. [CrossRef]
16. Stolarz, M.; Żuk, M.; Król, E.; Dziubińska, H. Circumnutation Tracker: Novel software for investigation of circumnutation. *Plant Methods* **2014**, *10*, 24. [CrossRef]
17. Schuster, J.; Engelmann, W. Circumnutations of Arabidopsis thaliana Seedlings. *Biol. Rhythm. Res.* **1997**, *28*, 422–440. [CrossRef]
18. Bastien, R.; Meroz, Y. The Kinematics of Plant Nutation Reveals a Simple Relation between Curvature and the Orientation of Differential Growth. *PLoS Comput. Biol.* **2016**, *12*, e1005238. [CrossRef]
19. Porat, A.; Tedone, F.; Palladino, M.; Marcati, P.; Meroz, Y. A General 3D Model for Growth Dynamics of Sensory-Growth Systems: From Plants to Robotics. *Front. Robot. AI* **2020**, *7*, 89. [CrossRef]
20. Gallentine, J.; Wooten, M.B.; Thielen, M.; Walker, I.D.; Speck, T.; Niklas, K. Searching and Intertwining: Climbing Plants and GrowBots. *Front. Robot. AI* **2020**, *7*, 118. [CrossRef]
21. Heikkila, J.; Silven, O. A four-step camera calibration procedure with implicit image correction. In Proceedings of the IEEE Computer Society Conference on Computer Vision and Pattern Recognition, San Juan, PR, USA, 17–19 June 1997; pp. 1106–1112.
22. Scaramuzza, D.; Martinelli, A.; Siegwart, R. A Toolbox for Easily Calibrating Omnidirectional Cameras. In Proceedings of the IEEE/RSJ International Conference on Intelligent Robots and Systems, Benjing, China, 9–15 October 2006; pp. 5695–5701. [CrossRef]
23. Urban, S.; Leitloff, J.; Hinz, S. Improved wide-angle, fisheye and omnidirectional camera calibration. *ISPRS J. Photogramm. Remote. Sens.* **2015**, *108*, 72–79. [CrossRef]
24. Zhang, Z. A flexible new technique for camera calibration. *IEEE Trans. Pattern Anal. Mach. Intell.* **2000**, *22*, 1330–1334. [CrossRef]
25. Lucas, B.D.; Kanade, T. An iterative image registration technique with an application to stereo vision. In Proceedings of the 7th Int. Joint Conf. on Artificial Intelligence (IJCAI), Vancouver, BC, USA, 24–28 April 1981; pp. 674–679. Available online: https://www.ri.cmu.edu/pub_files/pub3/lucas_bruce_d_1981_2/lucas_bruce_d_1981_2.pdf (accessed on 21 June 2021).
26. Hartley, R.; Zisserman, A.; Faugeras, O. *Multiple View Geometry in Computer Vision*; Cambridge University Press (CUP): Cambridge, UK, 2004.
27. Koo, T.K.; Li, M.Y. A Guideline of Selecting and Reporting Intraclass Correlation Coefficients for Reliability Research. *J. Chiropr. Med.* **2016**, *15*, 155–163. [CrossRef]

28. Castiello, U.; Dadda, M. A review and consideration on the kinematics of reach-to-grasp movements in macaque monkeys. *J. Neurophysiol.* **2019**, *121*, 188–204. [CrossRef] [PubMed]
29. Trewavas, A. The foundations of plant intelligence. *Interface Focus* **2017**, *7*, 20160098. [CrossRef]
30. Sustaita, D.; Pouydebat, E.; Manzano, A.; Abdala, V.; Hertel, F.; Herrel, A. Getting a grip on tetrapod grasping: Form, function, and evolution. *Biol. Rev.* **2013**, *88*, 380–405. [CrossRef]
31. Klein, B.G.; Deich, J.D.; Zeigler, H. Grasping in the pigeon (*Columba livia*). *Behav. Brain Res.* **1985**, *18*, 201–213. [CrossRef]
32. Whishaw, I.Q. An endpoint, descriptive, and kinematic comparison of skilled reaching in mice (Mus musculus) with rats (Rattus norvegicus). *Behav. Brain Res.* **1996**, *78*, 101–111. [CrossRef]
33. Jeannerod, M. The Timing of Natural Prehension Movements. *J. Mot. Behav.* **1984**, *16*, 235–254. [CrossRef]
34. Putz, F.E.; Holbrook, N.M. Biomechanical studies of vines. In *The Biology of Vines*; Cambridge University Press (CUP): Cambridge, UK, 1992; pp. 73–98.
35. Gianoli, E. The behavioural ecology of climbing plants. *AoB PLANTS* **2015**, *7*, plv013. [CrossRef]
36. Ceccarini, F.; Guerra, S.; Peressotti, A.; Peressotti, F.; Bulgheroni, M.; Baccinelli, W.; Bonato, B.; Castiello, U. Speed–accuracy trade-off in plants. *Psychon. Bull. Rev.* **2020**, *27*, 966–973. [CrossRef]
37. Ceccarini, F.; Guerra, S.; Peressotti, A.; Peressotti, F.; Bulgheroni, M.; Baccinelli, W.; Bonato, B.; Castiello, U. On-line control of movement in plants. *Biochem. Biophys. Res. Commun.* **2020**. [CrossRef]
38. Rowe, N.; Isnard, S.; Gallenmüller, F.; Speck, T. Diversity of Mechanical Architectures in Climbing Plants: An Ecological Perspective. In *Ecology and Biomechanics: A Mechanical Approach to the Ecology of Animals and Plants*; CRC Press: Boca Raton, FL, USA, 2006; pp. 36–59.

animals

MDPI

Article

Exploring the Cognitive Capacities of Japanese Macaques in a Cooperation Game

Ryan Sigmundson [1,†], Mathieu S. Stribos [2,†], Roy Hammer [2], Julia Herzele [3], Lena S. Pflüger [3,4] and Jorg J. M. Massen [2,3,*]

1 Department of Philosophy, University of Vienna, 1010 Vienna, Austria; ryansigmundson@gmail.com
2 Animal Behaviour and Cognition, Department of Biology, Utrecht University, 3584 CH Utrecht, The Netherlands; m.s.stribos@gmail.com (M.S.S.); roy.hammer@live.nl (R.H.)
3 Austrian Research Center for Primatology, 9570 Ossiach, Austria; herzele.julia@gmail.com (J.H.); lena.pflueger@univie.ac.at (L.S.P.)
4 Department of Behavioral and Cognitive Biology, University of Vienna, 1090 Vienna, Austria
* Correspondence: jorgmassen@gmail.com
† These authors contributed equally.

Simple Summary: Experiments using animal models are often conducted to explore the cognitive capacities of different species and to shed light upon the evolution of behavior and the mind that shapes it. Investigating the cognitions and motivations involved in cooperation is one such area that has attracted attention in recent years. As experiments examining these abilities in natural settings are underrepresented in the literature, our study was conducted in a setting closely resembling the natural environment of the study species so as to retain the social factors that help shape these behaviors. In our experiments, Japanese macaques needed to work together to simultaneously pull two loops in order to release food rewards onto a central platform. Over the course of the experiment, the macaques in our study came to make fewer attempts at the cooperative task when no potential partner was present. Furthermore, following an unequal division of the rewards, macaques receiving lesser rewards were more likely to express stress-related and aggressive behavior. Together, these results suggest that the Japanese macaques in our study understood the importance of having a partner in the cooperative task, paid attention to the relative value of the reward they received from the task and became distressed if their reward was inferior to that of another.

check for updates

Citation: Sigmundson, R.; Stribos, M.S.; Hammer, R.; Herzele, J.; Pflüger, L.S.; Massen, J.J.M. Exploring the Cognitive Capacities of Japanese Macaques in a Cooperation Game. *Animals* **2021**, *11*, 1497. https://doi.org/10.3390/ani11061497

Academic Editors: Lucia Regolin and Maria Loconsole

Received: 15 April 2021
Accepted: 20 May 2021
Published: 21 May 2021

Abstract: Cooperation occurs amongst individuals embedded in a social environment. Consequently, cooperative interactions involve a variety of persistent social influences such as the dynamics of partner choice and reward division. To test for the effects of such dynamics, we conducted cooperation experiments in a captive population of Japanese macaques (*Macaca fuscata*, N = 164) using a modified version of the loose-string paradigm in an open-experiment design. We show that in addition to becoming more proficient cooperators over the course of the experiments, some of the macaques showed sensitivity to the presence of potential partners and adjusted their behavior accordingly. Furthermore, following an unequal reward division, individuals receiving a lesser reward were more likely to display aggressive and stress-related behaviors. Our experiments demonstrate that Japanese macaques have some understanding of the contingencies involved in cooperation as well as a sensitivity to the subsequent reward division suggestive of an aversion to inequity.

Keywords: inequity aversion; ecological validity; field experiments; loose-string paradigm; partner choice; animal cognition

1. Introduction

Humans may stand out when it comes to intensive cooperation [1], but we are far from alone in possessing these abilities. Examples of cooperation exist throughout the animal kingdom in behavioral domains as diverse as predator defense [2], group hunting [3], cooperative breeding [4] and coalition formation [5]. The study of animal cooperation has flourished in recent years (for a recent review, see [6]), but many open questions remain regarding the cognitive mechanisms that promote and maintain it. Discussion of the subject has been impeded by the numerous definitions of cooperation existing across the literature, so to avoid confusion, throughout this paper we will use the definition, "all interactions or series of interactions that, as a rule (or 'on average'), result in a net gain for all participants" [7].

The experimental study of animal cooperation has been beset with uncertainty when it comes to determining what level of task understanding and intentionality is actually possessed by the animals engaging in it [8]. Many cooperative tasks could be completed through sheer coincidence by two animals acting independently, i.e., acting apart together (cf. [7]). Coincidental cooperation aside, successful cooperation could itself occur through merely responding to the presence of a conspecific, through attending to their actions or through a full-on understanding of their intentions. Different cognitions underlie these forms of cooperation, and it is for this reason that extreme care must be taken in designing experiments to differentiate between them.

The setup most frequently used to investigate cooperation in animals was originally developed by Crawford [9] and later modified into the loose-string paradigm by Hirata [10,11]. In this paradigm, two animals must coordinate their actions to pull two ends of a string simultaneously in order to move food rewards close enough to be accessed. If they mistime their attempt or make an attempt in the absence of a partner, the string comes loose and precludes further attempts. The paradigm has proved effective for testing a range of species such as chimpanzees (*Pan troglodytes*) [12], macaques (*Macaca sylvanus; Macaca fuscata; Macaca fascicularis*) [13–15], marmosets (*Callithrix jacchus*) [16], corvids (*Corvus frugilegus; Corvus corax*) [17,18], elephants (*Elephas maximus*) [19], wolves (*Canis lupus*) [20], hyenas (*Crocuta crocuta*) [21], otters (*Pteronura brasiliensis; Aonyx cinerea*) [22], parrots (*Nestor notabilis; Ara glaucogularis; Psittacus erithacus; Eupsittula aurea*) [23–27] and dolphins (*Tursiops truncatus*) [28], to name but a few.

Even though the design of the loose-string paradigm makes uncoordinated attempts conspicuous, it is still possible for animals to succeed at the task by virtue of arriving to it simultaneously. To counter this, some studies have taken the design a step further and introduced a delay before releasing a prospective cooperative partner to ensure that active coordination is necessary to complete the task [17–22,24–29]. Success then necessitates waiting behavior on the part of the initially arriving animal, which at minimum indicates understanding that a partner is required to complete the task.

In addition to task-understanding, another, and more general, pitfall of cooperation is that it is open to exploitation by so-called "free riders"—those who would reap the benefits of cooperative interactions while contributing little effort themselves [30]. For cooperation to have evolved in a species, it needs to consistently provide payoffs for those engaging in it. This requires some form of behavioral strategy to avoid interacting with individuals who reliably exploit others. The aversive reaction to inequity in reward distribution, termed inequity aversion (IA), is theorized to play just such a role and may have evolved alongside cooperation [31,32]. According to this theory, individuals compare their own efforts and benefits to that of their partner and judge whether the outcome is equitable or "fair". A negative assessment may be expressed through protest, agonistic behavior, unwillingness to continue cooperating or partner switching. These IA-mediated reactions may shield cooperation from free-riders, in which case the mechanism may act as a stabilizer in cooperative interactions [31,32]. IA can be divided into disadvantageous IA, in which the actor is on the detrimental side of the distribution, and advantageous IA, in which the actor is on the beneficial side [32]. As disadvantageous IA is much more common amongst animals [32,33], we use the term IA to refer specifically to this form.

The tendency to cooperate with individuals who are neither kin nor mate has emerged as a strong predictor of the existence of IA in a species [31]. Japanese macaques (*Macaca fuscata*) are known to form coalitions with non-kin for the purpose of social support during conflicts [34], making them excellent subjects to investigate this in. Even though the species does not habitually cooperate in the foraging domain, a study of the closely related long-tailed macaques (*Macaca fascicularis*) demonstrated that IA is not domain specific and thus can be generalized from the social to the foraging domain [35]. It is therefore likely that IA will be present in Japanese macaques and detectable in a food-rewarded task. Moreover, previous work established the capacity to succeed at an experimental cooperative task in Japanese macaques [14]. The same study, however, also found considerable variation in the success rates of different populations at the task that could be linked to differences in social tolerance between those two wild populations [14].

We tested cooperation in a population of Japanese macaques kept under conditions resembling those typical of their natural environment [36] in an enclosure large enough to justify use of the term semi-free (cf. [37–41]), which is how we will refer to their conditions from this point onwards. Our first experiment aimed to demonstrate that this particular population is capable of cooperation, then to further examine this capacity by establishing whether individuals understood that having a partner was necessary for their own success. To accomplish this, we employed a cooperation task that required two individuals to coordinate their pulling of opposing loops in order to release food rewards to a central platform. We predicted that individuals would increasingly come to take into account the presence of a partner when making attempts as the experiment progressed and they had opportunity to learn the contingencies of the task. This would be reflected behaviorally by a change in the proportion of attempts to pull made in the presence of a potential partner as well as by an increase in the frequency of waiting behavior, i.e., not pulling until a potential partner has arrived.

Our second experiment used the same setup, but with the introduction of an unequal reward condition to investigate the presence of IA, or at least an "emotional" response that may be a precursor of it [6], in the species. Here we predicted that individuals would display a higher rate of stress-related and aggressive behaviors when they had received a comparatively smaller reward than their cooperative partner.

A strategy of reacting to cumulative outcomes over a period of time rather than each instance of disadvantageous outcome leads to greater tolerance of accidental and unintentional unequal outcomes, thereby protecting potentially valuable cooperative bonds. Only if an individual consistently experiences disadvantageous outcomes over multiple collaborations should it act against it [31]. For this reason, we further predicted that individuals would show more frequent stress-related and aggressive behaviors when confronted with an accumulation of disadvantageous outcomes rather than only one.

2. Materials and Methods

2.1. Study Subjects and Housing

Our study was conducted at the Affenberg Landskron (Affenberg Zoobetriebsgesellschaft mbH), in Carinthia, Austria, where a population of Japanese macaques (*Macaca fuscata*) resides under semi-free conditions in a ±40,000 m^2 enclosure located in a coniferous forest. We classify this population as semi-free because they live in a large enclosure with conditions resembling their natural environment, living space and group size [36], while having the opportunity to naturally form social groups. The population originates from Minoo City, Japan, and was introduced to the area in 1996. At the beginning of data collection for these studies, the enclosure contained 164 individuals living in a single group. The population consisted of 79 adult females (>3.5 years of age), 51 adult males (>4.5 years of age), 24 juveniles and 10 infants. Experiment 1 was conducted from December 2018 to April 2019, and Experiment 2 was conducted from January 2020 to May 2020.

The macaques receive food twice daily with the first feeding occurring between 9:00am and 11:00am. Provisioned food consists of various fruits and vegetables, as well as wheat. The natural vegetation of the area provides the monkeys with additional foraging opportunities. Water is available ad libitum at any time from a natural stream that goes through the enclosure. Testing occurred between regular feeding sessions and did not involve subjecting the macaques to deprivation states.

2.2. Ethical Note

The experiments took place in a wooden hut within the enclosure. The wooden hut and experimental apparatus had been used in a previous experiment [42], so the animals had already had an opportunity to habituate to its presence. Open doorframes were located at each corner of the hut to allow individuals to enter and leave the area at will. The macaques were neither actively selected, nor separated from the group. Data were collected on an opportunistic basis, with active participants and group size in the experimental area varying considerably across sessions. All participation of the macaques in our experiments was thus uncoerced.

Since all our experiments were non-invasive, the study complied with the Austrian Law (§ 2. Federal Law Gazette number 501/1989) and the Code for Best Practices in Field Primatology and received oversight from and was authorized by the internal board of the Austrian Research Center for Primatology. No invasive research or experimental procedures requiring ethics approval according to the European Directive 2010/63 were performed. Our studies adhered to the American Society of Primatologists' principles for the ethical treatment of primates, and all applicable international, national and institutional guidelines for the care and use of animals were followed.

2.3. Apparatus

The apparatus used in our study (both Experiment 1 and 2) was a modified version of the string-pulling paradigm pioneered by Crawford [9] and later refined into the loose-string paradigm by Hirata [10,11]. The body of the apparatus consisted of a wooden platform with a clear lockbox mounted on top, in which food could be placed (Figure 1). A copper pipe that housed a wire ending in a loop extended from two opposite sides of the lockbox to an area in front of the apparatus. Each loop could be pulled to release one of the two pins within the lockbox. The floor of the lockbox contained a trapdoor mechanism that would drop open when both pins holding it in place were released. The experimenter was able to place food in the lockbox through a door in the top, but the only way for the monkeys to access the food was to spring the trapdoor by pulling both loops simultaneously. Successful release of the trapdoor resulted in the enclosed food being dropped onto the wooden platform below. The platform was positioned approximately 1 m away from the loops and could be approached from all directions, allowing for better observation of the impact of social dynamics on reward division. During both experimental phases, the loops were positioned two meters apart so that no single individual could reach both loops simultaneously (Figure 1).

2.4. Individual Training

Training consisted of two phases and was carried out from October to November 2018. Throughout the training period, individuals could succeed in releasing the food without a partner (see below). This allowed them to form an association between the apparatus and food acquisition while giving them the opportunity to learn how to operate it individually. For an individual to be considered trained, they needed to successfully complete at least three trials split between two separate days. Sessions were conducted up to three times daily and consisted of 10 trials. At the beginning of each trial, the experimenter locked two pieces of food within the apparatus before moving away to allow the monkeys clear access to the apparatus. Successful trial completion entailed pulling the active loop(s) to trigger the release of food. Once the food rewards were retrieved, the setup procedure was repeated until completion of the 10th trial marked the end of the session. During the first phase of training, only one loop was required to trigger the trapdoor and release the food ("single-sided pull" form), i.e., only one of the two pins holding the trapdoor was in place. The side actively required to operate the device was switched between the right and left side over the course of this phase to discourage perseveration of operation on only one side. This phase of training continued for 42 sessions.

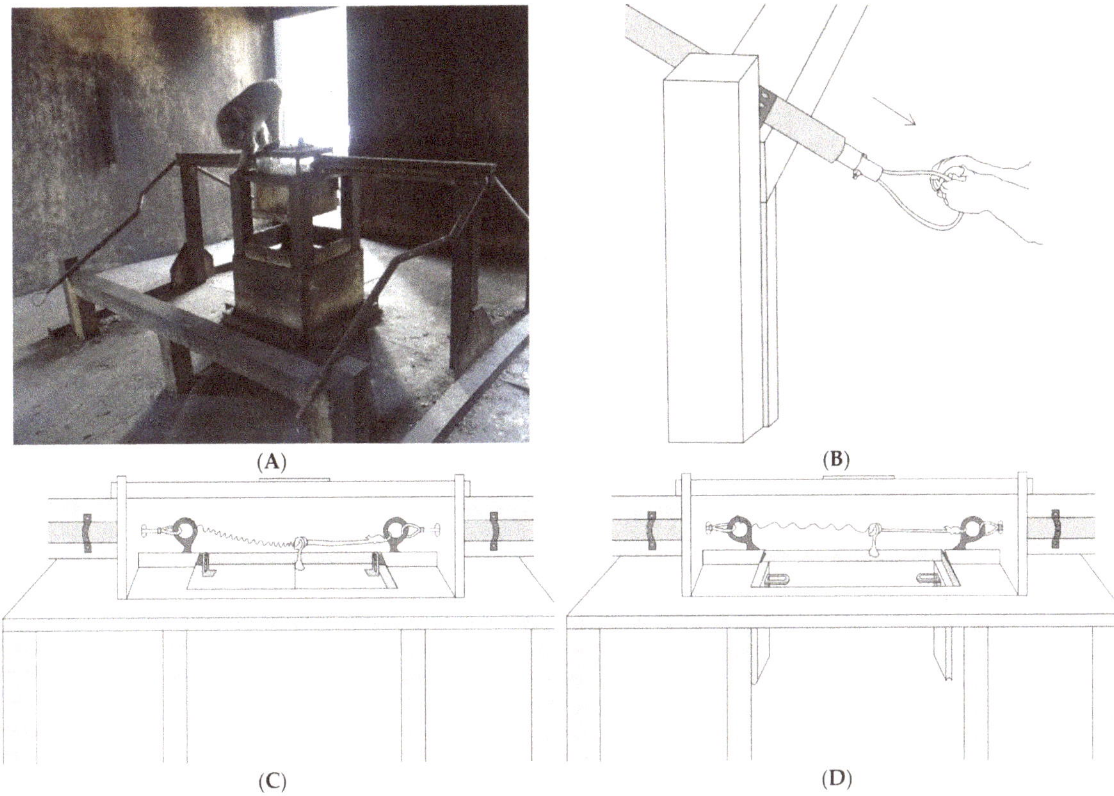

Figure 1. (**A**) Photo of the apparatus used in the study. Food rewards remain visible but inaccessible in the plexiglass lockbox until simultaneous pulling of the loops positioned at the end of the copper pipes triggers the opening of the trapdoor, allowing the rewards to fall onto the platform below. (**B**) Close-up illustration of one of the loops. (**C**) Close-up illustration of the lockbox in the closed position. A latch mechanism holds the trapdoor in place until both loops are pulled simultaneously, releasing the latch mechanism and allowing the trapdoor to fall open. (**D**) Close-up illustration of the lockbox in the open position. The trapdoor has fallen open after release of the latch mechanism.

During the second phase, which ran for 36 sessions, simultaneous pulling of both loops was required but the loops were positioned close enough together so that a single monkey could reach both loops at once (20 cm apart). The "simultaneous pull" form of the device was used only with monkeys who had met training criteria in the previous phase. If potential new pullers began manipulating the loops, the apparatus was switched back to the "single-sided pull" form of the previous phase. The "single-sided pull" form of the device was occasionally used with experienced monkeys as well as on an as-needed basis to maintain a high level of motivation.

By the end of the training period, 11 monkeys had reached training criteria on the "single-sided pull" form of the device, but no monkey had met the training criteria on the "simultaneous pull" form. Due to time-constraints, we decided to nonetheless proceed with Experiment 1. Note that Experiment 1 specifically tested for the monkey's understanding of the cooperative task and allowed further learning of the contingencies of the set-up (see below). We ran an additional training phase consisting of 51 sessions in February 2019 to address a motivational issue that developed during the initial testing sessions.

Two additional monkeys reached "single-sided pull" training criteria during this second training phase, bringing the total number of trained monkeys to 13. A further two learnt to use the apparatus during the testing phase of Experiment 1 without having participated in the training sessions.

The first 10 sessions of Experiment 1 were treated as a transitionary period for the dyadic form of the apparatus, with the experimenter acting as cooperative partner for any persistently interested individuals when no other potential partner was present. This was done to give individuals an opportunity to learn the necessity of a cooperative partner in the early stages of the setup and decrease the likelihood that they would lose interest in the apparatus before acquiring an effective cooperative partner. All of these sessions were, however, excluded from later analyses.

3. Experiment 1

The purpose of Experiment 1 was to investigate whether participants were capable of understanding the task outcomes associated with the cooperation paradigm. Special attention was given to demonstrating whether they understood the necessity of a partner. We ran a total of 126 testing sessions between December 2018 and April 2019. During these sessions, the loop handles were positioned two meters apart so that no single monkey could reach both loops simultaneously.

3.1. Procedure

Sessions consisted of 10 trials and were conducted up to three times daily, the first of which was conducted at least one hour after the morning feeding. In each trial, two pieces of food were locked into the apparatus to serve as rewards. For this purpose, we variously used apples, bananas, tomatoes and pineapples, all of which were known to be preferred foods for the monkeys [37]. The combination and presentation order of rewards was randomized. After loading the device, the experimenter moved to the edge of the research hut or left it entirely so as not to obstruct the monkeys' access to the device or affect their behavior. As the monkeys were well habituated to the presence of humans, we anticipated the presence of the experimenter would have very little impact on their behavior during the experiment. If the food items were successfully retrieved from the apparatus, the experimenter returned to repeat the setup procedure until the session had ended. If a single dyad was monopolizing the device, defined as succeeding in six successive trials, the researcher would leave the experimental area for a short period of time (5–10 min) to encourage those individuals to leave and provide a new dyad with an opportunity to use the apparatus.

A trial was terminated if no monkey was successful in operating the device within 15 min unless there were monkeys actively interested in the device at that time in which case the trial continued until they succeeded or lost interest in the apparatus. If there was only one monkey actively interested in the device at the designated termination time, the experimenter attempted to act as a cooperative partner to give the individual an opportunity to gain experience with the device and to ensure that the behavior of persistent individuals was occasionally reinforced (see also section Training above). If a session continued for 1.5 h without reaching 10 trials, the session was terminated.

3.2. Measures: Collection and Coding

All sessions were fully recorded using cameras (Sony Handycam HDR-CX130E) mounted in protective boxes in two opposite corners of the research hut. The cameras captured two different frontal angles of the experiment, ensuring that the entire apparatus was within view. Trials were coded for identity of the pulling individuals, number of pulls (for definition see below), number and identity of the individuals in the surrounding area, whether a potential partner was positioned in front of the opposite loop at the time of the attempt, and whether the pulling individual waited for a partner to arrive before pulling. An attempt was defined as "any manipulation of the loop using enough force to result in movement of the pin to the open position". Multiple pulls occurring within a 5-s span were aggregated and counted as a single attempt. A "potential partner" was defined as any individual who had previously succeeded in cooperation or met the first training criteria. An attempt was classified as a "wait" if it was preceded by the eventual puller delaying their pulling behavior for at least 15 s upon arriving to the area in front of the apparatus until such a time that a potential partner entered the area. All trials were coded live and afterwards from the video recordings. A second rater independently recoded 15% of the videos, and inter-rater agreement was excellent for participant identity (Cohen's kappa = 0.98), potential partner presence (Cohen's kappa = 0.97) and attempt (Cohen's kappa = 0.91), and good for waiting (Cohen's kappa = 0.71).

3.3. *Analyses*

To evaluate whether the Japanese macaques understood the specifics of the task, we ran three separate binomial generalized linear mixed models (GLMMs) with a logit link function on (a) whether individuals became increasingly proficient at cooperating, as reflected by the likelihood of an attempt to be successful, (b) whether individuals became more sensitive to the presence of a partner at the opposite loop, as measured by the likelihood of making an attempt with a potential partner present, and (c) whether waiting behavior increased in likelihood over time. Session number was entered as a fixed effect and number of attempts per individual was added as a random effect to control for repeated measurements. To correct for the alternative explanation that making an attempt with a partner present became more likely over the course of the experiment because the research hut increased in popularity, the number of individuals in the area was added to model b) as a fixed effect. We compared all full models to null models containing only the random effects. All statistical analyses were carried out using R statistical software (version 3.6.1) [43] with α set at 0.05. GLMMs were run using the lme4 package [44].

3.4. *Results*

Over the course of the experiment, we conducted 1165 trials spread across 126 sessions. Of these trials, 708 were successful, resulting in an overall success rate of 60.8% (708/1165). Twelve monkeys succeeded in cooperating in at least one instance (see Table 1). Four individuals who succeeded at least once in the training setup never engaged in cooperation and were excluded from further analyses. Of the cooperating individuals, only two were male. The age of participants ranged from 2 to 11, with an average of 6.1 (SD = 2.8). Participants had an average of 118 successes (SD = 183.7) and an average of 3.3 partners (SD = 2.8). Overall, these individuals combined into 20 unique cooperative dyads (Supplementary Table S1). Three of those dyads were composed of kin, and those three dyads accounted for 64.5% of all successful co-operations. Cooperative dyads had a mean of 35.4 successes (SD = 67.2).

Table 1. Summary of results by individual.

Individual	Age *	Sex	Partners	Total Successes
James	2	Male	11	635
Ingrid	10	Female	3	230
Pippi	2	Female	6	203
Kate	7	Female	3	177
Finn	5	Male	4	78
Alina	11	Female	2	58
Herta	6	Female	3	14
Montana	8	Female	1	10
Krato	5	Female	2	5
Sandra	7	Female	3	3
Julia	6	Female	1	2
Kurtney	4	Female	1	1

*: Age in years as of the beginning of the experiment.

Of the 126 sessions that were run, 10 sessions comprising the transitionary phase from the training to the experimental phase were excluded from all analyses as were an additional 12 sessions where problems with the videos made coding impossible. The co-operations that occurred during these sessions are still included in the descriptive summary tables.

Over the course of the experiment, individuals became more proficient at operating the apparatus with the likelihood of successful attempts increasing as a function of session (GLMM: Estimate = 0.398, z = 10.3, p = <0.001; Figure 2). This model explained the data significantly better than the respective null model ($\chi^2(1)$ = 107.07, p = < 0.001).

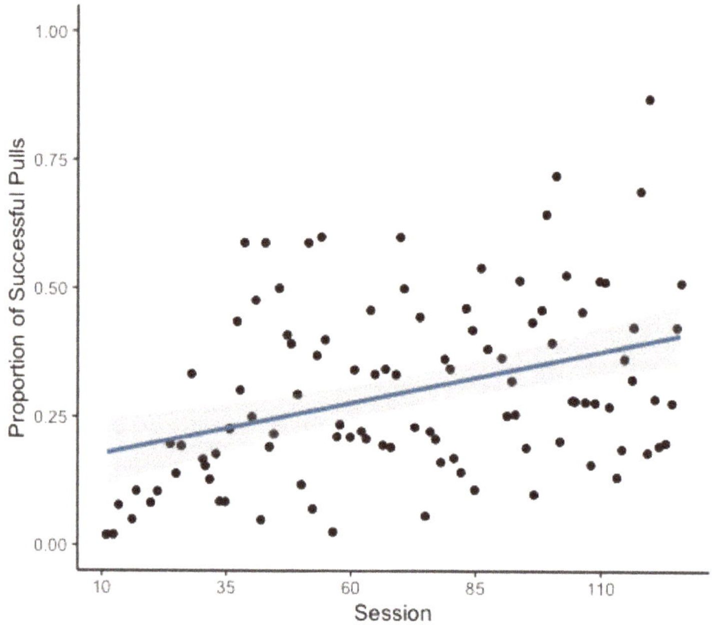

Figure 2. Scatterplot of proportion of successful pulls by session with line of best fit overlaid. Greyed area indicates standard error.

The second model revealed a significant decrease in the likelihood of attempts being made in the absence of a potential partner over the course of the experiment (GLMM: Estimate = 0.155, z = 3.666, p = <0.001; Figure 3). This model explained the data significantly better than the respective null model ($\chi^2(1)$ = 13.538, p = <0.001).

The final model revealed that the likelihood of waiting behavior did increase over the course of the experiment as a function of session (GLMM: Estimate = 0.378, z = 4.0, p = <0.001; Figure 4). This model explained the data significantly better than the respective null model ($\chi^2(1)$ = 16.205, p = <0.001). Examination of total instances of waiting on the individual level revealed that waiting was used as a strategy by only a few individuals (Supplemental Figure S1), suggesting that the main effect of waiting results from the behavior of only a minority of the participating individuals.

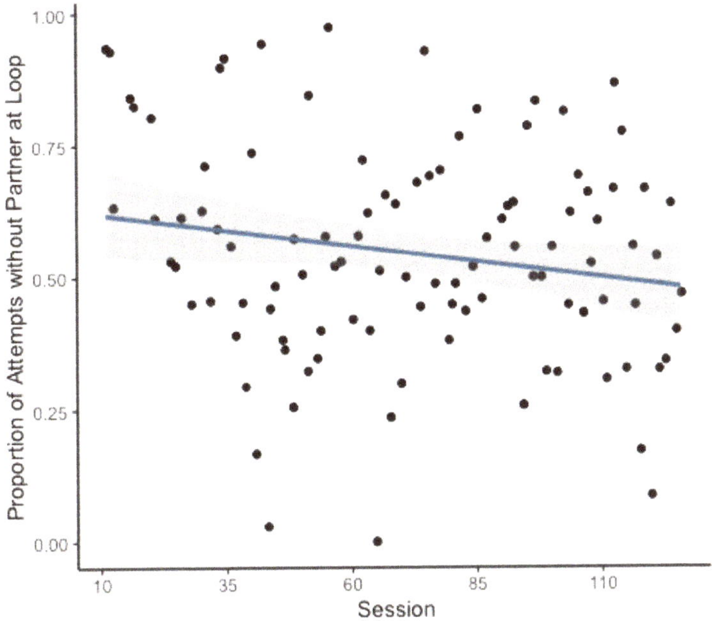

Figure 3. Scatterplot of overall proportion of attempts without a potential partner in front of the opposite loop. Line of best fit overlaid with surrounding greyed area indicating standard error.

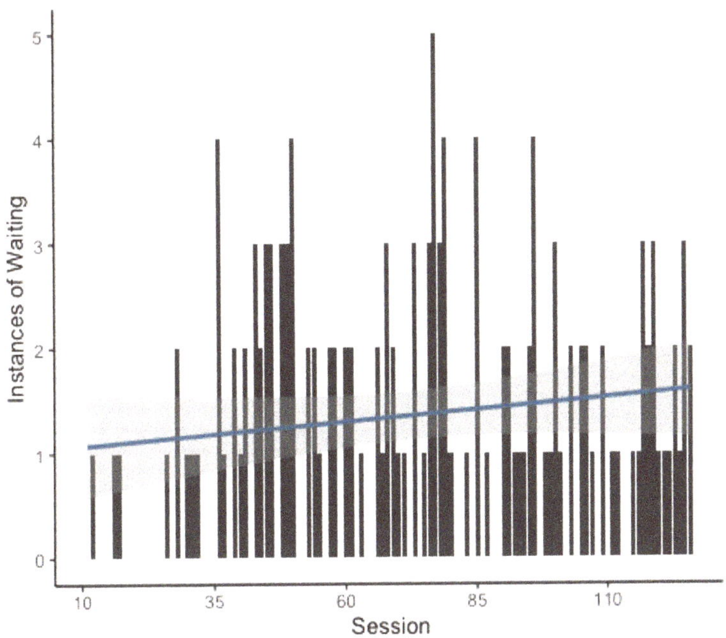

Figure 4. Bar graph showing instances of waiting per session with line of best fit overlaid. Greyed area indicates standard error.

4. Experiment 2

The purpose of Experiment 2 was to assess whether the Japanese macaques involved in our study were sensitive to differing reward equity outcomes. To evaluate this, the food rewards used in the experiment were modified to create a condition where one participant always received a lesser reward than the other. A post-reward observation period was added to assess whether unequal divisions had any effect on the subsequent behavior of participants. For this experiment we ran a total of 119 sessions between January and May 2020.

4.1. Procedure

Other than the addition of the new reward division condition, the procedure used in Experiment 2 was identical to that of Experiment 1. The only further exceptions were that the experimenter under no condition acted as a cooperative partner for the monkeys and the trial termination time was shortened to 10 min.

As in Experiment 1, two food rewards were paired together in every trial. Both bananas and zucchinis were used as rewards, with reward type always matched within the trial. In Condition A, the two food rewards were sliced into segments 1/9th the size of the entire food item. In Condition B, a 1/9th-sized segment was paired with a 1/3rd-sized segment to ensure that the cooperating monkeys would receive unequal rewards. It should be noted that reward theft was always a possibility, so receiving no reward was a further possible outcome for the participating monkeys in either condition. Sessions alternated between Condition A and Condition B, such that each new session marked a change in condition.

4.2. Collection and Coding

At the beginning of each session, the experimenter moved into a corner and began voice recording (using an Olympus WS-510M voice recorder) any stress-related or aggressive behaviors occurring within the research hut using an ad libitum/behavioral sampling method [45,46]. Stress-related behaviors included scratch, yawn and body shake, while prototypical examples of aggressive behaviors were bite, chase and lunge (for a full list, see Supplementary Table S2). The reward division outcomes of each trial as well as any refusals to act as a cooperation partner were also noted. Reward division outcomes were classified in the following way: reward equivalent to that of the other recipient (Outcome A), reward smaller than that of the other recipient (Outcome B), reward larger than that of the other recipient (Outcome C) or no reward due to theft (Outcome D). A refusal was defined as "any instance where a focal animal that had previously cooperated in the session with an individual currently waiting near a loop does not act to cooperate with that individual for at least 10 s". However, such a refusal was only observed 7 times over the course of 272 successful trials and was therefore excluded from future analyses.

To control for whether seeing an alternative reward available affected later behavior, whether an individual looked into the food box prior to cooperation was also recorded. Such instances were defined by an individual climbing up the apparatus to the plexiglass food box and looking inside. Sessions were video recorded using the same camera setup as Experiment 1, but in this experiment audio recordings from a handheld device served as the primary behavioral record so as to allow the experimenter greater mobility.

Following successful cooperation, post-cooperation behavior was recorded of one, or if possible, both members of the cooperative dyad. If one of the cooperating monkeys left the research hut following a successful trial, the session was paused, and a three-minute behavioral sampling period modelled after de Waal and Yoshihara's post-conflict matched-control method [47] was conducted. In our modified form of their method, receiving an unequal reward in relation to a partner (Condition B) was treated as the conflict, whereas an equal reward distribution among the cooperation partners (Condition A) was treated as the control. A further difference was that monkeys were observed for three minutes instead of five because previous studies using the post-conflict method with Japanese macaques have found that the majority of noteworthy post-conflict behaviors in the species occur within the first two minutes following the conflict [48,49]. Furthermore, instead of using an observation of the same monkey on the following day as a matched control, our study compared the average behavior following an unequal outcome with the average behavior following an equal outcome.

If the cooperating dyad did not leave the research area after a successful cooperation, the next trial began immediately so as to maintain the interest of participants. If any of the remaining participants failed to show interest in the new trial, their behavior was categorized as post-cooperative behavior for three minutes following the successful cooperation.

4.3. Analyses

The initial data set was heavily inflated by zeroes, so measures expressing stress and aggression were additionally coded as binomial scores representing whether or not the behavior occurred (yes/no; with 1 as yes) and a hurdle approach was chosen for the analysis. The hurdle approach is a multi-step procedure in which an initial analysis deals with zero-inflated count data and a subsequent analysis examines only those cases with a positive count. Our initial hurdle began by comparing across reward division outcomes the likelihood of any stress-related or aggressive behaviors occurring within the post-cooperation observation period using a binomial GLMM with a logit link function. The second hurdle took only those observation periods where stress or aggression actually occurred and explored whether the intensity with which those behaviors occurred varied systematically in accordance with reward division outcome using a linear GLMM with a logit link function. The term "intensity" in this paper refers to the frequency of expression of the behavior of interest within a specified time period. Focal animal, partner, food type and observer were added to the models as random effects. Both models examined separately the first minute of data and the first three minutes of data following a successful cooperation to investigate whether any behavioral response was more pronounced immediately following the event. An estimated marginal means (EMM) function was used to compare the likelihood of expressing stress and aggression across outcomes.

A post-hoc analysis was done to see whether an individual that inspected the box before cooperating had a higher likelihood of expressing stress-related or aggressive behaviors after experiencing Outcome B. We used a binomial GLMM with a logit link function to test for the effect of looking in the box on the likelihood of expressing stress-related and aggressive behavior. The variables focal animal, food type and observer were added as random effects.

To determine whether the behaviors observed during the post-cooperation observation period were products of only the last trial outcome or an accumulation of all previous trials, another binomial GLMM with a logit link function was carried out to control for repeated measures. To check for a pattern each outcome was assigned a value: A, B, C and D, respectively, 1, 0, 2, −1. These values are arbitrarily assigned numbers used to make a distinction between the outcomes and are based on the cooperator's direct personal food reward gains. Each trial had a value determined by the last outcome, and a cumulative value of the last and all previous outcomes of the individual during that session. Subsequently, the likelihood of stress-related and aggressive behaviors occurring within three minutes after the outcome was tested for the influence of the value of the last outcome and the "cumulative value of all previous trials" as a fixed effect. Consecutively, the same models were carried out with either only "cumulative value of all previous trials" or "direct value" as a fixed effect. In all models, focal animal was added as a random effect. ANOVA-based comparisons, using the Aikake information criterion (AIC) [50], between these models and null models were performed to see which model best explains the results. Since there were no cumulative values of previous trials for first trials per individual, only successful trials in which the individual had also succeeded in the previous trial were included in these analyses.

All statistical analyses were carried out using R statistical software [43] with α set at 0.05. GLMMs were run using the lme4 package [44] and EMM comparisons made use of the emmeans package [51].

4.4. Results

Over the course of the experiment, 119 sessions and 1059 trials were conducted, 272 of which were successful, resulting in an overall success rate of 25.7%. Seven monkeys succeeded in at least one cooperation, with four of those individuals accounting for 96% of all co-operations. Of the seven participants, six were returning monkeys from Experiment 1 while the seventh was a newly interested monkey who had shown no previous aptitude.

Breaking down the reward division outcomes observed across both conditions, in 41% of trials the cooperating monkey received an equal reward to that of their partner (Outcome A), in 11% of trials they received a smaller reward (Outcome B), in 12% of trials they received a larger reward (Outcome C), and in 41% of cases their reward was stolen (Outcome D).

4.4.1. Stress-Related Behavior

We found a significant effect of the outcome of reward division on the likelihood of expressing stress-related behavior in the 3-min period following a cooperation (Table 2). Specifically, the cooperators were more likely to express stress-related behavior after Outcome B than Outcome A in the 3-min period following a cooperation (Estimate = −1.179, z.ratio = -2.888, p = 0.02), more likely to express stress-related behavior in Outcome D compared to Outcome A (Estimate = −1.296, z.ratio = −4.241, p = <0.001) and more likely to express stress-related behavior after Outcome D than Outcome C (Estimate = −1.254, z.ratio = −2.576, p = 0.049; Figure 5). This model explained the data significantly better than the respective null model ($\chi^2(3) = 24.938$, p = <0.001). We further found that looking in the box had no impact on the increased likelihood of expressing stress-related behavior following a personally disadvantageous inequitable division (Estimate = 0.263, z-ratio = 0.399, p = 0.690).

Table 2. Results of the GLMM testing the effect of outcome on likelihood of expressing stress-related behaviors within 3 min after cooperation.

Outcome	Estimate	Standard Error	z-Value	*p*-Value
A	−2.111	0.305	−6.918	<0.001
B	1.179	0.408	2.888	0.004
C	0.042	0.516	0.081	0.935
D	1.296	0.306	4.241	<0.001

The same analysis was carried out using only the first minute of stress-related behavior following a cooperation, but no significant difference was found between the different reward outcomes.

The second hurdle, investigating the frequency of stress-related behavior when it occurred, found no significant difference in the intensity of stress-related behaviors (when the animals showed at least some stress-related behavior) following the different outcomes.

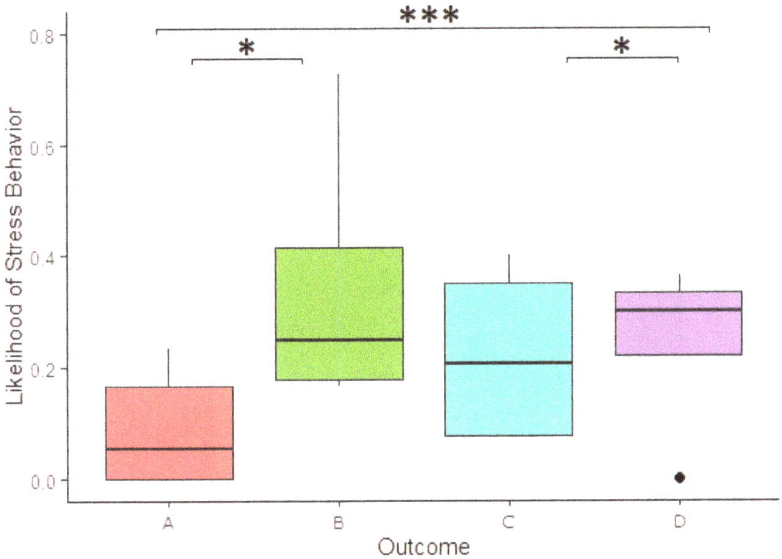

Figure 5. Likelihood of stress-related behaviors across outcomes. A = reward equivalent to that of other, B = reward smaller than that of other, C = reward larger than that of other, D = no reward due to theft. Boxplots show the median (solid line), 25th–75th percentile (box) and the largest and smallest value (whiskers). Dots reflects outliers. * indicates a difference at the $p < 0.05$ level and *** indicates a difference at the $p < 0.001$ level.

4.4.2. Aggression

We found a significant effect of the outcome of reward division on the likelihood of expressing aggressive behavior in the 3-min period following a cooperation (Table 3). Specifically, a higher likelihood of expressing aggressive behavior after Outcome A compared to Outcome D in the 3-min period following a cooperation (Estimate = −1.189, z.ratio = −3.008, p = 0.014) as well as higher likelihood of expression in Outcome A compared to Outcome D (Estimate = −1.191, z.ratio = −4.09, p = <0.001; Figure 6). This model explained the data significantly better than the respective null model ($\chi^2(3) = 20.735$, p = <0.001). We further found that looking in the box had no impact on the increased likelihood of expressing aggressive behavior following an inequitable division (Estimate = −0.105, z-ratio = −0.173, p = 0.862).

Table 3. Results of the GLMM testing the effect of outcome on the likelihood of expressing aggressive behavior within 3 min after cooperation.

Outcome	Estimate	Standard Error	z-Value	*p*-Value
A	−1.865	0.311	−6.008	>0.001
B	1.189	0.395	3.008	0.003
C	0.467	0.427	1.093	0.275
D	1.191	0.291	4.090	>0.001

The same pattern was present using only the first minute of data following successful cooperation, with a higher likelihood of aggressive behaviors again found in Outcome B compared to Outcome A (Estimate = −1.224, z.ratio = −3.094, p = 0.011) as well as higher likelihoods of aggressive behaviors in Outcome D compared to that of Outcome A (Estimate = −1.169, z.ratio = −3.983, p = <0.001), even after null model comparison ($\chi^2(3) = 20.421$, p = <0.001).

The second hurdle, investigating the frequency of aggression, when it occurred, found no significant difference in the intensity of aggressive behaviors following the different outcomes.

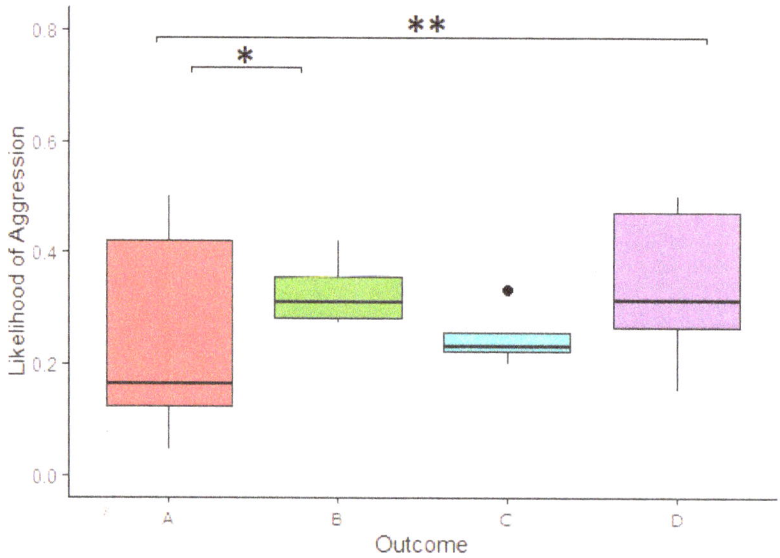

Figure 6. Likelihood of aggressive behaviors across outcomes. A = reward equivalent to that of other, B = reward smaller than that of other, C = reward larger than that of other, D = no reward due to theft. Boxplots show the median (solid line), 25th-75th percentile (box) and the largest and smallest value (whiskers). Dots reflects outliers. * indicates a difference at the $p < 0.05$ level and ** indicates a difference at the $p < 0.01$ level.

4.4.3. Cumulative Effects

As for comparing the models of the effect of cumulative value of all previous trials on the likelihood of expressing stress-related and aggressive behavior, we ran binomial GLMMs with logit link functions including and excluding the effect of cumulative reward outcome. The GLMM testing the effect of the value of the last trial on likelihood of stress excluding cumulative outcome value as a fixed effect revealed a significant effect of the last trial (Estimate = −0.698, z-ratio = −4.200, p = <0.001). The GLMM including cumulative outcomes as a fixed effect did not reach significance (Estimate = 0.097, z-ratio = 1.284, p = 0.199) nor did it when it was tested as a single fixed effect, i.e., without the value of the last trial (Estimate = -0.0833, z-value = −1.312, p = 0.189). For the likelihood of aggressive behavior, the effect of value of the last trial without the cumulative value of all previous trials as a fixed effect was significant (Estimate = −0.463, z-ratio = -3.384, p = <0.001), while the GLMM including cumulative outcomes as a fixed effect did not reach significance (Estimate = 0.0120, z-ratio = 0.177, p = 0.860), nor did it when it was tested as a single fixed effect (Estimate = −0.094, z-value = −1.634, p = 0.102). ANOVA-based comparisons revealed that the model where stress likelihood was tested as a result of the value of the last trial without additional fixed effects had the lowest AIC (265.64) compared to when only the effect of all previous trials was tested (AIC = 284.48), when the effect of all previous trials and that of the last were tested (AIC = 265.99) and the null model (AIC = 284.23). The same applied to the likelihood of aggression, where the model with effect of only the last trial had an AIC of 328.48. For comparison, the model testing only the effect of all previous trials had an AIC of 337.97, while the model including both previous trials and the last trial as fixed effects had an AIC of 330.45 and the null model had an AIC of 338.69. This led us to the conclusion that the likelihood of post-cooperation stress-related and aggressive behavior was better explained by the outcome of the last trial than it was by the cumulative value of all previous trials.

5. Discussion

Our study used a variation of the loose-string paradigm [10,11] to examine cognitive abilities associated with cooperation in Japanese macaques. Even though it has been previously demonstrated that the species is capable of succeeding in a cooperative task [14], we took this a step further by demonstrating that they are attentive to the presence of a partner. In addition, we combined our cooperation task with an examination of the reaction to unequal payoffs thereafter. Embedding our experiments within a setting resembling the natural environment of the species allowed for an examination of cooperation and inequity aversion (IA) within an intact social environment. This allowed us to place IA in the context of natural reward division following cooperation, thus more closely replicating the conditions that should accompany its occurrence in nature [52]. Our study used post-cooperation observation of stress-related and aggressive behaviors to investigate the potential existence of IA in the species, and succeeded in documenting a pattern of behavior consistent with the predicted profile of disadvantageous IA. That being said, we did not provide evidence of the adoption of IA-associated behavioral strategies such as future refusal to cooperate or partner switching (cf. [53]). Future studies may wish to focus specifically on how the negative states suggested by the behavioral profile we observed in our study may translate into strategies that could conceivably result in consistent positive fitness outcomes.

5.1. Sensitivity to Partner Presence

The primary aim of our first experiment was to demonstrate that Japanese macaques recognize the importance of their partner during a cooperative task. Demonstrating such an understanding is an essential step in arguing a case for intentional cooperation on the part of the macaques. We assessed this by examining changes in the proportion of attempts made in the presence of a potential partner as well as changes in the frequency of waiting behaviors. Our analyses demonstrated a behavioral shift in the predicted direction for both features, with a decrease in the proportion of attempts made without a potential partner present and an increase in the frequency of waiting behavior. This led us to the conclusion that at least some of the individuals involved in our experiment developed an understanding of the contingencies of the task, including the necessity of having a partner, though they may not have understood the role that their partner played in the task. This corresponds to "presence-dependent cooperation" in the framework set out by Albiach-Serrano [8], though the level of understanding in our study population could have extended beyond this.

Our experiment further demonstrated that individuals became increasingly proficient with more experience, with successful attempts increasing in relative frequency to unsuccessful attempts as the experiment progressed. Even though it is not a strong argument for task understanding in and of itself, when taken in conjunction with the increasing specificity in which participating monkeys made attempts, it bolsters the case for task understanding in at least some individuals. Increasing relative frequency of successful attempts is suggestive of greater coordination with cooperative partners.

Even though increasing specificity of pulling was evident in the case of most regular cooperators, this does not necessarily indicate that all individuals who showed this pattern understood the cooperative nature of the task. For a dyad to give the appearance of deliberate cooperation, only one member of that dyad needs to be actively coordinating with the other. Increasing specificity of pulling as we measured it may just as easily indicate that another individual became adept at coordinating with them, rather than that they became proficient at coordinating with others. Used as a proxy for task understanding, increasing specificity of pulling needs to be interpreted with caution when taken in isolation. There was a significant main effect for an increase in the frequency of waiting behavior as well, but closer examination revealed that this behavioral strategy was only frequently used by three individuals (Supplementary Figure S1). Given that all but two of the dyads that formed contained at least one of these individuals, it is a distinct possibility that all active coordination throughout the experiment rooted from them. It is worth noting that these three individuals were also the youngest monkeys involved in the experiment. Nevertheless, active coordination even in the case of only a few individuals still demonstrates a capacity for task understanding and deliberate cooperation in the species, thus serving as proof of concept.

Imposing a delay in partner release has been effectively used to demonstrate waiting behavior during cooperation tasks in a variety of species including Asian elephants (*Elephas maximus*) [19], capuchin monkeys (*Cebus apella*) [54], cotton-top tamarins (*Saguinus oedipus*) [55], chimpanzees (*Pan troglodytes*) [12], spotted hyenas (*Crocuta crocuta*) [21], domestic dogs (*Canis familiaris*) [56], orangutans (*Pongo pygmaeus*) [57], coral trout (*Plectropomus leopardus*) [58] and keas (*Nestor notabilis*) [24], but the practical constraints of the semi-free setting used in our study prevented us from designing our own experiment in this way. Even though naturally occurring waiting was observed in a few of our monkeys, our argument for sensitivity to partner presence rests more heavily on the change in the proportion of attempts made in the presence of a partner.

When it comes to behavioral inhibition, macaque species have been demonstrated to underperform in comparison to other primates [59,60], which may have impaired their capacity to wait for a partner. Indeed, only one of the Japanese macaques in Kaigaishi, Nakamichi, and Yamada's previous study of the species showed any indication of using waiting as a strategy [14]. Evans and Beran, in their study of the closely related rhesus macaque (*Macaca mulatta*), found that the species had a particularly difficult time maintaining self-control over a delay when an impulsive response option was continually available [60], as it was in the present study. An absence of waiting due to a lack of inhibitory ability in a species does not necessarily mean that the individuals involved do not have some understanding that a partner is necessary to complete the task, it may merely reflect the behavioral constraints intrinsic to the species. It was with this in mind that in addition to examining waiting behavior, we chose to emphasize the more subtle measure of change in proportion of attempts as a measure of sensitivity to partner presence in our own study.

An alternative explanation for the increase in the relative frequency of pulling behaviors in the presence of others is the possibility of social facilitation, a well-documented phenomenon in which a dominant behavioral response is amplified in the presence of others [61]. Our argument against the social facilitation explanation is as follows: firstly, the effect of social facilitation should have been more or less constant across all individuals who came to associate pulling behavior with reward, but this was not the case. Some individuals that were highly active in the experiment did not increase at all in relative frequency of pulling behaviors in the presence of others. One of the most successful individuals even showed a decrease by this metric over the course of the experiment, which is more in line with our assertion that only a few individuals gained insight into the cooperative nature of the apparatus and adjusted their behavior accordingly. Furthermore, as training sessions took place before the experimental phase began, pulling behavior should have already been the dominant response at the beginning of the experiment and remained the dominant response throughout. Any effect of social facilitation should therefore have remained constant.

As for why pulling in the absence of a partner never fully extinguished, the fact that our experimental apparatus automatically reset itself after failed attempts likely played a large part. In the typical loose-string paradigm, an attempt made in the absence of a partner results in an inability to further participate in the experiment until the apparatus is reset by an attendant. With the apparatus used in our experiment, repeated attempts could be made without precluding the possibility of future attempts should a partner appear. In practical terms, this meant that the only cost of making an attempt on the device was the small energy input required of the attempt. This cost was likely low enough that the pressures acting toward the extinction of indiscriminate pulling were not strong enough to extinguish that behavior outright. The benefit of designing our experiment in this way is that it allowed for individuals not met with immediate success to continue exploratory behavior with no ill effect until a potential partner appeared and gave them an opportunity to learn the appropriate reward contingency. This was well-suited to the natural setting of our study, as long periods of time may pass before a potential partner arrives.

5.2. Partial Evidence for Inequity Aversion

Our second experiment explored whether Japanese macaques react aversively following cooperation that has resulted in an unequal reward outcome. Our results demonstrate that the subjects in our study were more likely to display stress-related and aggressive behaviors after receiving a disadvantageous reward outcome in comparison to another individual (Outcome B) than when receiving an equally sized reward, but in this condition also equal to that of their partner (Outcome A). The higher likelihood of expressing stress-related and aggressive behavior in the post-cooperation period following Outcome B does indicate that a personally disadvantageous outcome following cooperation incites aversive behaviors, which may be caused by IA. The lack of a difference in the likelihood of expressing stress-related behavior and aggression between Outcome C, i.e., when the subject received the larger of the two rewards, and Outcome A, suggests that this cannot be explained by the frustration of individuals over the fact that they received a lesser reward after having received the larger reward previously [62–64]. We cannot completely exclude, however, the possibility that the reaction of the Japanese macaques was due to the violation of expectation of receiving the visible larger rewards [65–70]. Even though the available rewards were visible before reward division occurred, any interested macaque had to move directly up to the plexiglass food box to see them. We included whether this occurred in our analysis and found that it had no impact on the increased likelihood of expressing stress-related or aggression following an inequitable division. Consequently, we concluded that a social comparison of rewards was most likely responsible for the pattern that we observed.

Engelman et al. presented another explanation for disadvantageous IA in the form of the "social disappointment hypothesis" [71], in which an individual rejects a reward because it is disappointed in the provider for not supplying a better reward. Their findings suggest that reward rejection may be based on a social interaction with the researcher, rather than a comparison with the reward of the conspecific, because they only found rejections of those rewards when provided by a researcher and not when provided by a machine. Even though a researcher was present during the experiments in the current study, our method distinguishes itself from the researcher condition of the study of Engelman et al. [71] in that in our experiment, the researcher did not distribute the food rewards. Instead, the rewards became simultaneously available to all individuals after joint completion of the dyadic task. It must be noted, however, that the macaques could see the researcher loading the device with food. Nevertheless, as the specifics of the reward division were left to the macaques, they could at most be disappointed in the experimenter for loading the apparatus with unequal food rewards. Such inter-conspecific negotiation tasks with minimal researcher interference are suggested by Oberliessen and Kalenscher to control for the "social disappointment hypothesis" [72].

Finally, a comparison of the outcomes is not the only important factor in fairness sensitivity—comparison of the efforts is also of relevance [35,73–76]. One could argue that, in the context of the current study, a thief has a lower effort input than a cooperator. It is then of some relevance that the macaques in our study showed no difference in their likelihood to react aversively whether a thief or their cooperative partner obtained the second reward (Supplementary Note S1). This suggests that macaques may not base their aversive reaction on a comparison of effort when obtaining a reward. However, the comparison between the role of thief and cooperation partner was not one of the premeditated goals of this study, and there are methods better suited for comparing effort-reward (in)equity [35,74–76]. Future studies that aim to research effort (in)equity should take advantage of these more direct methods.

In general, the proximate mechanism underlying IA are still poorly understood, but frustration and anger, as proxied here by stress behavior and aggression, seem potential candidates for an affective component involved in IA. Consequently, our findings provide tentative support for the theory that an aversion for inequity, potentially mediated by affective responses [6], is necessary to stabilize cooperation [77–80] and that cooperation and a sense for fairness have coevolved [31,77–79].

Our experiment distinguishes itself from the bulk of animal inequity studies in that subjects perform a cooperation task instead of conducting a version of the impunity game, which often translates into a token-exchange experiment where the subject must exchange a token with a researcher or machine in order to receive a reward (Capuchin monkeys (*Cebus apella*) [53,74,81,82], crows (*Corvus corone*) and ravens (*Corvus corax*) [75], chimpanzees (*Pan troglodytes*) [71,83], keas (*Nestor notabilis*) [24], Goffin's cockatoos (*Cacatua goffiniana*) [76]). As it has been argued that IA has coevolved with cooperation [31,77–79], our cooperation experiment thus more closely resembles the context that may have created the selection pressures leading to the evolution of IA. Nevertheless, the use of more artificial set-ups may provide clearer indications of IA. In the impunity game, a subject can choose whether to continue performing a rewarded task after seeing a conspecific receive a greater reward for the same task (though their own refusal has no impact on whether that conspecific receives future rewards). Refusing to continue the task is seen as a potential indicator of IA in the species. In our current study, refusal to continue cooperating meant that neither the refuser nor their partner were able to obtain a reward, making our experimental setup more closely resemble that of the ultimatum game [84]. Despite the potential impact it could have had on an inequitable partner, refusal to cooperate was seldom observed in our experiment. It should be noted however, that as a reaction an individual could also simply leave the cooperation device. Yet, it was impossible to distinguish the different potential reasons for an individual to leave the cooperation device in our open setting. Nevertheless, this reaction should be investigated in more detail in future studies that also try to embed an examination of IA in a more natural cooperative situation.

The current study found that the likelihood of expressing stress-related behavior or aggression did not increase with the number of previous trials in which a monkey had personally experienced inequity. As our results suggest that macaques base their reaction only on the last reward-division outcome, our second hypothesis must be rejected. Brosnan suggests that individuals should react aversively to personal inequity primarily when it becomes a systematic outcome [31]. This sort of reaction profile allows for some tolerance of the accidental inequities that happen frequently in natural settings, which serves to maintain cooperative bonds that have overall equal payoffs. However, reacting to cumulative inequity rather than direct inequity may be reserved for species that cooperate in the food domain, where differences in obtained food after a cooperative task are more readily quantifiable. Even though IA in its basic form may not be domain specific [35], the cognitive accounting mechanisms required of cumulative IA may be. This ability would not necessarily be of use in the social domain, where outcomes are not as readily measurable. This may lead species that cooperate in only the social domain to employ a more cognitively simplistic mechanism such as only considering the immediately previous outcome. Japanese macaques may fall into this category, as they cooperate in the social domain and have an extensive network of kin [85] but are not known to hunt together or share food.

5.3. *Limitations and Future Directions*

An added obstacle when conducting research in a natural setting is the difficulty of obtaining a high rate of participation in the population [86]. Even though there are many advantages to conducting experiments in an intact social environment, these experimental designs come with the disadvantage of potential active interference from other group members and monopolization of the experiment by high-ranking individuals. This may be especially relevant in the study of socially despotic species such as the Japanese macaques used in our study. Furthermore, the alternative foraging opportunities available in a natural setting may reduce the appeal of food-rewarded experiments. Even though our study population contained 164 monkeys at the beginning of our first experiment, only 12 of them participated in it. Participation in our second experiment dropped down to seven monkeys, with only four of them accounting for 96% of all co-operations. Due to this fact, our conclusions need to be interpreted with caution. Future studies conducting research in a natural setting may wish to consult Cronin et al.'s recommendations for countering the reduced participation typical of such studies [86].

The monkeys in our study tended to have very few partners, which may be a further effect of the low level of social tolerance typical of the species under investigation [87]. Most individuals cooperated with a very small proportion of the other participants, and some of the monkeys who reached training criteria during the training phase never came to cooperate with any others despite a continued interest in the experiment. The juveniles in our experiment tended to have a much higher number of partners, constituting a notable exception to this pattern. This may be due to the greater social tolerance characteristic of juveniles in primate species [88]. Even though our sample size was too small to draw any meaningful conclusions regarding partner preference, the dyadic success rates we observed were suggestive of a strong preference for kin-kin or juvenile-juvenile dyads, as is expected of a socially despotic and nepotistic species. Future studies making use of a larger sample size may be able to shed further light on partner preference in socially despotic species such as the Japanese macaque.

Our limited sample in Experiment 2 further prevented us from investigating the impact of rank on reaction to reward division. Half of the monkeys involved in that experiment were sub-adult males, which occupy a particularly transient position in the hierarchy system of Japanese macaques [89–91]. All active dyads in this experiment included at least one of these males, so an analysis of the impact of rank on reaction to inequity could not be performed. This is nevertheless an interesting avenue of study, as past experiments have demonstrated that dominance plays a large role in access to food sources [92–95], and future studies may wish to revisit this subject in the species.

Even though we did make a case for "presence-dependent cooperation" in our study, we did not go so far as to investigate "action-dependent cooperation", the next step in demonstrating fully intentional cooperation [8]. Past studies have used partner-directed glancing behavior as an indicator of attentiveness to partner action [54,57,96]. Our own study design could potentially be adapted to take glancing behavior into account using the same approach we used in making a case for sensitivity to partner presence. In addition to tracking the actions of a cooperation partner, active signaling may also contribute to deliberate coordination. Japanese macaques are extremely vocal animals [97], and it would be of further interest to investigate whether they use these vocalizations to recruit cooperation partners or otherwise coordinate their cooperative efforts.

6. Conclusions

We demonstrated not only that Japanese macaques are capable of succeeding in a cooperation task, but that they are also attentive to the presence of a partner and will actively adjust their attempt rate to match this. Over the course of the first experiment, the proportion of attempts made in the absence of a partner decreased and a subset of the participants learnt to wait for a partner when none was present. Our second experiment found that individuals had a higher likelihood to express stress-related and aggressive behaviors after receiving a lesser reward than their partner in the cooperation task when compared to a control condition where both individuals received an equivalent reward. This matches the pattern of behavior predicted by disadvantageous IA, suggesting a possible sensitivity to inequity in the species. A major strength of our study was that it combined a cooperation paradigm with a measure of IA within an intact social environment to better replicate food division dynamics that would be found in nature. This design leant a high degree of ecological validity to our experiments, which may be of particular importance for studies examining socially bound phenomena such as cooperation and IA.

Supplementary Materials: The following 5 files are available online at https://www.mdpi.com/article/10.3390/ani11061497/s1, File 1: containing Table S1, a summary of results by dyad, Figure S1. instances of waiting by individual and Table S2. definitions of behavioral variables recorded during experiment 2; File 2: Data for experiment 1; File 3: R-code for Experiment 1; File 4: Data for experiment 1; and File 5: R-code for Experiment 2.

Author Contributions: Conceptualization, R.S., M.S.S. and J.J.M.M.; methodology, R.S., M.S.S. and J.J.M.M.; validation, J.H.; formal analysis, R.S., M.S.S. and J.J.M.M.; investigation, R.S., M.S.S. and R.H.; data curation, R.S. and M.S.S.; writing—original draft preparation, R.S. and M.S.S.; writing—review and editing, J.J.M.M.; supervision, L.S.P. and J.J.M.M. All authors have read and agreed to the published version of the manuscript.

Funding: This research received no external funding.

Institutional Review Board Statement: Since all our experiments were non-invasive, the study complied with the Austrian Law (§ 2. Federal Law Gazette number 501/1989) and the Code for Best Practices in Field Primatology and received oversight from and was authorized by the internal board of the Austrian Research Center for Primatology. No invasive research or experimental procedures requiring ethics approval according to the European Directive 2010/63 were performed. Our studies adhered to the American Society of Primatologists' principles for the ethical treatment of primates and all applicable international, national and institutional guidelines for the care and use of animals were followed.

Data Availability Statement: The data presented in this study are available in the supplementary materials (Files 2 and 4).

Acknowledgments: We are grateful to the Affenberg Zoobetriebsgesellschaft mbH, and specifically to its directors Peter and Svenja Gaubatz for allowing us to do our research there. We also thank all animal caretakers for their good care of our study animals and Max Dorner for his help with the building of the apparatus. Additionally, we would like to thank Pia Böhm for her training with regard to the behavior of the macaques. Finally, we would like to thank two anonymous reviewers for their comments on an earlier draft of the manuscript.

Conflicts of Interest: The authors declare no conflict of interest.

References

1. Burkart, J.M.; Allon, O.; Amici, F.; Fichtel, C.; Finkenwirth, C.; Heschl, A.; Huber, J.; Isler, K.; Kosonen, Z.K.; Martins, E.; et al. The evolutionary origin of human hyper-coooperation. *Nat. Commun.* **2014**, *5*, 4747. [CrossRef]
2. Krams, I.; Krama, T.; Igaune, K.; Mänd, R. Experimental evidence of reciprocal altruism in the pied flycatcher. *Behav. Ecol. Sociobiol.* **2008**, *62*, 599–605. [CrossRef]
3. Boesch, C.; Boesch, H. Hunting behavior of wild chimpanzees in the Taï National Park. *Am. J. Phys. Anthropol.* **1989**, *78*, 547–573. [CrossRef]
4. Clutton-Brock, T. Breeding together: Kin selection and mutualism in cooperative vertebrates. *Science* **2002**, *296*, 69–72. [CrossRef] [PubMed]
5. Dugatkin, L.A. Animal cooperation among unrelated individuals. *Naturwissenschaften* **2002**, *89*, 533–541. [CrossRef] [PubMed]
6. Massen, J.J.M.; Behrens, F.; Martin, J.S.; Stocker, M.; Brosnan, S.F. A comparative approach to affect and cooperation. *Neurosci. Biobehav. Rev.* **2019**, *107*, 370–387. [CrossRef] [PubMed]
7. Noë, R. Cooperation experiments: Coordination through communication versus acting apart together. *Anim. Behav.* **2006**, *71*, 1–18. [CrossRef]
8. Albiach-Serrano, A. Cooperation in primates: A critical, methodological review. *Interact. Stud.* **2015**, *16*, 361–382. [CrossRef]
9. Crawford, M.P. The cooperative solving of problems by young chimpanzees. *Comp. Psychol. Monogr.* **1937**, *14*, 1–88.
10. Hirata, S. Cooperation in chimpanzees. *Hattatsu* **2003**, *95*, 103–111.
11. Hirata, S.; Fuwa, K. Chimpanzees (*Pan troglodytes*) learn to act with other individuals in a cooperative task. *Primates* **2007**, *48*, 13–21. [CrossRef] [PubMed]
12. Melis, A.P.; Hare, B.; Tomasello, M. Chimpanzees recruit the best collaborators. *Science* **2006**, *311*, 1297–1300. [CrossRef]
13. Molesti, S.; Majolo, B. Cooperation in wild Barbary macaques: Factors affecting free partner choice. *Anim. Cogn.* **2016**, *19*, 133–146. [CrossRef] [PubMed]

14. Kaigaishi, Y.; Nakamichi, M.; Yamada, K. High but not low tolerance populations of Japanese macaques solve a novel cooperative task. *Primates* **2019**, *60*, 421–430. [CrossRef] [PubMed]
15. Stocker, M.; Loretto, M.C.; Sterck, E.H.; Bugnyar, T.; Massen, J.J.M. Cooperation with closely bonded individuals reduces cortisol levels in long-tailed macaques. *R. Soc. Open Sci.* **2020**, *7*, 191056. [CrossRef]
16. Martin, J.S.; Koski, S.E.; Bugnyar, T.; Jaeggi, A.V.; Massen, J.J.M. Prosociality, social tolerance and partner choice facilitate mutually beneficial cooperation in common marmosets, *Callithrix jacchus*. *Anim. Behav.* **2021**, *173*, 115–136. [CrossRef]
17. Seed, A.M.; Clayton, N.S.; Emery, N.J. Cooperative problem solving in rooks (*Corvus frugilegus*). *Proc. R. Soc. B* **2008**, *275*, 1421–1429. [CrossRef] [PubMed]
18. Massen, J.J.M.; Ritter, C.; Bugnyar, T. Tolerance and reward equity predict cooperation in ravens (*Corvus corax*). *Sci. Rep.* **2015**, *5*, 15021. [CrossRef] [PubMed]
19. Plotnik, J.M.; Lair, R.; Suphachoksahakun, W.; de Waal, F.B.M. Elephants know when they need a helping trunk in a cooperative task. *Proc. Natl. Acad. Sci. USA* **2011**, *108*, 5116–5121. [CrossRef] [PubMed]
20. Marshall-Pescini, S.; Schwarz, J.F.L.; Kostelnik, I.; Virányi, Z.; Range, F. Importance of a species' socioecology: Wolves outperform dogs in a conspecific cooperation task. *Proc. Natl. Acad. Sci. USA* **2017**, *114*, 11793–11798. [CrossRef] [PubMed]
21. Drea, C.M.; Carter, A.N. Cooperative problem solving in a social carnivore. *Anim. Behav.* **2009**, *78*, 967–977. [CrossRef]
22. Schmeltz, M.; Duguid, S.; Bohn, M.; Völter, C.J. Cooperative problem solving in giant otters (*Pteronura brasiliensis*) and Asian small-clawed otters (*Aonyx cinerea*). *Anim. Cogn.* **2017**, *20*, 1107–1114. [CrossRef]
23. Schwing, R.; Jocteur, E.; Wein, A.; Noë, R.; Massen, J.J.M. Kea cooperate better with sharing affiliates. *Anim. Cogn.* **2016**, *19*, 1093–1102. [CrossRef]
24. Heaney, M.; Gray, R.D.; Taylor, A.H. Keas perform similarly to chimpanzees and elephants when solving collaborative tasks. *PLoS ONE* **2017**, *12*, e0169799. [CrossRef] [PubMed]
25. Tassin de Montaigu, C.; Durdevic, K.; Brucks, D.; Krasheninnikova, A.; von Bayern, A. Blue-throated macaws (*Ara glaucogularis*) succeed in a cooperative task without coordinating their actions. *Ethology* **2020**, *126*, 267–277. [CrossRef]
26. Peron, F.; Rat-Fischer, L.; Lalot, M.; Nagle, L.; Bovet, D. Cooperative problem solving in African grey parrots (*Psittacus erithacus*). *Anim. Cogn.* **2011**, *14*, 545–553. [CrossRef] [PubMed]
27. Ortiz, S.T.; Castro, A.C.; Balsby, T.J.S.; Larsen, O.N. Problem-solving in a cooperative task in peach-fronted conures (*Eupsittula aurea*). *Anim. Cogn.* **2020**, *23*, 265–275. [CrossRef]
28. Yamamoto, C.; Kashiwagi, N.; Otsuka, M.; Sakai, M.; Tomonaga, M. Cooperation in bottlenose dolphins: Bidirectional coordination in a rope-pulling task. *PeerJ* **2019**, *7*, e7862. [CrossRef]
29. Massen, J.J.M.; Schaake, W.A.A.; Bugnyar, T. A comparison of cooperative cognition in corvids, chimpanzees, and other animals. In *Chimpanzees in Context*; Hopper, L., Ross, S., Eds.; University of Chicago Press: Chicago, IL, USA, 2020; pp. 368–390.
30. Rand, D.G.; Nowak, M.A. Human cooperation. *Trends Cogn. Sci.* **2013**, *17*, 413–425. [CrossRef] [PubMed]
31. Brosnan, S.F. A hypothesis of the co-evolution of cooperation and responses to inequity. *Front. Neurosci.* **2011**, *5*, 43. [CrossRef]
32. Brosnan, S.F.; de Waal, F.B.M. Evolution of responses to (un)fairness. *Science* **2014**, *346*, 1251776. [CrossRef] [PubMed]
33. Brosnan, S.F.; de Waal, F.B.M. Fairness in animals: Where to from here? *Soc. Just. Res.* **2012**, *25*, 336–351. [CrossRef]
34. Chapais, B.; Girard, M.; Primi, G. Non-kin alliances, and the stability of matrilineal dominance relations in Japanese macaques. *Anim. Behav.* **1991**, *41*, 481–491. [CrossRef]
35. Massen, J.J.M.; van den Berg, L.M.; Spruijt, B.M.; Sterck, E.H.M. Inequity aversion in relation to effort and relationship quality in long-tailed macaques (*Macaca fascicularis*). *Am. J. Primatol.* **2012**, *74*, 145–156. [CrossRef] [PubMed]
36. Fooden, J.; Aimi, M. Systematic review of Japanese macaques, *Macaca fuscata* (Gray, 1870). *Fieldiana Zool.* **2005**, *104*, 1–198. [CrossRef]
37. Setchell, J.M.; Abbott, K.M.; Gonzalez, J.; Knapp, L.A. Testing for post-copulatory selection for major histocompatibility complex genotype in a semi-free-ranging primate population. *Am. J. Primatol.* **2013**, *75*, 1021–1031. [CrossRef]
38. Naud, A.; Chailleux, E.; Kestens, Y.; Bret, C.; Desjardins, D.; Petit, O.; Ngoubangoye, B.; Sueur, C. Relations between spatial distribution, social affiliations and dominance hierarchy in a semi-free mandrill population. *Front. Psychol.* **2016**, *7*, 612. [CrossRef]
39. Ottoni, E.B.; Izar, P. Capuchin monkey tool use: Overview and implications. *Evol. Anthropol.* **2008**, *17*, 171–178. [CrossRef]
40. Blomquist, G.E.; Sade, D.S.; Berard, J.D. Rank-related fitness differences and their demographic pathways in semi-free-ranging rhesus macaques (*Macaca mulatta*). *Int. J. Primatol.* **2011**, *32*, 193–208. [CrossRef]
41. Trapanese, C.; Bey, M.; Tonachella, G.; Meunier, H.; Masi, S. Prolonged care and cannibalism of infant corpse by relatives in semi-free-ranging capuchin monkeys. *Primates* **2020**, *61*, 41–47. [CrossRef]
42. Armster, S.I. The effect of different food rewards on performance in a string pulling task and subsequent theft avoidance in Japanese macaques (*Macaca fuscata*). Master's Thesis, University of Vienna, Vienna, Austria, 2018.
43. R Core Team. R: A Language Environment for Statistical Computing. Version 3.6.1. R Foundation for Statistical Computing: Vienna, 2019. Available online: https://www.R-project.org/ (accessed on 13 March 2021).
44. Bates, D.; Maechler, M.; Bolker, B.; Walker, S. lme4: Linear mixed-effects models using Eigen and S4. R Package Version 2015. Volume 1, pp. 1–7. Available online: https://CRAN.R-project.org/package=lme4 (accessed on 13 March 2021).
45. Altman, J. Observational study of behavior: Sampling methods. *Behaviour* **1974**, *49*, 227–267. [CrossRef] [PubMed]
46. Martin, P.; Bateson, P. *Measuring Behaviour: An Introductory Guide*, 3rd ed.; Cambridge University Press: Cambridge, UK, 2007.
47. de Waal, F.B.M.; Yoshihara, D. Reconciliation and redirected affection in rhesus monkeys. *Behaviour* **1983**, *85*, 224–241. [CrossRef]

48. Majolo, B.; Ventura, R.; Koyama, N. Postconflict behavior among male Japanese macaques. *Int. J. Primatol.* **2005**, *26*, 321–336. [CrossRef]
49. Kutsukake, N.; Castles, D.L. Reconciliation and variation in postconflict stress in Japanese macaques (*Macaca fuscata fuscata*): Testing the integrated hypothesis. *Anim. Cogn.* **2001**, *4*, 259–268. [CrossRef]
50. Zuur, A.F.; Ieno, E.N.; Walker, N.J.; Saveliev, A.A.; Smith, G.M. Mixed effects models and extensions in ecology with R. In *Statistics for Biology and Health*; Gail, M., Krickeberg, K., Samet, J.M., Tsiatis, A., Wong, W., Eds.; Springer: New York, NY, USA, 2009; pp. 323–341.
51. Lenth, R.V. Emmeans: Estimated Marginal Means, aka Least-Squares Means. R Package Version 1.5.4. 2021. Available online: https://CRAN.R-project.org/package=emmeans (accessed on 13 March 2021).
52. Brosnan, S.F.; Bshary, R. On potential links between inequity aversion and the structure of interactions for the evolution of cooperation. *Behaviour* **2016**, *153*, 1267–1292. [CrossRef]
53. Brosnan, S.F.; de Waal, F.B.M. Monkeys reject unequal pay. *Nature* **2003**, *425*, 297–299. [CrossRef] [PubMed]
54. Mendres, K.A.; de Waal, F.B.M. Capuchins do cooperate: The advantage of an intuitive task. *Anim. Behav.* **2000**, *60*, 523–539. [CrossRef] [PubMed]
55. Cronin, K.A.; Kurian, A.V.; Snowdon, C.T. Cooperative problem solving in a cooperatively breeding primate (*Saguinus oedipus*). *Anim. Behav.* **2005**, *69*, 133–142. [CrossRef] [PubMed]
56. Ostojić, L.; Clayton, N.S. Behavioural coordination of dogs in a cooperative problem-solving task with a conspecific and a human partner. *Anim. Cogn.* **2014**, *17*, 445–459. [CrossRef] [PubMed]
57. Chalmeau, R.; Lardeaux, K.; Brandibas, P.; Gallo, A. Cooperative problem solving by orangutans (*Pongo pygmaeus*). *Int. J. Primatol.* **1997**, *18*, 23–32. [CrossRef]
58. Vail, A.L.; Manica, A.; Bshary, R. Fish choose appropriately when and with whom to collaborate. *Curr. Biol.* **2014**, *24*, 791–793. [CrossRef] [PubMed]
59. Amici, F.; Aureli, F.; Call, J. Fission-fusion dynamics, behavioral flexibility, and inhibitory control in primates. *Curr. Biol.* **2008**, *18*, 1415–1419. [CrossRef]
60. Evans, T.A.; Beran, M.J. Delay of gratification and delay maintenance by rhesus macaques (*Macaca mulatta*). *J. Gen. Psychol.* **2007**, *134*, 199–216. [CrossRef]
61. Zajonc, R.B. Social facilitation. *Science* **1965**, *149*, 269–274. [CrossRef] [PubMed]
62. Roma, P.G.; Silberberg, A.; Ruggiero, A.M.; Suomi, S.J. Capuchin monkeys, inequity aversion, and the frustration effect. *J. Comp. Psychol.* **2006**, *120*, 67–73. [CrossRef]
63. Silberberg, A.; Cescimbene, L.; Addessi, E.; Anderson, J.R.; Visalberghi, E. Does inequity aversion depend on a frustration effect? A test with capuchin monkeys (*Cebus apella*). *Anim. Cogn.* **2009**, *12*, 505–509. [CrossRef]
64. Hopper, L.M.; Lambeth, S.P.; Schapiro, S.J.; Brosnan, S.F. Social comparison mediates chimpanzees' responses to loss, not frustration. *Anim. Cogn.* **2014**, *17*, 1303–1311. [CrossRef] [PubMed]
65. Wynne, C.D.L. Fair refusal by capuchin monkeys. *Nature* **2004**, *428*, 140. [CrossRef]
66. Dubreuil, D.; Gentile, M.S.; Visalberghi, E. Are capuchin monkeys (*Cebus apella*) inequity averse? *Proc. R. Soc. B* **2006**, *273*, 1223–1228. [CrossRef] [PubMed]
67. Bräuer, J.; Call, J.; Tomasello, M. Are apes really inequity averse? *Proc. R. Soc. B* **2006**, *273*, 3123–3128. [CrossRef]
68. Bräuer, J.; Call, J.; Tomasello, M. Are apes inequity averse? New data on the token-exchange paradigm. *Am. J. Primatol.* **2009**, *71*, 175–181. [CrossRef]
69. Neiworth, J.J.; Johnson, E.T.; Whilock, K.; Greenberg, J.; Brown, V. Is a sense of inequity an ancestral primate trait? Testing social inequity in cotton top tamarins (*Saginus oedipus*). *J. Comp. Psychol.* **2009**, *123*, 10–17. [CrossRef] [PubMed]
70. Sheskin, M.; Ashayeri, K.; Skerry, A.; Santos, L.R. Capuchin monkeys (*Cebus apella*) fail to show inequality aversion in a no-cost situation. *Evol. Hum. Behav.* **2014**, *35*, 80–88. [CrossRef]
71. Engelmann, J.M.; Clift, J.B.; Herrmann, E.; Tomasello, M. Social disappointment explains chimpanzees' behavior in the inequity aversion task. *Proc. R. Soc. B* **2017**, *284*, 20171502. [CrossRef] [PubMed]
72. Oberliessen, L.; Kalenscher, T. Social and non-social mechanisms of inequity aversion in non-human animals. *Front. Behav. Neurosci.* **2019**, *13*, 133. [CrossRef]
73. Adams, J.S. Towards an understanding of inequity. *J. Abnorm. Soc. Psych.* **1963**, *67*, 422–436. [CrossRef]
74. van Wolkenten, M.; Brosnan, S.F.; de Waal, F.B. Inequity responses of monkeys modified by effort. *Proc. Natl. Acad. Sci. USA* **2007**, *104*, 18854–18859. [CrossRef]
75. Wascher, C.A.F.; Bugnyar, T. Behavioral responses to inequity in reward distribution and working effort in crows and ravens. *PLoS ONE* **2013**, *8*, e56885. [CrossRef]
76. Laumer, I.B.; Massen, J.J.M.; Wakonig, B.; Lorck-Tympner, M.; Carminito, C.; Auersperg, A.M.I. Tentative evidence for inequity aversion to unequal work-effort but not to unequal reward distribution in Goffin's cockatoos. *Ethology* **2020**, *126*, 185–194. [CrossRef]
77. Fehr, E.; Fischbacher, U. Social norms and human cooperation. *Trends Cogn. Sci.* **2004**, *8*, 185–190. [CrossRef] [PubMed]
78. Tomasello, M. *A Natural History of Human Morality*; Harvard University Press: Cambridge, UK, 2016.
79. Henrich, J. *The Secret of our Success: How Culture is Driving Human Evolution, Domesticating our Species, and Making Us Smarter*; Princeton University Press: Princeton, NJ, USA, 2015.

80. Yamamoto, S.; Takimoto, A. Empathy and fairness: Psychological mechanisms for eliciting and maintaining prosociality and cooperation in primates. *Soc. Just. Res.* **2012**, *25*, 233–255. [CrossRef]
81. Fontenot, M.B.; Watson, S.L.; Roberts, K.A.; Miller, R.W. Effects of food preferences on token exchange and behavioural responses to inequality in tufted capuchin monkeys, *Cebus apella*. *Anim. Behav.* **2007**, *74*, 487–496. [CrossRef]
82. Brosnan, S.F.; Talbot, C.; Ahlgren, M.; Lambeth, S.P.; Schapiro, S.J. Mechanisms underlying responses to inequitable outcomes in chimpanzees, *Pan troglodytes*. *Anim. Behav.* **2010**, *79*, 1229–1237. [CrossRef]
83. Brosnan, S.F.; Hopper, L.M.; Richey, S.; Freeman, H.D.; Talbot, C.F.; Gosling, S.D.; Lambeth, S.P.; Schapiro, S.J. Personality influences responses to inequity and contrast in chimpanzees. *Anim. Behav.* **2015**, *101*, 75–87. [CrossRef]
84. Thaler, R.H. Anomalies: The ultimatum game. *J. Econ. Perspect.* **1988**, *2*, 195–206. [CrossRef]
85. Hill, D.A. The effects of demographic variation on kinship structure and behavior in cercopithecines. In *Kinship and Behavior in Primates*; Chapais, B., Berman, C.M., Eds.; Oxford University Press: New York, NY, USA, 2004; pp. 132–150.
86. Cronin, K.A.; Jacobson, S.L.; Bonnie, K.E.; Hopper, L.M. Studying primate cognition in a social setting to improve validity and welfare: A literature review highlighting successful approaches. *PeerJ* **2017**, *5*, e3649. [CrossRef] [PubMed]
87. Thierry, B. Covariation of conflict management patterns across macaque species. In *Natural Conflict Resolution*, 1st ed.; Aureli, F., de Waal, F.B.M., Eds.; University of California Press: Berkeley, CA, USA, 2000; pp. 106–128.
88. Pereira, M.E.; Fairbanks, L.A. *Juvenile Primates: Life History, Development and Behavior*; University of Chicago Press: Chicago, IL, USA, 2002.
89. Kawai, M. On the rank system in a natural group of Japanese monkey (I). *Primates* **1958**, *1*, 111–130. [CrossRef]
90. Sugiyama, Y. Life history of male Japanese monkeys. *Adv. Stud. Behav.* **1976**, *7*, 255–284. [CrossRef]
91. Suzuki, S.; Hill, D.A.; Sprague, D.S. Intertroop transfer and dominance rank structure of nonnatal male Japanese macaques in Yakushima, Japan. *Int. J. Primatol.* **1998**, *19*, 703–722. [CrossRef]
92. Furuichi, T. Interindividual distance and influence of dominance on feeding in a natural Japanese macaque troop. *Primates* **1983**, *24*, 445–455. [CrossRef]
93. Whitten, P.L. Diet and dominance among female vervet monkeys (*Cercopithecus aethiops*). *Am. J. Primatol.* **1983**, *5*, 139–159. [CrossRef] [PubMed]
94. Ekman, J.B.; Lilliendahl, K. Using priority to food access: Fattening strategies in dominance-structured willow tit (*Parus montanus*) flocks. *Behav. Ecol.* **1993**, *4*, 232–238. [CrossRef]
95. Gerald, M.S. The finding of an inverse relationship between social dominance and feeding priority among pairs of unfamiliar adult male vervet monkeys (*Cercopithecus aethiops sabaeus*). *Primates* **2002**, *43*, 127–132. [CrossRef] [PubMed]
96. Chalmeau, R.; Gallo, A. What chimpanzees (*Pan troglodytes*) learn in a cooperative task. *Primates* **1996**, *37*, 39–47. [CrossRef]
97. Itani, J. Vocal communication of the wild Japanese monkey. *Primates* **1963**, *4*, 11–66. [CrossRef]

animals

MDPI

Article

Carrion Crows and Azure-Winged Magpies Show No Prosocial Tendencies When Tested in a Token Transfer Paradigm

Lisa Horn [1,*], Jeroen S. Zewald [2], Thomas Bugnyar [1] and Jorg J. M. Massen [1,2]

1 Department of Behavioral and Cognitive Biology, University of Vienna, Althanstraße 14, 1090 Vienna, Austria; thomas.bugnyar@univie.ac.at (T.B.); j.j.m.massen@uu.nl (J.J.M.M.)
2 Animal Behavior and Cognition, Department of Biology, Utrecht University, Padualaan 8, 3584 CH Utrecht, The Netherlands; j.s.zewald@students.uu.nl
* Correspondence: lisa.horn@univie.ac.at

Simple Summary: Sharing your possessions or donating them to others, as well as helping others, can be summarized under the term, "prosocial behavior". Recently, researchers have become interested in whether, and in which situations, animals help and share. In this study, we trained carrion crows and azure-winged magpies—two bird species that have previously been found to donate food to their group members—to exchange objects, called "tokens", with an experimenter for food. We then tested whether the birds would provide these tokens to their group members when they themselves did not have access to the experimenter, but their group members did. We found, however, that there were very few attempted and/or successful token transfers between the birds, suggesting that they were not prosocial in this situation. We argue that the carrion crows and azure-winged magpies might not have fully understood the value of the tokens, either as placeholders for food or as "currency" with which food could be obtained. This limited understanding might have prevented them from exhibiting prosocial behavior in the current study. Therefore, we advocate the use of simpler and more naturalistic paradigms to study prosocial behaviors, such as providing food or resources to others, in a broader range of species.

Abstract: To study the evolution of humans' cooperative nature, researchers have recently sought comparisons with other species. Studies investigating corvids, for example, showed that carrion crows and azure-winged magpies delivered food to group members when tested in naturalistic or simple experimental paradigms. Here, we investigated whether we could replicate these positive findings when testing the same two species in a token transfer paradigm. After training the birds to exchange tokens with an experimenter for food rewards, we tested whether they would also transfer tokens to other birds, when they did not have the opportunity to exchange the tokens themselves. To control for the effects of motivation, and of social or stimulus enhancement, we tested each individual in three additional control conditions. We witnessed very few attempts and/or successful token transfers, and those few instances did not occur more frequently in the test condition than in the controls, which would suggest that the birds lack prosocial tendencies. Alternatively, we propose that this absence of prosociality may stem from the artificial nature and cognitive complexity of the token transfer task. Consequently, our findings highlight the strong impact of methodology on animals' capability to exhibit prosocial tendencies and stress the importance of comparing multiple experimental paradigms.

Keywords: cooperation; prosociality; instrumental helping; comparative methods; corvid; *Corvus corone*; *Cyanopica cyana*

check for updates

Citation: Horn, L.; Zewald, J.S.; Bugnyar, T.; Massen, J.J.M. Carrion Crows and Azure-Winged Magpies Show No Prosocial Tendencies When Tested in a Token Transfer Paradigm. *Animals* **2021**, *11*, 1526. https://doi.org/10.3390/ani11061526

Academic Editors: Lucia Regolin and Maria Loconsole

Received: 30 April 2021
Accepted: 16 May 2021
Published: 24 May 2021

1. Introduction

Prosocial behavior (i.e., voluntary behavior that benefits another individual at no gain for and no or low costs to the actor [1]) has long been regarded as one of the most important foundations of human social interactions. Various hypotheses have been formulated to explain the evolution of prosocial behavior in our lineage, proposing, for example, a high social tolerance [2] or allomaternal care [3] as important prerequisites for the emergence of prosociality. While prosocial behavior relates to the actions of a specific actor [1], it is clearly linked to and has been shown to promote cooperation [4], and it may therefore play an important role in the evolution of cooperation in general [3]. Growing evidence of prosocial tendencies in non-human animals has recently led researchers to widen their focus and gain a comparative perspective on prosociality [1]. Understanding the presence and prevalence of, as well as the motivations for, prosocial behavior in non-human animals has therefore been the aim of a host of studies in the fields of behavioral biology and comparative psychology in the last three decades (for a review, see [1,5]). Most studies on prosocial behavior have so far been conducted in non-human primates, whereas there have been far fewer attempts to study prosociality in other mammals (for an exception, see [6,7] for some positive results in dogs and rats, respectively) or other classes of animals, such as birds. While giving a comprehensive review of all species whose prosocial behavior has been investigated in recent years is beyond the scope of this article, it is important to note that researchers have used a multitude of different approaches to investigate animal prosociality, ranging from observations of natural behavior to complex experimental paradigms designed to probe the cognitive mechanisms guiding animals' prosocial choices. We will therefore first give an overview of different methodological approaches to studying prosociality and subsequently review the existing evidence of prosocial behavior in corvids and parrots, two bird taxa that are renowned for their sophisticated socio-cognitive skills.

Providing food to others is often regarded as a relatively common form of prosocial behavior in non-human animals that can be easily observed in naturalistic or semi-naturalistic settings. In such studies, one individual typically receives one or multiple food item(s), and it is observed whether the individual provides some or all of this food to other group members [8]. However, discerning the underlying motivations for transferring food (e.g., voluntary provisioning vs. harassment avoidance) can be difficult when only observing naturally occurring interactions [8]. By using restricted setups, in which only one donor animal has access to food that can be provided to a recipient in an adjacent compartment, and by including control conditions (e.g., no recipient present; recipient already has food), researchers have been able to investigate the motivations for providing food in these situations more closely [9]. Similarly, in a relatively simple experimental setup, the so-called group service paradigm, individuals can choose whether or not to operate an apparatus installed in their regular enclosure, which results in food becoming available to their group members but not to themselves [3,10,11]. One or two control conditions (e.g., empty control: no food available; blocked control: access to food is blocked) ascertain that operating the apparatus in the prosocial test condition is aimed at benefitting the group members and does not reflect the absence of sufficient inhibitory control.

Another common way to investigate prosocial behavior in non-human animals is through prosocial choice tasks. In these tasks, a donor animal and a recipient animal are placed in adjacent compartments. The donor is given a choice between two trays or containers with different reward combinations, one of which typically contains food rewards for both the donor and the recipient (i.e., prosocial choice), while the other contains only a food reward for the donor (i.e., selfish choice; see [1] for an overview of other possible reward combinations and control conditions). The donor chooses by either pulling a tray within reach or opening a container, thereby making the specific reward combination available to both individuals. The advantage of prosocial choice tasks is that they can be used to discern the exact conditions under which an individual is willing to share with another. However, understanding the task contingencies often poses a cognitive challenge for the tested individuals, and the visibility of food can be distracting for the donors, resulting in limited attention being paid to the consequences of the donor's choice for the recipient [1].

One way to circumvent the distracting aspect of having the donor animal deal directly with food rewards is to use tokens. A token is an inherently non-valuable object that acquires an associative value upon exchange with the experimenter—typically by exchanging the token for a food reward [12]. Tokens have thus been shown to become placeholders for actual food rewards [13]. One way to use tokens is in token choice tasks, which are analogous to the abovementioned prosocial choice tasks. In this case, the donor first learns to distinguish between two tokens that result in different reward combinations for the donor and the recipient (e.g., prosocial token vs. selfish token [14]). Instead of pulling a tray or opening a container, the donor animal's choice of a specific token indicates the reward combination that will be delivered by the experimenter. Consequently, such paradigms involve an extra step for the donor: instead of seeing the reward distribution in front of them and making a physical choice, the donors first have to acquire the correspondence between a specific token and a specific reward distribution and then use this knowledge to make their choice between different reward distributions. Therefore, learning the task contingencies in token choice paradigms is challenging.

Somewhat cognitively simpler than token choice tasks are token transfer paradigms. Here, the donor animal is provided with tokens, each of which can be exchanged for one food item. Therefore, the token may become a placeholder for a food reward, instead of a specific reward distribution. In the crucial prosocial test condition, the donor itself cannot use these tokens to obtain food, either because the donor received tokens that would only result in a food reward when another individual exchanges them with the experimenter [15] or because the donor's access to the exchanging experimenter is blocked [16]. In both cases, the donor has the possibility to transfer the tokens to one or several recipient animals in an adjacent compartment, which would then be able to exchange the tokens with the experimenter for a food reward. The latter paradigm, in which the donor cannot access the exchanging experimenter, has been argued to resemble instrumental helping tasks. In such tasks, a donor is in possession of a tool but does not have access to an apparatus that can be operated with this tool, while a recipient has access to the apparatus but does not possess the tool [17]. Tool transfers from the donor to the recipient have been regarded as prosocial behavior [1]. Using actual tools in an experimental setup, however, restricts the possibilities of testing prosocial tendencies in this way to those animals that habitually use tools or, at the very least, understand the use of tools. Overall, there is a great variety of experimental setups to test prosociality in animals, each with their own benefits and downsides, which need to be tailored to the specific question and the (cognitive) constraints of the studied species, particularly when investigating prosocial behavior outside of the primate order [1].

In birds, for example, our understanding of the prevalence and extent of prosocial behavior is relatively incomplete due to the lack of studies, although several bird species show the social and cognitive prerequisites for prosociality. Large-brained and long-living bird species, such as corvids and parrots, exhibit sophisticated socio-cognitive abilities, such as social learning [18–20], perspective taking [21–23], and cooperation [24–28], among other abilities (for a review see [29]). Corvid and parrot species typically form monogamous pair bonds and food sharing between bonded partners, as well as food provisioning to dependent offspring by both parents is common [30]. Moreover, in cooperatively breeding bird species (e.g., carrion crows [31], azure-winged magpies [32], and eclectus parrots [33]), other group members—kin as well as non-kin—provide food to the breeding female and her offspring [34]. Systematic observations of food provisioning during free interactions revealed that multiple corvid species and one parrot species actively provided food to conspecifics, regardless of bond status, sex, and kinship [35–37]. In restricted setups, where only one donor bird received food, pinyon jays and azure-winged magpies actively provided food to their conspecifics [38,39]. The azure-winged magpies even took into account whether the potential recipients had access to food of their own or not and preferentially gave their own food to the group members, when no food was available in the recipients' compartment [39]. Similarly, tool-using Goffin's cockatoos were found to regularly provide tools to their group members [40]. Certain individuals even paid attention to the specific apparatus that the recipient was faced with and preferentially provided functional tools over non-functional tools [40].

Prosocial choice tasks in corvids and parrots have so far delivered mixed results, sometimes even within the same species. In a two-way choice task, pinyon jays preferred to deliver food to a conspecific over delivering it to an empty cage, but only when the tray also contained a reward for the donor itself (i.e., prosociality) and not when there was nothing to gain for the donor (i.e., altruism) [41]. Eurasian jackdaws were tested in a paradigm that required them to remove the lid from one of two transparent boxes [42]. One box contained a food reward for both the donor and the recipient, while the other contained a reward only for the donor. The jackdaws did provide food to their partners, but only when the recipients first showed interest in the side where food was available, meaning that the donors' prosocial choices could likely be explained by social or stimulus enhancement [42]. In a similar setup, however, common ravens were completely indifferent to the gains of a conspecific recipient and chose randomly between the prosocial and selfish options [43]. Similarly, in a two-alternative prosocial choice task ravens did not preferentially pull a tray with a food reward on the recipient's side, compared to another tray in front of an empty control compartment, but mostly stopped pulling the trays altogether when they received nothing for themselves [44]. However, in these prosocial choice studies, it was not clear how acquiring the task contingencies, the visibility of food, and the donors' attention influenced prosocial performance [1]. Horn and colleagues [45] tested eight corvid species in the group service paradigm, in which the birds were tested in their regular social groups and could choose to provide food to their group members by landing on a simple seesaw apparatus. The authors found high rates of prosocial food deliveries in three corvid species (azure-winged magpies, carrion crows, and one group of New-Caledonian crows). Other species (e.g., common ravens) delivered very little food to their group members. The authors argued that the variation of prosocial provisioning across species could be traced to sex-specific positive effects of cooperative breeding and colonial nesting [45].

Despite the fact that tokens have been used successfully to test economic decision-making processes in corvids and parrots (e.g., delay of gratification [46–48]), the results of prosocial token choice tasks, in which donors chose between tokens that indicated prosocial or selfish reward combinations, have so far indicated a limited understanding of the task contingencies in two tested parrot species (African grey parrots [49,50]; kea [51]). Similarly, in a token transfer experiment, in which ravens and crows learned to discriminate between self-value tokens, partner-value tokens, and no-value tokens, there were very few transfers between the individuals, and the pattern of the token transfers indicated an equally limited understanding [52]. In a simpler token transfer task, however, in which donors had the chance to transfer tokens to recipients once they themselves were prevented from exchanging the tokens, African grey parrots spontaneously and voluntarily transferred tokens to the recipients [53]. They did so significantly more when they could truly benefit a partner than in two control conditions, where either no recipient was present in the adjacent compartment or where the recipient did not have the possibility to exchange tokens with the experimenter [53]. Blue-headed macaws [53] and ravens [54], on the other hand, did not show any strong evidence of prosocial behavior when tested in the same paradigm. These findings illustrate that there seems to be considerable species variation in the propensity to engage in prosocial behavior, though they can only be addressed by applying comparable methodologies across species. Additionally, when investigating a complex phenomenon like prosociality, it seems essential to use multiple experimental paradigms within the same species but, most importantly, to also explore the validity of these paradigms for those species.

Therefore, in this study, we decided to use the simple token transfer paradigm that has already been used in common ravens, African grey parrots and blue-headed macaws [53,54] to investigate prosocial tendencies in two corvid species that are promising candidates for exhibiting prosocial behavior: carrion crows (*Corvus corone*) and azure-winged magpies (*Cyanopica cyana*). Cooperative breeding, including food provisioning for breeding females and offspring by kin and non-kin helpers, has been documented in both species in the wild [31,32]. Both species exhibited high rates of prosocial food deliveries when tested in the group service paradigm [45,55]. Additionally, azure-winged magpies actively provided food to their group members in a naturalistic food provisioning experiment [39]. Consequently, we predicted that, if the animals understand the contingencies of this specific task, as it was demonstrated in African grey parrots, the crows and magpies would transfer tokens to their group members once their own access to the exchanging experimenter was blocked. To probe the underlying motivations for transferring tokens and to control for the effects of motivation, as well as of social and stimulus enhancement, we additionally tested each individual in three control conditions (i.e., non-social control, social control, and motivational control).

2. Materials and Methods

2.1. Subjects, Housing, and Ethical Considerations

We tested seven captive carrion crows (3F/4M) and four captive azure-winged magpies (3F/1M). In terms of appearance, the crows were either carrion crows or hybrids of carrion and hooded crows, reflecting the hybridization belt in Europe. Both crow species have highly similar life histories and are often considered to be subspecies [56]. All of the crows were adults (i.e., between 3–8 years old) and hand-raised. At the start of this study, five of the crows (Daisy, Paula, Peppi, Caruso, Saul) had been kept in a group for more than two years, with three additional crows that never learned to successfully exchange tokens but were present as potential recipients (Juno, Signore, Soukie). Caruso and Juno left the group in order to establish a breeding pair in January 2016. In February 2017, Walter and Willi joined the group from another research station (Konrad Lorenz Research Station for Behaviour and Cognition, Grünau, Austria), where they had previously participated in multiple other behavioral studies. The crows were housed in a social group setting at the Haidlhof Research Station, Bad Vöslau, Austria. The aviary comprised a large outdoor part (12 × 9 × 5 m) and two adjacent roofed experimental compartments (3 × 4 × 5 m each), where the experiment was conducted. During the whole period the crows were fed a diverse diet containing meat, milk products, cereal, vegetables, and fruit twice a day. The experiment was conducted prior to the first feeding of the day with a high-quality food reward (little pieces of Frolic® (Mars, Inc.; McLean, VA, USA dog food). The crows had ad libitum access to water in each compartment of the aviary at all times.

The four azure-winged magpies were siblings from the same nest, 4 years old at the time of testing, and hand-raised. The magpies were housed in a social group setting at the Animal Care Facility of the Department of Behavioral and Cognitive Biology, the University of Vienna, Vienna, Austria. The outdoor aviary (4.3 × 3 × 3m) was partially covered with a semi-transparent roof and consisted of two equally sized compartments that were separated by a central wire mesh partitioning. The two compartments could be closed off with two sliding doors made from wire mesh. The magpies were fed a diverse diet of different fruits, insects, seeds, meat, and egg. The experiment was conducted prior to the first feeding of the day with a high-quality food reward (mealworms). The magpies had ad libitum access to water and pellets ("Beo komplet", NutriBird®, Versele-Laga; Deinze, Belgium) in each compartment of the aviary at all times.

All subjects participated in the experiment voluntarily. The study followed the Guidelines for the Use of Animals [57] in accordance with the national legislation. The study was reviewed and approved by the ethical board of the Faculty of Life Sciences, the University of Vienna (case number: 2016-018).

2.2. Training

Prior to the experiment, the subjects were trained to be isolated from the group in each of the compartments of the respective aviaries and to exchange tokens through the wire mesh of the aviary with an experimenter (L.H. and J.S.Z.) for a piece of food. We chose different tokens for each species due to the size difference between the two species. For the carrion crows, the tokens were green plastic bottle tops (diameter: 30 mm, weight: 3 g) and for the azure-winged magpies, they were white plastic rings (diameter: 16 mm, weight: 0.3 g). Exchanging was shaped using positive reinforcement with a food reward and verbal praise. First, we started to train the birds to exchange in a group setting in several different areas around the respective aviaries. Subsequently, they were trained to exchange when isolated from the group and *only* at the site where a small table was placed outside the aviary. The subjects reached the criterion to participate in the experiment when they were able to transport and then transfer 10 tokens placed at a large distance from the site of the exchange table (e.g., in the other compartment for the carrion crows, in the back of the same compartment for the azure-winged magpies, etc.) within 10 min in a minimum of 4 training sessions (see Supplementary Video S1; the video shows the donor's exchange behavior in the motivation control condition, where exchange with the experimenter is possible, and illustrates the motivation and training level in token exchange in the donors of both species).

The training and subsequent experiments with the carrion crows were conducted between September 2015 and September 2016 with Daisy, Paula, Peppi, Caruso, and Saul and between April 2017 and September 2017 with Walter and Willi. The individual training to criterion took 3–11 months. The training and subsequent experiments with the azure-winged magpies were conducted between July 2020 and October 2020. The individual training to criterion took 2 months. The birds were never trained to transfer tokens through the wire mesh partitioning between the two compartments that were used in the experimental setup (see description below; red dashed lines in Figure 1).

2.3. Experimental Setup

We replicated the experimental setup of experiment 1 of Massen and colleagues [54] in a group setting, instead of a dyadic setting. For the carrion crows, the donor was separated from the group in one of the two experimental compartments and was tested in four different experimental conditions. In each of these conditions, the donor received 10 tokens, i.e., these were placed into the compartment in which the donor was residing. In the **test condition** (Figure 1a), the group—not including the donor—had access to the adjacent compartment, and the exchange table was placed in front of that compartment. Therefore, the donor did not have the opportunity to exchange tokens with the experimenter, but if the donor transferred a token to the adjacent compartment, the other birds could subsequently exchange the token for a food reward. The setup in the **non-social control condition** (Figure 1b) was the same as in the test, but the group did not have access to the adjacent compartment. Therefore, there was nobody present to receive the transferred tokens. With this condition, we aimed to control for the possibility that the donor might transfer tokens to the other compartment in the test condition simply to get the tokens as close to the exchange table as possible, rather than actually sharing them. In the **social control condition** (Figure 1c), the group had access to the adjacent compartment, but there was no exchange table. Therefore, even if the donor transferred a token to the adjacent compartment, the other birds could not exchange the token with the experimenter. This condition was implemented to control for potential playfulness, i.e., transferring tokens to the group members as play, rather than with the aim of helping them obtain the food reward. Finally, in the **motivation control condition** (Figure 1d), the group had access to the adjacent compartment, but the exchange table was placed in front of the donor's compartment. Therefore, the donor itself had the opportunity to exchange tokens for food rewards with the experimenter.

For the azure-winged magpies, the experimental setup was identical to the carrion crows in the **test condition** (Figure 1e), the **social control condition** (Figure 1g), and the **motivation control condition** (Figure 1h). The **non-social control condition** was modified, as there were only two compartments in the aviary, so having an empty compartment was not possible. Therefore, the exchange table was placed next to the aviary on the side of the donor's compartment at a distance of 40 cm (see Figure 1f). The other three magpies were in the second compartment.

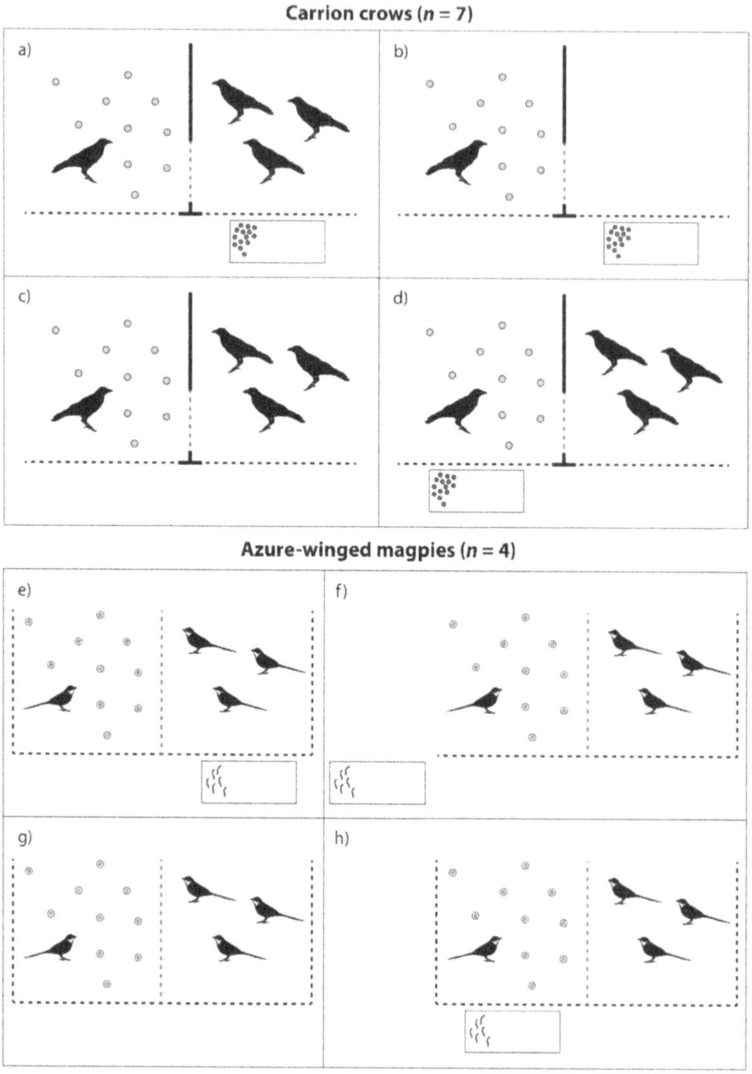

Figure 1. Schematic representation of the experimental setup in the four different conditions for the carrion crows (**a**) test; (**b**) non-social control; (**c**) social control; and (**d**) motivation control and for the azure-winged magpies (**e**) test; (**f**) non-social control; (**g**) social control; and (**h**) motivation control. The red dashed line indicates the wire mesh partitioning separating the two experimental compartments. The blue dashed line indicates the wire mesh of the aviary opposite the other compartment, near to where the exchange table was placed in the non-social control condition for the azure-winged magpies. The dimensions in the drawing are not to scale.

2.4. Experimental Procedure

Each trial started with the subject receiving 10 tokens and lasted until the 10 tokens were transferred or exchanged or until a maximum of 10 min. Each subject was tested in two sessions, each session comprising one trial of each of the four experimental conditions (i.e., 8 trials in total). The two sessions were conducted on two different days. The sequence of conditions was semi-randomized, with the stipulation that it was different in each of the subject's sessions. Each subject was tested once in the left compartment and once in the right compartment, and the first location was counterbalanced across birds.

The azure-winged magpies started to lose motivation to participate in the experiment after the first session (see results section). Therefore, with the magpies, we added 3 days of token training and a second round of two test sessions for each subject. During the second round, we conducted the experimental sessions in the morning before the first feeding of the day and conducted additional individual training sessions every day in the afternoons.

All experiments were recorded simultaneously using two high-definition camcorders (Canon LEGRIA® HD, Canon Inc., Tokyo, Japan; Panasonic HC-X909, Panasonic Co., Osaka, Japan).

2.5. Data Collection and Analysis

The donors' behavior was scored live by L.H. and J.S.Z., and the scores were confirmed from the video recordings. We scored how many tokens were transferred through the wire mesh partitioning separating the two experimental compartments (red dashed lines in Figure 1). In the non-social control with the azure-winged magpies, we additionally scored whether any tokens were transferred through the outer wire mesh of the aviary opposite the other compartment, near to where the exchange table was placed in this condition (blue dashed line in Figure 1f). Further, we scored all unsuccessful attempts to transfer tokens in the same locations mentioned above (i.e., pushing a token against the wire mesh partitioning without transferring it through or placing a token close to the wire mesh, on the rim of the wire mesh frame, or in a crack or below the wire mesh frame). Finally, since many corvid species have been found to cache valuable food items or objects [58–60], and ravens cached many of the tokens when tested in a similar paradigm [54], we scored how many tokens each individual donor cached in each condition. For the donor scores, we took each variable's sum across the sessions of a specific condition.

Additionally, the recipients' behavior was scored from the video recordings. We were unable to reliably identify individual recipients from the video recordings in the case of the azure-winged magpies due to the small size of their identifying color rings and the distance and resolution of the camera. Therefore, for both species, we coded the duration that one or more recipient bird(s) spent within one body length distance of the wire mesh partitioning, where they could have received a token (red dashed lines in Figure 1), in each donor's test and social control conditions. For only the carrion crows, we also coded each individual recipient's presence close to the wire mesh partitioning, and these results are presented in File S1, Table S1. Further, for both species, we scored the total frequency of begging calls and of all other vocalizations given by all recipients in each donor's test and social control conditions. Additionally, in each donor's test condition, we coded the duration that one or more recipient bird(s) spent in front of the exchange table in their compartment. For only the carrion crows, we also coded each individual recipient's presence in front of the exchange table (see File S1, Table S1). The recipient behavior scores were always averaged across the sessions of a specific condition for each donor.

All statistical analyses were carried out with the pooled data from both species. Due to the second round of testing, which we added only for the azure-winged magpies, we ran each analysis on two datasets that differed in terms of the data included from the magpies: The first dataset contained only the magpies' scores across the first two sessions, which were carried out using exactly the same procedure as those employed in the case of the carrion crows. The second dataset contained the magpies' scores across all four sessions. The data from the carrion crows were the same in both datasets (full raw data on the recipient behavior is available in File S1). To assess whether the donors were prosocial, we compared the number of tokens they transferred and attempted to transfer to the adjacent compartment in the three conditions, in which they could not exchange the tokens with the experimenter themselves (i.e., test, non-social control, and social control), using Friedman tests and pair-wise Wilcoxon tests, with Holm-Bonferroni correction for multiple testing. Similarly, we compared the number of cached tokens in those three conditions with a Friedman test. To assess whether the recipients displayed more attention-getting behavior in the test condition, in which they had the possibility to exchange tokens with the experimenter, than in the social control, in which they did not, we compared the duration spent in proximity to the wire mesh partitioning, the number of begging calls, and the number of all other vocalizations across these two conditions using a Wilcoxon test. Additionally, we compared the duration spent in proximity to the wire mesh partitioning and the duration spent at the exchange table in the test condition to assess where the recipients spent more time. All analyses were performed with RStudio (R version 3.6.0).

3. Results

3.1. Donor Behavior

The motivation control conditions showed that the carrion crow donors were highly motivated to participate in the experiment. They exchanged all available tokens in both sessions. All subjects except for two birds in one session each (Caruso, S1; Paula, S2) did so within the first two minutes of the trial (mean time to exchange all 10 tokens ± SD = 103 s ± 84 s). The azure-winged magpie donors started to lose motivation to participate in the experiment after the first session: two birds exchanged no tokens for food rewards in the motivation control condition in the second session (BB8, Poe), and one bird exchanged only one token (Rey). After introducing additional training sessions, in the second round, the azure-winged magpie donors exchanged all available tokens in the motivation control conditions (mean time to exchange all 10 tokens ± SD = 155 s ± 187 s).

Over the course of the experiment, the carrion crow donors only transferred 2 tokens to the adjacent compartment (Walter, Willi; both in the non-social control; see Table 1). Most attempts to transfer a token were observed in the non-social control ($n = 7$), followed by the test ($n = 5$) and the social control ($n = 1$). Only 1 token was transferred by an azure-winged magpie donor (Poe). The bird transferred the token in the non-social control through the outer wire mesh of the aviary, near to where the exchange table was placed and opposite to the compartment containing the group members (Figure 1f). Most of the azure-winged magpies' attempts to transfer a token were observed in the non-social control ($N_{Round1} = 4$; $N_{Round2} = 8$), followed by the test ($N_{Round1} = 1$; $N_{Round2} = 1$) and the social control ($N_{Round1} = 1$). Across the two species, there was a significant difference in the number of attempts between the three conditions, both when considering only the first round of the azure-winged magpies (Friedman test: $n = 11$, $X^2 = 6.348$, $p = 0.042$) and when considering the magpies' attempts in both rounds ($X^2 = 11.455$, $p = 0.003$; see Table 1). However, pair-wise comparisons did not reveal significant differences between any of the conditions when considering only the first round of the azure-winged magpies. Only when using the magpies' pooled data from both rounds were there significantly more attempts in the non-social control than in the social control (Wilcoxon test: $n = 11$, $W = 2.5$, $p = 0.034$). Importantly, however, there was no significant difference between the test and any of the control conditions.

Table 1. Name and sex of the donor, total number of attempted transfers, successful transfers, and exchanges, as well as the total number of caches in the 4 different conditions. Soc C = social control, Non-Soc C = non-social control, Mot C = motivation control. Note that the donor itself performed exchanges in the motivation control. * All of the attempted transfers and transfers listed here occurred at the outer wire mesh of the aviary, opposite to the other compartment and near to where the exchange table was placed in this condition (blue dashed line in Figure 1f).

Carrion Crows ($n = 7$)									
Donor	Sex	Total # Attempts/Transfers/Exchanges				Total # Caches			
		Test	Soc C	Non-Soc C	Mot C	Test	Soc C	Non-Soc C	Mot C
Daisy	F	0/0/0	0/0/–	1/0/–	0/0/20	1	0	0	0
Paula	F	0/0/0	0/0/–	1/0/–	0/0/20	0	1	1	0
Peppi	F	0/0/0	0/0/–	1/0/–	0/0/20	1	0	0	0
Caruso	M	2/0/0	1/0/–	0/0/–	0/0/20	6	6	2	0
Saul	M	0/0/0	0/0/–	0/0/–	0/0/20	0	2	2	0
Walter	M	2/0/0	0/0/–	2/1/–	0/0/20	1	0	0	0
Willi	M	1/0/0	0/0/–	2/1/–	0/0/20	0	2	0	0
Total		**5/0/0**	**1/0/–**	**7/2/–**	**0/0/140**	**9**	**11**	**5**	**0**

Table 1. *Cont.*

Azure-Winged Magpies (n = 4)—Round 1									
		Total # Attempts/Transfers/Exchanges				Total # Caches			
Donor	Sex	Test	Soc C	Non-Soc C *	Mot C	Test	Soc C	Non-Soc C	Mot C
BB8	F	1/0/0	1/0/–	1/0/–	0/0/10	1	0	0	0
Poe	F	0/0/0	0/0/–	0/0/–	0/0/10	0	0	0	0
Rey	F	0/0/0	0/0/–	0/0/–	1/0/11	0	0	0	0
Kylo	M	0/0/0	0/0/–	3/0/–	0/0/20	0	0	1	0
Total		**1/0/0**	**1/0/–**	**4/0/–**	**1/0/51**	**1**	**0**	**1**	**0**
Azure-Winged Magpies (n = 4)—Round 2									
		Total # Attempts/Transfers/Exchanges				Total # Caches			
Donor	Sex	Test	Soc C	Non-Soc C *	Mot C	Test	Soc C	Non-Soc C	Mot C
BB8	F	0/0/0	0/0/–	2/0/–	0/0/20	0	0	1	0
Poe	F	1/0/0	0/0/–	2/1/–	0/0/20	0	0	0	0
Rey	F	0/0/0	0/0/–	4/0/–	0/0/20	0	0	4	0
Kylo	M	0/0/0	0/0/–	0/0/–	0/0/20	2	0	0	0
Total		**1/0/0**	**0/0/–**	**8/1/–**	**0/0/80**	**2**	**0**	**5**	**0**

Further, there was no significant difference in how many tokens were cached for both species combined between the three conditions, both when considering only the first round of the azure-winged magpies (Friedman test: n = 11, X^2 = 0.667, p = 0.717) and when considering the magpies' caches across both rounds (X^2 = 0.839, p = 0.658; see Table 1). The carrion crow donors pushed five tokens through the front wire mesh of the donors' own compartment when no exchange table was present, which did not lead to a food reward (test: n = 3, social control: n = 1, non-social control: n = 1). The azure-winged magpie donors never showed this behavior.

3.2. Recipient Behavior

Combining both species, there was no significant difference in how much time the recipients spent close to the wire mesh partitioning between the test and the social control, both when considering only the first round of the azure-winged magpies (Wilcoxon test: n = 11, W = 37, p = 0.765; Figure 2) and when considering the magpies' proximity to the wire mesh in both rounds (W = 43, p = 0.413; see Supplementary Information File S1, Figure S1 for a detailed depiction of the azure-winged magpie behavior in both rounds). Correspondingly, there was no significant difference in the number of begging calls (only first round: W = 2, p = 0.789; in both rounds: W = 3, p = 1.000) or all other vocalizations (only first round: W = 4.5, p = 0.498; in both rounds: W = 4, p = 0.419) between these two conditions.

The recipients of both species spent significantly more time close to the exchange table than close to the wire mesh partitioning in the test condition, when considering only the first round of the azure-winged magpies (W = 66, p = 0.001). This difference was a non-significant trend, when considering the azure-winged magpies' behavior averaged from both rounds (W = 54, p = 0.067; see also Supplementary Information File S1).

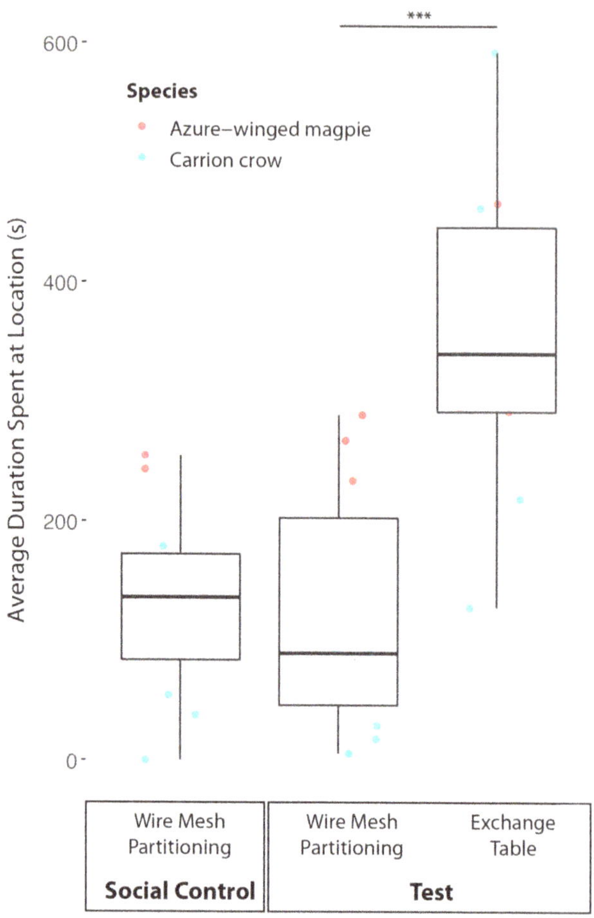

Figure 2. Duration that one or more recipients spent at the given locations averaged from the first two sessions of each donor individual per condition. The box plots represent the medians (horizontal lines), inter-quartile ranges (boxes), as well as minima and maxima (whiskers). All of the data are represented with dots. The dot colors indicate the species according to the legend in the top left corner. *** $p \leq 0.001$.

4. Discussion

In this experiment, we found no evidence for prosocial behavior in captive carrion crows and azure-winged magpies, when tested in a simple token transfer paradigm. In total, there were very few instances when the donors attempted to transfer a token (e.g., pushing it against the wire mesh partitioning or placing it close to the wire mesh), and only three tokens were transferred successfully. All of the transfers occurred in the non-social control condition and were transferred either to an empty compartment or through the outer wire mesh of the aviary, where no potential recipients were present in both cases. This result contradicts our main hypothesis that the crows and magpies would behave prosocially towards their group members, as has been observed in the wild [31,32] and demonstrated experimentally in captive birds in the group service paradigm for both species [45,55] and in a naturalistic food provisioning experiment with azure-winged magpies [39]. Given that prosocial behavior has been shown to depend on the context and/or on the specific group, when tested in other non-human animals (e.g., chimpanzees [11,61,62]), it is possible that the subjects were not motivated to act prosocially in the context of the current experiment.

It is, however, always difficult to interpret null results, because apart from an actual lack of a prosocial disposition, they may stem from a flaw in the experimental setup, its applicability to the specific species, and/or a lack of understanding of the procedure on the part of the subjects. In our case, for example, the possibility to transfer tokens through the wire mesh partitioning to the adjacent compartment where the potential recipients were located might not have been apparent to the donors due to a lack of experience. In order to get a spontaneous reaction in the test and to avoid shaping the birds' behavior, we never let them transfer tokens through that particular wire mesh partitioning during training. In a similar experiment with capuchin monkeys, the subjects were trained to both receive and transfer tokens through the partitioning separating them from their recipients prior to the test [16]. This might have been the reason why the rate of token transfers exhibited by the capuchins was higher than in our current study. However, the three birds that managed to transfer a token successfully in a control trial in our study did not continue to transfer tokens in subsequent test trials, suggesting that even with the experience of transferring tokens through the wire mesh partitioning, they "chose" not to do so in the crucial test condition. In another study investigating crows' and ravens' prosociality with a token transfer paradigm involving tokens of different value, the birds had ample experience with exchanging tokens with the experimenter through the partitioning separating them from their potential recipients. Despite this, the rates of token transfers between the birds were also very low for both species [52]. Therefore, it is unlikely that the lack of experience of transferring tokens through the wire mesh partitioning alone prevented the crows' and magpies' prosocial behavior in the current study. Nevertheless, it might be beneficial to provide experience with receiving and transferring tokens in all potential transfer locations in future studies, which would potentially increase the occurrence of the behavior. Control conditions and knowledge probes would still allow for the disentangling of different motivations for transferring the tokens during the experiment.

Scrutinizing the pattern of attempted transfers, however, might help to resolve the donors' motivations for transferring the tokens. There was a significant difference in the number of attempted transfers between the conditions, with the greatest number of attempts occurring in the non-social control condition. For both species, the attempted transfers in this condition occurred through the wire mesh that separated the donors from the area where the exchange table was placed, but no group members were present. For the azure-winged magpies, this meant attempting to transfer tokens outside of the aviary, rather than through the wire mesh partitioning to the other compartment, where potential recipients would have been present. In contrast, only one transfer attempt was made by each species in the social control condition, in which potential recipients would have been present but there was no exchange table. Therefore, it is likely that the pattern of attempted token transfers observed in the current study reflected the birds' attempts to move the tokens closer to the out-of-reach exchange table, rather than their willingness to provide the tokens to their group members. Similar arguments have been made when analyzing the patterns of token transfers in capuchin monkeys [16] and blue-headed macaws [53] in a similar token transfer paradigm.

The results obtained in the current study seem to indicate that the birds did not fully understand the quasi-symbolic nature of the token as a placeholder for food. Instead, they may have acquired an association between the action of exchanging an item with the experimenter and the corresponding food reward during training. While the birds never received food for transferring objects other than the tokens in the course of the study, the birds sometimes pushed other objects through the wire mesh when no tokens were immediately available. Similar observations have been made in ravens [54] and capuchin monkeys [63,64] and have been attributed to an intrinsic motivation to exchange tokens (e.g., that the exchange behavior is performed for its own sake [64]). Therefore, it is likely that the donors failed to attribute a specific value to the tokens. Even capuchin monkeys that were trained to successfully discriminate between three different tokens representing different food qualities showed poorer judgments of quantity when they had to choose between the tokens than when they had to choose between the actual food rewards [65]. Based on these findings, Beran and Parrish [13] argue that the use of tokens likely produces a high cognitive load, which prevents a true equivalence between the tokens and the represented food reward in non-human animals. Interestingly, the only bird species that has so far been found to express prosocial behavior by transferring tokens to a partner is the African grey parrot [53]. Through a long-running language acquisition project, this species became known for its symbolic abilities, such as acquiring labels for objects and categories, as well as expressing an understanding for quantity and similarity/difference concepts [66]. Therefore, it is possible that sophisticated symbolic abilities, which have so far only been demonstrated in very few non-human animal species, are a necessary prerequisite for successfully using tokens to investigate other abilities, such as prosocial behavior.

Another indication that the donors did not understand the value of the tokens is that they only rarely cached the tokens when they themselves did not immediately have the possibility to exchange them with the experimenter (azure-winged magpies: 2.3% of the tokens; carrion crows: 6%). In contrast, common ravens cached about half of the tokens when tested in a similar paradigm [54]. The ravens' behavior was interpreted as storing the valuable tokens for a potential future opportunity to exchange them with the experimenter, thereby fulfilling an important cognitive prerequisite for using tokens as a form of "currency" [13]. However, object manipulation and caching are important for ravens from an early age [67] and have been found to be generally more frequent in ravens than in crows [68]. To our knowledge, object caching has not yet been documented in azure-winged magpies. Therefore, the substantial differences obtained between the two studies could also be attributed to species-typical differences in object caching propensity. Nevertheless, the scarcity of caching behavior in the carrion crows and azure-winged magpies is a strong indication that the donors did not attribute a high value to the tokens and did not understand them as placeholders for food. This argument is corroborated by our analysis of the recipients' behavior. The potential recipients did not show more attention-getting behavior (e.g., begging, close proximity, etc.) towards the donors when they could have used the tokens to exchange them for food (i.e., in the test condition) than when there was no possibility to exchange the tokens (i.e., in the social control condition). By contrast, azure-winged magpies in a naturalistic food provisioning experiment begged more when they were in need of food and a group member had food that could be provided [39]. During the test condition in the current study, the recipients actually spent more time in front of the exchange table—despite their lack of exchangeable tokens—than close to the donor, who could have provided them with tokens. It is possible that the recipients were anticipating a possibility to exchange tokens with the experimenter in this location, although the duration spent in front of the exchange table decreased in the azure-winged magpies when they were tested in a second round (see Supplementary Information File S1, Figure S1). Nevertheless, this is another indication that the birds were more focused on the action of exchanging with the experimenter than on the tokens as valuable objects.

5. Conclusions

Using a token transfer paradigm, we found no evidence of prosociality in carrion crows and azure-winged magpies. We propose that the absence of prosocial token transfers in the current study may have stemmed from the complexity of the token transfer task and the required quasi-symbolic representation of the tokens. The extensive training required in this task also makes this paradigm less generalizable across contexts (e.g., when comparing captive birds with individuals in the wild). Consequently, using experimental paradigms that are more intuitive and put less cognitive load on the subject appear more promising for the investigation of prosocial behavior in corvids and other non-human animal species in general. Naturalistic food provisioning experiments, for example, allow animals to express their species-specific behavior, while well-thought-out control conditions can still help to probe the underlying motivations for providing the food [39] and can even allow us to investigate complex socio-cognitive mechanisms in the tested animals, such as understanding the desire states of conspecifics [69]. Simple experimental setups, such as the group service paradigm, which do not require individuals to be separated from the group, and where behavioral contingencies are obvious to the birds, are an equally promising avenue for the investigation of prosocial behavior [45,55]. Both paradigms can be easily adapted to other bird species or other non-human animal taxa (e.g., primates [3,4,10,11]) and could even be adapted to testing individuals in the wild (for an example of using the group service paradigm on corvids living in the wild, see [45]). In conclusion, while the current study fails to show prosocial tendencies in carrion crows and azure-winged magpies, when tested in a token transfer paradigm, our findings highlight the strong impact of methodology on whether animals are able to exhibit prosocial tendencies, as well as the importance of using valid comparative methods and systematic replications across different species.

Supplementary Materials: The following are available online at https://www.mdpi.com/article/10.3390/ani11061526/s1, Supplementary Information File S1: Supplementary recipient behavior results; Supplementary Video S1: Videos from the motivation control condition, showing one donor of each species and illustrating the donors' training level in token exchange; Supplementary Data File S1: Data on the recipient behavior used in the main analysis.

Author Contributions: Conceptualization, L.H. and J.J.M.M.; Data Collection, L.H. and J.S.Z.; Formal Analysis and Visualization, L.H.; Writing—Original Draft Preparation, L.H.; Writing—Review and Editing, L.H., J.S.Z., T.B. and J.J.M.M.; Funding Acquisition, L.H., J.S.Z., T.B. and J.J.M.M. All authors have read and agreed to the published version of the manuscript.

Funding: This study was funded by a Marie Jahoda grant from the University of Vienna to L.H. and an Erasmus+ Higher Education Traineeship to J.S.Z. The study received additional financial support from the Austrian Science Foundation (FWF) grants to J.J.M.M. (P26806) and T.B. (Y366-B17, W1234-G17).

Institutional Review Board Statement: The study was conducted according to the guidelines of the Declaration of Helsinki and the Guidelines for the Use of Animals, in accordance with national legislation. The study was reviewed and approved by the ethical board of the Faculty of Life Sciences, the University of Vienna (case number: 2016-018).

Data Availability Statement: Data on the donor behavior are available in full in Table 1 of the main manuscript. Raw data on the recipient behavior used in the main analysis are available in Supplementary Data File S1. Data on individual carrion crow recipient behavior are available in Table S1 of Supplementary Information File S1.

Acknowledgments: We thank Selina Schöndorfer and Sara Vutirakis for their help with coding the videos and the animal care staff at both research facilities. Additionally, we thank Kate Malone for proofreading the manuscript.

Conflicts of Interest: The authors declare no conflict of interest. The funders had no role in the design of the study; in the collection, analyses, or interpretation of data; in the writing of the manuscript, or in the decision to publish the results.

References

1. Marshall-Pescini, S.; Dale, R.; Quervel-Chaumette, M.; Range, F. Critical issues in experimental studies of prosociality in non-human species. *Anim. Cogn.* **2016**, *19*, 679–705. [CrossRef]
2. Hare, B. Survival of the Friendliest: *Homo sapiens* Evolved via Selection for Prosociality. *Annu. Rev. Psychol.* **2017**, *68*, 155–186. [CrossRef]
3. Burkart, J.M.; Allon, O.; Amici, F.; Fichtel, C.; Finkenwirth, C.; Heschl, A.; Huber, J.C.; Isler, K.; Kosonen, Z.K.; Martins, E.; et al. The evolutionary origin of human hyper-cooperation. *Nat. Commun.* **2014**, *5*, 1–9. [CrossRef]
4. Martin, J.; Koski, S.; Bugnyar, T.; Jaeggi, A.; Massen, J. Prosociality, social tolerance and partner choice facilitate mutually beneficial cooperation in common marmosets, *Callithrix jacchus*. *Anim. Behav.* **2021**, *173*, 115–136. [CrossRef]
5. Cronin, K.A. Prosocial behaviour in animals: The influence of social relationships, communication and rewards. *Anim. Behav.* **2012**, *84*, 1085–1093. [CrossRef]

6. Quervel-Chaumette, M.; Dale, R.; Marshall-Pescini, S.; Range, F. Familiarity affects other-regarding preferences in pet dogs. *Sci. Rep.* **2016**, *5*, 18102. [CrossRef]
7. Hernandez-Lallement, J.; Van Wingerden, M.; Marx, C.; Srejic, M.; Kalenscher, T. Rats prefer mutual rewards in a prosocial choice task. *Front. Neurosci.* **2015**, *8*, 443. [CrossRef] [PubMed]
8. Stevens, J.R. The selfish nature of generosity: Harassment and food sharing in primates. *Proc. R. Soc. B Boil. Sci.* **2004**, *271*, 451–456. [CrossRef] [PubMed]
9. De Waal, F.B. Attitudinal reciprocity in food sharing among brown capuchin monkeys. *Anim. Behav.* **2000**, *60*, 253–261. [CrossRef]
10. Burkart, J.M.; Van Schaik, C. Group service in macaques (*Macaca fuscata*), capuchins (*Cebus apella*) and marmosets (*Callithrix jacchus*): A comparative approach to identifying proactive prosocial motivations. *J. Comp. Psychol.* **2013**, *127*, 212–225. [CrossRef]
11. van Leeuwen, E.J.C.; DeTroy, S.E.; Kaufhold, S.P.; Dubois, C.; Schütte, S.; Call, J.; Haun, D.B.M. Chimpanzees behave prosocially in a group-specific manner. *Sci. Adv.* **2021**, *7*, eabc7982. [CrossRef]
12. Brosnan, S.F.; De Waal, F.B.M. Socially Learned Preferences for Differentially Rewarded Tokens in the Brown Capuchin Monkey (*Cebus apella*). *J. Comp. Psychol.* **2004**, *118*, 133–139. [CrossRef]
13. Beran, M.J.; Parrish, A.E. Non-human primate token use shows possibilities but also limitations for establishing a form of currency. *Philos. Trans. R. Soc. B Biol. Sci.* **2021**, *376*, 20190675. [CrossRef]
14. Horner, V.; Carter, J.D.; Suchak, M.; de Waal, F.B.M. Spontaneous prosocial choice by chimpanzees. *Proc. Natl. Acad. Sci. USA* **2011**, *108*, 13847–13851. [CrossRef]
15. Pelé, M.; Dufour, V.; Thierry, B.; Call, J.; Pelé, M. Token transfers among great apes (*Gorilla gorilla, Pongo pygmaeus, Pan paniscus,* and *Pan troglodytes*): Species differences, gestural requests, and reciprocal exchange. *J. Comp. Psychol.* **2009**, *123*, 375–384. [CrossRef]
16. Skerry, A.E.; Sheskin, M.; Santos, L.R. Capuchin monkeys are not prosocial in an instrumental helping task. *Anim. Cogn.* **2011**, *14*, 647–654. [CrossRef]
17. Yamamoto, S.; Humle, T.; Tanaka, M. Chimpanzees' flexible targeted helping based on an understanding of conspecifics' goals. *Proc. Natl. Acad. Sci. USA* **2012**, *109*, 3588–3592. [CrossRef]
18. Mioduszewska, B.; Schleuning, X.; Brunon, A.; O'Hara, M.; Auersperg, A.M.I.; Federspiel, I.G.; Von Bayern, A.M.P. Task Aspects Triggering Observational Learning in Jackdaws (*Corvus monedula*). *Anim. Behav. Cogn.* **2020**, *7*, 567–588. [CrossRef]
19. Auersperg, A.M.I.; Von Bayern, A.M.I.; Weber, S.; Szabadvari, A.; Bugnyar, T.; Kacelnik, A. Social transmission of tool use and tool manufacture in Goffin cockatoos (*Cacatua goffini*). *Proc. R. Soc. B Boil. Sci.* **2014**, *281*, 20140972. [CrossRef] [PubMed]
20. Fritz, J.; Kotrschal, K. Social learning in common ravens, *Corvus corax*. *Anim. Behav.* **1999**, *57*, 785–793. [CrossRef]
21. Bugnyar, T.; Stöwe, M.; Heinrich, B. Ravens, *Corvus corax*, follow gaze direction of humans around obstacles. *Proc. R. Soc. B Boil. Sci.* **2004**, *271*, 1331–1336. [CrossRef]
22. Peron, F.; Chardard, C.; Nagle, L.; Bovet, D. Do African grey parrots (*Psittacus erithacus*) know what a human experimenter does and does not see? *Behav. Process.* **2011**, *87*, 237–240. [CrossRef] [PubMed]
23. Bugnyar, T.; Reber, S.A.; Buckner, C. Ravens attribute visual access to unseen competitors. *Nat. Commun.* **2016**, *7*, 10506. [CrossRef]
24. Schwing, R.; Jocteur, E.; Wein, A.; Noë, R.; Massen, J.J.M. Kea cooperate better with sharing affiliates. *Anim. Cogn.* **2016**, *19*, 1093–1102. [CrossRef]
25. Péron, F.; Rat-Fischer, L.; Lalot, M.; Nagle, L.; Bovet, D. Cooperative problem solving in African grey parrots (*Psittacus erithacus*). *Anim. Cogn.* **2011**, *14*, 545–553. [CrossRef]
26. Asakawa-Haas, K.; Schiestl, M.; Bugnyar, T.; Massen, J.J.M. Partner Choice in Raven (*Corvus corax*) Cooperation. *PLoS ONE* **2016**, *11*, e0156962. [CrossRef]
27. Massen, J.J.M.; Ritter, C.; Bugnyar, T. Tolerance and reward equity predict cooperation in ravens (*Corvus corax*). *Sci. Rep.* **2015**, *5*, 15021. [CrossRef] [PubMed]
28. Seed, A.M.; Clayton, N.S.; Emery, N.J. Cooperative problem solving in rooks (*Corvus frugilegus*). *Proc. R. Soc. B Boil. Sci.* **2008**, *275*, 1421–1429. [CrossRef]
29. Lambert, M.L.; Jacobs, I.; Osvath, M.; Von Bayern, A.M. Birds of a feather? Parrot and corvid cognition compared. *Behaviour* **2019**, *156*, 505–594. [CrossRef]
30. Emery, N.J.; Seed, A.M.; Von Bayern, A.M.; Clayton, N.S. Cognitive adaptations of social bonding in birds. *Philos. Trans. R. Soc. B Biol. Sci.* **2007**, *362*, 489–505. [CrossRef]
31. Baglione, V.; Marcos, J.M.; Canestrari, D. Cooperatively Breeding Groups of Carrion Crow (*Corvus corone*) in Northern Spain. *Auk* **2002**, *119*, 790–799. [CrossRef]
32. Komeda, S.; Yamagishi, S.; Fujioka, M. Cooperative Breeding in Azure-Winged Magpies, *Cyanopica cyana*, Living in a Region of Heavy Snowfall. *Condor* **1987**, *89*, 835. [CrossRef]
33. Heinsohn, R.; Legge, S. Breeding biology of the reverse-dichromatic, co-operative parrot *Eclectus roratus*. *J. Zool.* **2003**, *259*, 197–208. [CrossRef]
34. Riehl, C. Evolutionary routes to non-kin cooperative breeding in birds. *Proc. R. Soc. B Boil. Sci.* **2013**, *280*, 20132245. [CrossRef] [PubMed]
35. Emery, N.J.; De Kort, S.R.; Von Bayern, A.M.; Clayton, N.S. The role of food- and object-sharing in the development of social bonds in juvenile jackdaws (*Corvus monedula*). *Behaviour* **2007**, *144*, 711–733. [CrossRef]

36. Scheid, C.; Schmidt, J.; Noë, R. Distinct patterns of food offering and co-feeding in rooks. *Anim. Behav.* **2008**, *76*, 1701–1707. [CrossRef]
37. Liévin-Bazin, A.; Pineaux, M.; Le Covec, M.; Gahr, M.; Bovet, D.; Bayern, A.M.P. Food sharing and affiliation: An experimental and longitudinal study in cockatiels (*Nymphicus hollandicus*). *Ethology* **2019**, *125*, 276–288. [CrossRef]
38. Duque, J.F.; Stevens, J.R. Voluntary food sharing in pinyon jays: The role of reciprocity and dominance. *Anim. Behav.* **2016**, *122*, 135–144. [CrossRef]
39. Massen, J.J.M.; Haley, S.M.; Bugnyar, T. Azure-winged magpies' decisions to share food are contingent on the presence or absence of food for the recipient. *Sci. Rep.* **2020**, *10*, 1–12. [CrossRef]
40. Laumer, I.B.; Massen, J.J.M.; Boehm, P.M.; Boehm, A.; Geisler, A.; Auersperg, A.M.I. Goffin cockatoos show flexible targeted helping in a tool transfer task. *PLoS ONE*. in review.
41. Duque, J.F.; Leichner, W.; Ahmann, H.; Stevens, J.R. Mesotocin influences pinyon jay prosociality. *Biol. Lett.* **2018**, *14*, 20180105. [CrossRef]
42. Schwab, C.; Swoboda, R.; Kotrschal, K.; Bugnyar, T. Recipients Affect Prosocial and Altruistic Choices in Jackdaws, *Corvus monedula*. *PLoS ONE* **2012**, *7*. [CrossRef]
43. Di Lascio, F.; Nyffeler, F.; Bshary, R.; Bugnyar, T. Ravens (*Corvus corax*) are indifferent to the gains of conspecific recipients or human partners in experimental tasks. *Anim. Cogn.* **2012**, *16*, 35–43. [CrossRef] [PubMed]
44. Lambert, M.L.; Massen, J.J.; Seed, A.M.; Bugnyar, T.; Slocombe, K.E. An 'unkindness' of ravens? Measuring prosocial preferences in Corvus corax. *Anim. Behav.* **2017**, *123*, 383–393. [CrossRef]
45. Horn, L.; Bugnyar, T.; Griesser, M.; Hengl, M.; Izawa, E.-I.; Oortwijn, T.; Rössler, C.; Scheer, C.; Schiestl, M.; Suyama, M.; et al. Sex-specific effects of cooperative breeding and colonial nesting on prosociality in corvids. *eLife* **2020**, *9*, 235–244. [CrossRef] [PubMed]
46. Dufour, V.; Wascher, C.A.F.; Braun, A.; Miller, R.; Bugnyar, T. Corvids can decide if a future exchange is worth waiting for. *Biol. Lett.* **2011**, *8*, 201–204. [CrossRef]
47. Koepke, A.E.; Gray, S.L.; Pepperberg, I.M. Delayed gratification: A grey parrot (*Psittacus erithacus*) will wait for a better reward. *J. Comp. Psychol.* **2015**, *129*, 339–346. [CrossRef] [PubMed]
48. Krasheninnikova, A.; Höner, F.; O'Neill, L.; Penna, E.; Von Bayern, A.M.P. Economic Decision-Making in Parrots. *Sci. Rep.* **2018**, *8*, 12537. [CrossRef] [PubMed]
49. Krasheninnikova, A.; Brucks, D.; Blanc, S.; von Bayern, A.M.P. Assessing African grey parrots' prosocial tendencies in a token choice paradigm. *R. Soc. Open Sci.* **2019**, *6*, 190696. [CrossRef]
50. Péron, F.; John, M.; Sapowicz, S.; Bovet, D.; Pepperberg, I.M. A study of sharing and reciprocity in grey parrots (*Psittacus erithacus*). *Anim. Cogn.* **2012**, *16*, 197–210. [CrossRef]
51. Heaney, M.; Bastos, A.P.M.; Gray, R.D.; Taylor, A.H. Are kea prosocial? *Ethology* **2020**, *126*, 176–184. [CrossRef]
52. Wascher, C.A.F.; Feider, B.; Bugnyar, T.; Dufour, V. Crows and common ravens do not reciprocally exchange tokens with a conspecific to gain food rewards. *Ethology* **2020**, *126*, 278–287. [CrossRef]
53. Brucks, D.; von Bayern, A.M. Parrots Voluntarily Help Each Other to Obtain Food Rewards. *Curr. Biol.* **2020**, *30*, 292–297.e5. [CrossRef]
54. Massen, J.J.M.; Lambert, M.; Schiestl, M.; Bugnyar, T. Subadult ravens generally don't transfer valuable tokens to conspecifics when there is nothing to gain for themselves. *Front. Psychol.* **2015**, *6*, 885. [CrossRef] [PubMed]
55. Horn, L.; Scheer, C.; Bugnyar, T.; Massen, J.J.M. Proactive prosociality in a cooperatively breeding corvid, the azure-winged magpie (*Cyanopica cyana*). *Biol. Lett.* **2016**, *12*, 20160649. [CrossRef]
56. Vijay, N.; Bossu, C.M.; Poelstra, J.W.; Weissensteiner, M.H.; Suh, A.; Kryukov, A.P.; Wolf, J.B.W. Evolution of heterogeneous genome differentiation across multiple contact zones in a crow species complex. *Nat. Commun.* **2016**, *7*, 13195. [CrossRef] [PubMed]
57. Guidelines for the treatment of animals in behavioral research and teaching. *Anim. Behav.* **2018**, *135*. [CrossRef]
58. De Kort, S.R.; Clayton, N.S. An evolutionary perspective on caching by corvids. *Proc. R. Soc. B Boil. Sci.* **2006**, *273*, 417–423. [CrossRef]
59. Berrow, S.D.; Kelly, T.C.; Myers, A.A. The mussel caching behaviour of Hooded Crows *Corvus cornix*. *Bird Study* **1992**, *39*, 115–119. [CrossRef]
60. Jacobs, I.F.; Osvath, M.; Osvath, H.; Mioduszewska, B.; Von Bayern, A.M.; Kacelnik, A. Object caching in corvids: Incidence and significance. *Behav. Process.* **2014**, *102*, 25–32. [CrossRef]
61. Schmelz, M.; Grueneisen, S.; Kabalak, A.; Jost, J.; Tomasello, M. Chimpanzees return favors at a personal cost. *Proc. Natl. Acad. Sci. USA* **2017**, *114*, 7462–7467. [CrossRef] [PubMed]
62. Brosnan, S.F. Understanding social decision-making from another species' perspective. *Learn. Behav.* **2017**, *46*, 101–102. [CrossRef] [PubMed]
63. Carpenter, C.R.; Locke, N.M. Notes on Symbolic Behavior in a Cebus Monkey (*Capucinus appella*). *Pedagog. Semin. J. Genet. Psychol.* **1937**, *51*, 267–278. [CrossRef]
64. Addessi, E.; Mancini, A.; Crescimbene, L.; Visalberghi, E. How Social Context, Token Value, and Time Course Affect Token Exchange in Capuchin Monkeys (*Cebus apella*). *Int. J. Primatol.* **2010**, *32*, 83–98. [CrossRef]

65. Addessi, E.; Mancini, A.; Crescimbene, L.; Padoa-Schioppa, C.; Visalberghi, E. Preference Transitivity and Symbolic Representation in Capuchin Monkeys (*Cebus apella*). *PLoS ONE* **2008**, 3, e2414. [CrossRef]
66. Pepperberg, I.M. Symbolic communication in the grey parrot. In *The Oxford Handbook of Comparative Evolutionary Psychology*; Vonk, J., Shackelford, T.K., Eds.; Oxford University Press: Oxford, UK, 2012; pp. 297–319.
67. Bugnyar, T.; Stöwe, M.; Heinrich, B. The ontogeny of caching in ravens, *Corvus corax*. *Anim. Behav.* **2007**, *74*, 757–767. [CrossRef]
68. Miller, R.; Bugnyar, T.; Pölzl, K.; Schwab, C. Differences in exploration behaviour in common ravens and carrion crows during development and across social context. *Behav. Ecol. Sociobiol.* **2015**, *69*, 1209–1220. [CrossRef]
69. Ostojić, L.; Shaw, R.C.; Cheke, L.G.; Clayton, N.S. Evidence suggesting that desire-state attribution may govern food sharing in *Eurasian jays*. *Proc. Natl. Acad. Sci. USA* **2013**, *110*, 4123–4128. [CrossRef]

animals

MDPI

Article

Where Is Ethology Heading? An Invitation for Collective Metadisciplinary Discussion

Nereida Bueno-Guerra

Department of Psychology, Pontifical Comillas University, 28049 Madrid, Spain; nbguerra@comillas.edu

Simple Summary: I analyzed the current state of Ethology (i.e., the study of animal behavior and cognition) from the researchers' perspective through an online questionnaire that was responded to by almost a hundred participants. Despite that the number of the respondents was small, it is interesting to highlight some of the results, since they align with some published analyses. First, not many basic concepts of the discipline, nor its denomination, reached homogeneous consensus. This is alarming because researchers need common vocabulary to communicate effectively between them. Second, despite the enormous biodiversity existent, the researchers seem to be biased towards studying primates, our own family species. Also, the apparatuses employed in the studies are not always adapted to the species' specific characteristics, so our conclusions about their behavior or cognition might be biased. Fortunately, the technology needed to conduct further studies already exists. However, there is not enough funding or collaboration with engineers to make it real. Establishing big scale networking, adopting some research principles such as transparency, and broadening gender and ethnic diversity in research teams may help in adopting new endeavors.

Abstract: Many factors can impact the advancement of scientific disciplines. In the study of animal behavior and cognition (i.e., Ethology), a lack of consensus about definitions or the emergence of some current events and inventions, among other aspects, may challenge the discipline's grounds within the next decades. A collective metadisciplinary discussion may help in envisioning the future to come. For that purpose, I elaborated an online questionnaire about the level of consensus and the researchers' ways of doing in seven areas: Discipline name and concepts, species, Umwelt, technology, data, networking, and the impact of sociocultural and ecological factors. I recruited the opinion of almost a hundred of colleagues worldwide ($N = 98$), both junior and seniors, working both in the wild and in the lab. While the results were pitted against the literature, general conclusions should be taken with caution and considered as a first attempt in exploring the state of the discipline from the researchers' perspective: There is no unanimity for the discipline's name; 71.4% of the researchers reported there is limited consensus in the definition of relevant concepts (i.e., culture, cognition); primate species still predominate in publications whereas the species selection criteria is sometimes based on fascination, chance, or funding opportunities rather than on biocentric questions; 56.1% of the apparatuses employed do not resemble species' ecological problems, and current tech needs would be solved by fostering collaboration with engineers. Finally, embracing the Open Science paradigm, supporting networking efforts, and promoting diversity in research teams may help in gathering further knowledge in the area. Some suggestions are proposed to overcome the aforementioned problems in this contemporary analysis of our discipline.

Keywords: Ethology; discipline; philosophy of science; Umwelt; networking

Citation: Bueno-Guerra, N. Where Is Ethology Heading? An Invitation for Collective Metadisciplinary Discussion. *Animals* **2021**, *11*, 2520. https://doi.org/10.3390/ani11092520

Academic Editors: Lucia Regolin and Maria Loconsole

Received: 30 April 2021
Accepted: 25 August 2021
Published: 27 August 2021

1. Introduction

Any ethologist will be familiar with how the four whys of Niko Tinbergen [1] revolutionized the questions and designs that ethologists could use. Indeed, the implications of his thoughts together with those of Konrad Lorenz and Carl von Frisch positioned Ethology at the highest level of human scientific recognition, which earned the Nobel

Prize in 1973. This social event became a turning point for the area. Even the laurates would have never expected such award: "Many of us have been surprised at the unconventional decision of the Nobel Foundation to award this year's prize 'for Physiology or Medicine' to three men who had until recently been regarded as 'mere animal watchers'", said Tinbergen in his Nobel Lecture [1]. Lorenz, during the banquette, reflected on the consequences of the award saying that "this trust goes so far that under certain circumstances world opinion about the importance of an entire branch of research can be influenced by this judgment" [2]. They were correct: The Nobel Prize established the discipline; new vocations were attracted; departments flourished across continents, and research grew. However, almost half of a century has gone by since then and new concepts, theories, methodologies, discoveries, inventions, and events have happened. Far from being viewed as threatening, these new knowledge and tools can nurture and empower the potential and the extent of disciplines [3]. How has Ethology embraced them? Has Ethology been consolidated as a discipline since the Nobel Prize? What are the further challenges to come? As one prominent ethologist stated, it is never a bad time to invite one to rethink the scope and the basic principles of animal behavior [4]. Therefore, the purpose of the present manuscript is to provide a collective look into Ethology by ethologists to produce an updated "metadisciplinary analysis" that may help in building a common and solid ground for the discipline's advancement.

To provide a broad analysis about the discipline but still produce a feasible abbreviated piece of research, the present text has been limited to seven areas: Concepts, species, Umwelt, technology, data, networking, and future. The selection of these areas has been grounded on three considerations: the principles of philosophy of science on how disciplines are consolidated, see [5]; the fact that new discoveries and inventions impact disciplines [3]; and the areas that previous analyses of Ethology have investigated (e.g., [6–8]). Next, the reasons that these seven areas were selected are detailed.

I Concepts: Each discipline receives a concrete name to be addressed in the United Nations Educational, Scientific and Cultural Organization (UNESCO) Nomenclature coding [9]. This name should be shared by researchers to demarcate the area of knowledge from others. If researchers choose different names for the discipline, eventually new methodologies, new experimental designs, new departments, working positions, and differentiated disciplines arise [10]. In the case of Ethology, Shettleworth [7] (Figure 4, p. 215) composed a picture of how there were different names given to the study of animal behavior and how they intersected each other. It is therefore timely to explore whether researchers nowadays find consensus on the name assigned to the discipline or not. Moreover, any given discipline is grounded on a body of theoretical concepts and assumptions shared by all the researchers, so comparisons and reviews can be done (see the Handbook of Comparative Psychology [11]). However, this has not always been the case of Ethology. To name two prominent examples, Levitis, Lidicker, and Freund [8] showed how there was no unanimous definition of "behavior" between behavioral biologists and a debate on how "cognition" should be framed has largely existed in Ethology (e.g., [7,12]). Moreover, the findings of other disciplines, such as the coined "plant intelligence" [13] or new techniques such as the genome editing CRISPR-Cas9 (Clustered Regularly Interspaced Short Palindromic Repeats and CRISPR-associated protein 9) [14] may challenge core ethologic concepts such as "intelligence" or "species". Therefore, it is timely to explore which concepts find more or less consensus in the discipline to apply for common agreements, if needed.

II Species: Any consolidated discipline has some specific object of study, which in the case of Ethology is (the behavior of) species. However, the question of which species should be studied and the reasons to study them is unsolved (e.g., [6,7,15,16]). As certain species selection criteria might bias the knowledge of the discipline (i.e., anthropomorphism can lead researchers to selectively study non-human primates over other species), for a discussion about the future of the discipline, it is interesting to explore the species that researchers are currently studying and their reasons to select them.

III Umwelt: Disciplines possess specific methodologies adapted to their object of study. In the case of Ethology, this is of special relevance because some animals (i.e., humans), with their species-specific perceptual and motor system (known as "Umwelt", [17]), are studying other animals (i.e., chosen species), with potentially different species-specific systems. Therefore, caution to employ ecologically valid apparatus and to consider the different factors that affects each species' Umwelt in the experimental designs is crucial [18,19]. Exploring whether the Umwelt of the species is being consistently considered by researchers in their experiments would reveal the health of the discipline.

IV Technology: Inventions can revolutionize an area of research [3], as it happened with the development of the microscope for different disciplines. In Ethology, the development of new devices, such as drones, OMICS (an abbreviation for all the biological disciplines whose names end in the suffix -omics, such as genomics), or neurobiology measurement apparatus, as well as the release of new machine techniques, such as big data analysis, may challenge how data are collected and which new questions the discipline can ask. Knowing which technology is currently used or missed may help to envisage the future methods of Ethology.

V Data: All scientific areas eventually provide data that need to be analyzed and published to become part of the discipline knowledge. However, the analysis techniques are not alien to fashions and the publication system is sometimes corrupted by how humans configure scientific paths (i.e., publish or perish) or disregard ethical principles (i.e., see discussions about the replication crisis in [20,21]). Getting to know how data are being analyzed in Ethology and how the different crises are being solved may illuminate the modifications to be done for the advancement of the discipline.

VI Networking: A decade ago, it was stated that "people publishing on comparative cognition are not just talking to each other" ([7], p. 212). Currently, within a progressively interconnected and globalized world, networking is easier than ever and collective collaboration might become the seed of future robust data. Therefore, analyzing whether researchers do engage in collaborative research and maintain sporadic contact with other colleagues is a first step to explore whether the 2009 statement is still accurate.

VII Future: Many environmental, cultural, and societal changes may permeate Ethology and affect the future of the discipline: Universal events such as climate change, the effects of human action over the environment (known as "Anthropocene"), the inclusion of ethnic minorities and queer diversity in research teams, or the reconsideration of ethical questions about animal welfare, to name a few. Starting a discussion about which of these changes are deemed as relevant by current researchers may help to understand how Ethology is preparing for the new years to come.

Finally, to make this analysis "metadisciplinary", the examinees of the aforementioned areas should be ethologists since their daily practice entails working with them. Therefore, the opinions of almost a hundred colleagues worldwide were collected (N = 98, from at least 40 different institutions and 14 nationalities) in which both junior and senior researchers participated. Among the latter, many of the currently well-known and leading researchers responded, so that both novice and experts could contribute their vision to this discussion. It is important to note that even though the total number of researchers devoted to Ethology is difficult to estimate, the recruited sample is unlikely to be representative enough of the whole professional group. Hence, the conclusions of this study must be taken with caution. However, this analysis can serve as the starting signal of a collective contemporary analysis of our discipline that may foster further fine-grained studies. Next, I will provide the characteristics of the sample to then go one by one for each of the seven areas providing a brief introduction, sharing the results of the sample's opinion, and discussing the main conclusions.

2. Materials and Methods

Two semi-structured questionnaires with questions regarding the seven areas mentioned above were distributed online. Both questionnaires were divided into six sections (see the full list of questions in the annex of Supplementary Materials): (1) "Demographic data": Continent of origin, continent of their academic affiliation, age (optional), years of experience, type of worksite (wild, free-range or lab), e-mail (optional); (2) "Our discipline": Name given to the discipline, level of consensus about some theoretical concepts; (3) "Species": The current species the researchers study versus the desired species to study, criteria for the species selection; (4) "Procedure": Awareness about the meaning of Umwelt; use of ecological apparatuses; the level of relevance attributed to some species-specific characteristics in any experimental design; the current and the potential uses of technology both in the field and in the lab; the statistical analysis techniques employed; (5) "Data Analysis": The level of impact assigned of some social, environmental, ethical, and academic factors to our area of study; and (6) "Past, present and future of our discipline": Assessment of some historical moments that contributed to the area or potential future challenges that the discipline will face and how much networking they did.

The first questionnaire was exploratory and mainly contained open questions (e.g., "How would you name our discipline?"). It served to extract the most repeated responses to configure a second questionnaire, which mainly contained multiple-choice questions (e.g., "How would you name our discipline? Please, choose only one even when many are related to each other or even if there is none that entirely conveys your preference"). The second questionnaire can be accessed here (shorturl.at/uKLRT accessed on 8 August 2021) and the full list of questions and responses can be checked in the Annex. Responses from both could be pooled after coding (see data sheet attached in Supplementary Materials, which can also be downloaded from the author's institutional repository here: http://hdl.handle.net/11531/59820, permanent link accessed on 8 August 2021). Both questionnaires were elaborated in Google Forms and participants provided written consent to participate. Only some questions were mandatory to avoid participants' frustration or abandonment, therefore, in the Result sections, the total number of respondents will slightly vary between the analyzed variables.

The questionnaire was distributed between February 2021 and March 2021 among the authors' colleagues both by email and text phone services (mainly primatologists); among the collaborators of a collective book about different species' Umwelt [19], who were mainly top-referent researchers of a wide array of animal taxa (e.g., ants, birds, and hyenas, to name a few); among different associations for the study of animal behavior (e.g., the Association for the Study of Animal Behaviour (ASAB), the Australasian Society for the Study of Animal Behaviour (ASSAB)); and among colleagues with active profiles in Twitter, asking all of them for dissemination. In an effort to avoid some potential Western bias, some overseas institutions (e.g., Kyoto University; Macquarie University), associations, and non-governmental organizations located in Africa, Asia, and Oceania working in the lab/in the field were also contacted. Therefore, the data collection technique was a non-probabilistic snowball method.

3. Sample

Sociodemographic: The sample is composed by 98 participants (round 1: 63, round 2: 35). Gender is not indicated because it was not included as a mandatory question. Despite the effort put in including underrepresented minorities, the sample nationality at birth (first data in the parenthesis) as well as the current institution of origin (second data in the parenthesis) were mainly from WEIRD (Western, educated, industrialized, rich and democratic) societies [22]: European (66.7%, $n = 58$; 69.4%, $n = 68$), and North American (24.1%, $n = 21$; 20.4%, $n = 20$). There was very little representation from South America, Asia, and Oceania (9.1%, $n = 8$; 10.2%, $n = 10$).

Education: Half of the participants have studied some Biology degree (55.1%, $n = 54$), followed by Psychology (24.5%, $n = 24$). Other studies were also reported, such as Philosophy, Anthropology, or Environmental Ecology.

Age and professional experience: Only 79 participants reported their age, which ranged from 21 to 79 years old with a mean of 39.8 years ($SD = 13.25$). With regards to professional experience, the sample is almost equally distributed among all levels. There is almost a quarter each of junior professionals (0 to 5 years of experience, 23.5%), young leaders (5 to 10 years, 31.6%), senior professionals (10 to 20 years, 21.4%), and experts (more than 20 years, 23.5%). For analysis purposes, the variable "professional experience" was dichotomized by pooling those researchers with 10 or less years of experience, called "juniors" and those with more than 10 years of experience, called "seniors".

Worksite: I considered three different worksites depending on the level of freedom that the studied animal had (lab, free-range, wild). From the total sample, the majority work at labs/zoos (45.9%, $n = 45$) or combine free-range/wild studies with experiments at labs/zoos (32.6%, $n = 32$). Only 21.5% of the sample works exclusively with non-human animals in their natural habitat or at natural reserves. For analysis purposes, the variable "worksite" was dichotomized into "lab" and "field", and it was coded as "lab" for those participants that reported working in the lab and those working in the lab and in free-range worksites and coded as "field" for those participants that reported working in the wild and those working in the wild and in free-range worksites. Those who marked all options ($n = 16$) were excluded.

4. Results

4.1. Concepts

> *"The scientist does not study nature because it is useful; he studies it because he delights in it, and he delights in it because it is beautiful. If nature were not beautiful, it would not be worth knowing, and if nature were not worth knowing, life would not be worth living"* (Henri Poincaré)

4.1.1. Discipline Name

All disciplines become renowned as scientific once they have a common name to be referred to and once there is some common pool of concepts that constitute the core of their set of research questions and studies [10]. Indeed, the six-digit UNESCO nomenclature does make an effort to nominate and separate all the different disciplines by a single name and code. This way, not only is the discipline and a particular "Universe corner" [23] recognized and differentiated among others, but also specific departments, positions, and funding options can emerge and consolidate its expansion [10]. Along the history of Science, many disciplines have undergone crucial denomination changes that led to the emergence of different methodologies and new ways of understanding concepts (e.g., Genetics [24]; Physics [25]). Until recently, there have been several definitions for the discipline whose central interest is getting to know more about the behavior and cognition of non-human (and human) animals [26], but the word "Ethology" has served as the wider umbrella term to embrace all. However, after the progressive emergence of different associations, conferences, and handbooks that chose related but different denominations to refer to this area of study, and after the emergence of the cognitive revolution during the 1960s, it may have happened that not all researchers would pick up the same name for the discipline (for the evolution of the discipline denomination, the interested reader may refer to elsewhere, e.g., [27]), with a subsequent risk to its consolidation. Thus, the participants were asked which preferred denomination would better embrace the studies they do. The response options included the terms of the famous Hinde's title [4] (Animal Behavior, Ethology, Comparative Psychology) plus two other terms (Animal Cognition and Behavioral Ecology), thus covering the names of some of the most reputed scientific journals in the field [6,7], and still limiting the responses to a few options.

The preferred denomination for the discipline among the whole sample ($N = 98$) was Animal Behavior (48%, $n = 47$), followed by Animal Cognition (26.5%, $n = 26$). Other less selected denominations included Ethology (13.3%, $n = 13$), Comparative Psychology ($n = 8$), Primatology ($n = 1$), or Behavioral ecology ($n = 1$), while two instances reported just "other". If we compare these results to the three UNESCO nomenclatures that refer to studies about non-human behavior (i.e., Ethology, Animal Behavior, and Comparative Psychology), we find that the one with higher level of analysis, namely, Ethology (2408), is not the most chosen one by this sample, but a lower-level area of analysis, namely, Animal Behavior (240102). This choice is relevant for considering the areas of research that grant applicants can be assigned to, for choosing journal denominations, and for organizing academic departments, since it seems that the term Ethology is becoming *démodé*. Potentially, the prevalence of English as the scientific communication language is replacing the Greek etymological origin (*ethos-*, behaviour; *-logos*, study) of the term. Adopting "animal behavior" as the new term for the discipline may be the solution, although it may feel awkward for cognitive scientists (because of the traditional behavioralist vs. cognitivist perspectives) and may not intuitively include humans in the discipline's scope (because of the traditional differentiation between human vs. non-human animals). Until that new term is widely agreed, Ethology was still used in the title of the present text for historical reasons. In the future, the scientific community should be open to find agreed-upon terms to refer to this area of study (and to potential subareas of study) since many factors crucial for its survival depend on that: Education offer, funding resources, academic positions, and platforms for knowledge dissemination.

The name of any discipline is related to the research questions that researchers pose, therefore the participants were also asked which underlying scientific interest they had when doing their research, providing three response options: Non-human species focus ("To know more about X non-human species"), non-human species comparison ("To know more about the differences/similarities between X and Y non-human species"), and human species focus ("To know more about the differences/similarities between X non-human species and humans"). Most researchers reported to have a non-human species focus (45%). The rest of the sample was similarly divided into a two non-human species comparison (28.6%) and human species focus (26.5%).

At least according to this sample, it seems that Ethology is not biased by some potential anthropocentric vision (namely, a majority of human species focus), but an intrinsic interest in getting to know the nature of non-human species prevails. Interestingly, whereas this vision was shared similarly between lab and field researchers ($N = 87$, $U = 568.500$, $p = 0.189$), the underlying researchers' motivation varied depending on the denomination they assigned to the discipline ($\chi^2 = 15.460$, df $= 6$, $p = 0.017$), as it can be seen in Table 1. Concretely, those who chose "Ethology" rarely conducted comparative studies to indagate more about humans and focused instead on non-human species *per se*, both individually or in comparison with another non-human species. Those who preferred the denomination "Animal Behavior" and "Animal Cognition" mostly preferred to focus on one non-human species, although many others were similarly divided between the other motivations. Finally, those who chose "Comparative Psychology" were mostly human-centered.

These results show that, despite that all the researchers felt called to participate in the present study because they did ethological research, they were not unanimous in how to refer to the discipline. Indeed, there are different names for the specialized journals, conferences, or associations (e.g., Comparative Cognition Society, Association for the Study of Animal Behavior, Animal Behavior Society, The International Society for Behavioral Ecology). Each denomination may attract different research motivations and, by extension, different methodologies, topics, funding sources, and researchers, with the potential risk of future disaggregation. The negative consequences of this may be ignoring much of the content associated with each discipline term [28], failing to compose a full picture about the common object of study, which is describing animal behavior and cognition to understand how they evolved. It seems therefore urgent to reach common consensus on how to refer to this field of study.

Table 1. Matrix with raw data relating the name of preference for the discipline and the underlying research motivation ($N = 94$).

Research Motivation	Ethology	Animal Behaviour	Animal Cognition	Comparative Psychology
Non-human species focus	6	25	12	1
Non-human species comparison	6	12	7	1
Human species focus	1	10	7	6

4.1.2. Theoretical Concepts

The next step for the consolidation of a discipline is harboring a common pool of agreed-upon terms to pull in the same direction and to favor dialogue between colleagues, otherwise the risk of diverging in conclusions as well as precluding comparisons or systematic reviews is very high. The proper use of terms can even radically change the perspective over a discipline, as it happened with the Avian Brain Nomenclature Consortium: "names have a powerful influence on the experiments we do and the way in which we think. For this reason, and in the light of new evidence about the function and evolution of the vertebrate brain, an international consortium of neuroscientists has reconsidered the traditional, 100-year-old terminology that is used to describe the avian cerebrum" ([29] p. 151).

Indeed, the "accuracy" in terminology was deemed by Kuhn [30] as one of the most relevant characteristics for good scientific theories, therefore a recurrent source of scientific conflict is the difference in the interpretation of the words that we use [31]. However, Ethology deals with many concepts in which, despite being covered in a single and accepted word, their interpretation is subject to controversy: Levitis, Lidicker, and Freund [8] investigated whether "behavior" had a common definition between researchers and found that "there was not a single question (. . .) that produced a unanimous consensus" (p. 107); consciousness (e.g., [32,33]) is said not to have an "universally accepted definition for the term" ([34], p. 210); the same happens with personality (e.g., [35]). It is true that Science advancement does not necessarily require entire homogeneity, but some "degree of consensus is the key" ([36], p. 260). Consensus is defined as "to have a working definition of behaviour that will be as operational and as essential as possible, thereby providing conceptual guidance as to where the boundaries are" ([8], p. 107). To explore the degree of consensus that participants had in core concepts in Ethology, they were firstly asked how much general level of consensus existed in their opinion in the discipline in general, and secondly, what level of consensus they attributed to some concrete concepts.

The majority of the sample reported that there is consensus only in a limited nucleus of theoretical concepts (71.4%, $n = 70$), and this opinion did not seem to be associated with the experience of the researchers ($N = 98$, $U = 1165.500$, $p = 0.839$) nor with the worksite where the participants conducted their studies ($N = 98$, $U = 636.000$, $p = 0.457$). The actual absence of consensus or at least the perception of a lack of consensus that this sample reported should be warning that Ethology might be suffering from some deep conceptual crisis. If these data were representative of the current total amount of ethologists, the conceptual crisis should be directly addressed with collective collaboration before we continue producing much data to avoid establishing little, separate, and independent knowledge.

Indeed, the consensus on core concepts is not homogeneous either, according to this sample. During the first round, there was an open question included about which concepts would find less or more consensus in the discipline. From the responses ($N = 63$), the concepts with higher mode were extracted, yielding eleven items: Cognition, social learning, associative learning, Tinbergen's questions, Evolution Theory, culture, tool use, emotion, linguistics, cooperation, and prosociality. In the second round, the researchers ($N = 35$) were asked about the degree of consensus on these eleven concepts using a Likert scale (ranging from 1 = little consensus to 3 = lot of consensus), see Table 2 for the results. Importantly, I ran two cross tables to explore the distribution of the responses according to the researchers' worksite and according to their years of experience, but neither factor yielded any significant differences.

Table 2. Level of perceived consensus in eleven relevant concepts in Ethology (obtained from qualitative responses in a previous survey, *N* = 63).

Concept	Min.	Max.	Mean	Standard Deviation
Evolution Theory	1	3	2.80	0.473
Tinbergen's questions	1	3	2.77	0.490
Associative Learning	1	3	2.63	0.598
Social Learning	1	3	2.20	0.584
Cooperation	1	3	2.14	0.550
Tool use	1	3	2.09	0.658
Prosociality	1	3	2.03	0.664
Cognition	1	3	1.94	0.639
Culture	1	3	1.77	0.547
Linguistics	1	3	1.74	0.657
Emotion	1	3	1.46	0.611

The theoretical concepts that showed more consensus between the researchers were the principles of the Evolution Theory (M = 2.80, SD = 0.473), the Tinbergen's questions (M = 2.77, SD = 0.490), and the definition of associative learning (M = 2.63, SD = 0.598). It seems that the discipline foundations are clearly established, or at least are deemed as commonly agreed. If this was true, it is good news because it provides solid soil to build new knowledge. Indeed, it is informative about the initial origins and emergence of our discipline (i.e., Darwin, Skinner, Tinbergen). The common ground of Ethology seems to be the fact that there are selective ecological, sexual, and interspecies pressures that affect the development of species; that there are four approaches to study animal behavior: Adaption, phylogeny, mechanism, ontogeny; and that behavior can be promoted or made extinct through reinforcement and punishments. Yet, the potential existence of consensus does not exclude the emergence of some integrative updates proposed in light of new findings and techniques. For example, on the 50th anniversary of Tinbergen's questions, Bateson and Laland [37] reviewed the relevance of Tinbergen's seminal legacy for Ethology and proposed to incorporate three more questions, such as "What do we currently understand as causation and function?", "How is the behaviour inherited?", and "How can the four questions be integrated?".

The theoretical concepts that reached only moderate consensus were those related to behaviors that entail more than one individual, such as social learning (M = 2.20, SD = 0.584), cooperation (M = 2.14, SD = 0.550), or prosociality (M = 2.03, SD = 0.664). Perhaps the level of mechanics of the researcher together with the ethogram of the species and how the concepts are operationalized in a given experiment are some factors that start distancing colleagues. The concepts that reached the least consensus were abstract entities such as cognition (M = 1.94, SD = 0.639), culture (M = 1.77, SD = 0.547), language (M = 1.74, SD = 0.657), and emotion (M = 1.46, SD = 0.611). These results seem representative of the current debates in the area. For example, Bräuer and colleagues [12] recently discussed how cognition should be reframed in Ethology to lose anthropocentrism and gain a more biocentric approach. However, even when it is conceivable that different interpretations of abstract concepts exist, we should consider facing this disagreement because these concepts lie at the frontier with other sister areas of knowledge, such as Psychology, Anthropology, Linguistics, or Artificial Intelligence, to name a few. Whereas this proximate distance between disciplines could be very fruitful in terms of interdisciplinary collaboration, it can also turn to become very dangerous if researchers, not finding a consensus in their own discipline, opted to arbitrarily borrow the formulations of other disciplines, thus adding even more conceptual and theoretical differences to the current disagreement. Tackling these conceptual differences is also very relevant for the discipline growth, because, as already anticipated in the introduction, new findings such as what is coined "plant intelligence" are challenging our traditional understanding of what underlying mechanisms favor the existence of cognition, which eventually has profound consequences in exploring how these capacities evolved.

Probably forecasting these barriers, some researchers already suggested in the final open question of the questionnaire the need for an "international colloquium to agree on definitions" or some "fertile cross-talk". Similar initiatives have already been done in the shape of special issues, conferences, or forums. Currently, the potential solution for this lack of consensus might be convoking some sort of World Consensus Conference to discuss different concepts, similar to what the Avian Brain Nomenclature Consortium did [29]. The present study could inspire a precise list of the concepts we should start with. One source of inspiration for how this world consensus meeting could be shaped might be the Strüngmann Forum Reports (e.g., [38] especially p. 67-on about social cognition). These are international periodical forums in which some hot topic is chosen, some concrete questions are formulated, and a considerable number of researchers with different perspectives and theoretical models meet for discussion to reach conclusions. Following this map route, but also opening the discussion to non-leading researchers as well as trying to favor the representation of minorities, gender, worksites, and years of experience in the participants, might be very fruitful. Otherwise, the vision of the discipline might be biased again. However, one of the main biases that Ethology is currently facing is found in the species we select to study.

4.2. Species

> *"There is grandeur in this view of life ... and that, whilst this planet has gone cycling on according to the fixed law of gravity, from so simple a beginning endless forms most beautiful and most wonderful have been and are being evolved"* (Charles Darwin)

If we study non-human animals, which species should we investigate to produce knowledge in the discipline? While the better integrative answer should be all of them, it is true that we lack enough resources (e.g., funding, staff, time, etc.) to do so. Therefore, we need to be selective to produce data and choose only an extremely narrow window of species within the current rich diversity. However, this *a priori* selection is already skewing the real diverse natural world we live in. Some authors have analyzed the species studied through looking at the research published in different relevant journals (see Figures 1 and 2 in [6]; Figures 1–3 in [7]; Figures 1–3 in [39]; Table 1 in [40]). They all conclude that Ethology started focusing on a poor diversity of species (i.e., albino rat, pigeon) to progressively incorporate more different taxa. However, the current predominance of non-human primates in publications is higher than other species: "the species represented shifted toward apes, monkeys, and humans" ([7], p. 212); "the reporting data about primates quadrupl[ed]" ([40], p. 1, see also Table 1 in p. 2). Cronin and colleagues ([41], abstract) also found that, at least in social cognition "while a wide range of species were studied, they were not equally represented, with 19% of the publications reporting data for chimpanzees". This primate focus to investigate social cognition was also detected by others some years before [42] and the specific predominance of chimpanzees within the studies with non-human primates led to the so-called "chimpocentrism" [15,43]. This might be informative of the anthropocentric bias underlying the species selection criteria of researchers, but other biases can also coexist: "a mixture of taxonomic prejudices, cultural aspects of behavioural ecology as a field, and of academia in general" ([39], abstract).

Therefore, getting to know the species studied and the selection criteria is essential because a potentially narrow-biased vision would be contributing to the foundations of our discipline. Thus, first, participants were asked which group of animals they were currently studying and why. Second, they were asked whether they would be interested in incorporating another group of animals in their research and, being affirmative, which group of animals would be and why. Six potential species selection criteria were provided: Model species for a research question, availability (including low-cost maintenance or easy access to big sample sizes), chance (including job opportunity, funding), fascination about the species, the species being understudied, or the species being "very skilled". Whereas the first reason would be grounded on answering some specific theoretical question, the existent of model species could also reduce the study of other species. "Availability" refers to the easiness of access and maintenance of the animal, thus skewing the study of other animals such as those in extreme conditions, whose study could illuminate much about how ecological pressures influence behavior or cognition. "Chance" would imply an opportunistic approach to Ethology, sustained by the direction that funding organisms already had chosen for the discipline, but that could also eventually turn into research vocation and generate new research questions. "Fascination" is a subjective reason, grounded in the particular interest of each researcher, which can eventually cause them to look to peculiar species. The "understudied species" reason may reflect an interest to expand our knowledge to other taxa. Finally, the reason framed as "the species being skilled" raises concern about whether "skilled" is a hidden anthropomorphism or, by contrast, whether "skilled" refers to explore non-human capacities that can expand our knowledge on how those capacities evolved.

The participants mostly worked with non-human primates (27.6%), followed by birds (21.4%), invertebrates (11.2%), dogs/wolves (9.2%), and rodents (7.1%) (see Figure 1). There was a minority of researchers that worked with non-avian reptiles, fish or marine animals, wild carnivores, and farm animals (i.e., horses, pigs, chickens). This sample seems to be intuitively representative of the total amount of publications in our area as stated above; however, it could also be providing a biased picture of the full discipline due to the non-probabilistic recruitment followed. The main reason that researchers (N = 98) reported some species to be eligible for their studies was being a model species to answer some research question (37.8%, n = 37). Thus, it seems that most of the researchers purposefully choose one species for a concrete scientific interest. However, the rest of the reasons reported by the participants distributed very similarly between availability, chance, fascination, unique skills, or being understudied. This variety of reasons should be taken with optimism, since its concurrence leads to including species diversity in further studies.

Interestingly, when the participants were asked whether they would like to incorporate some new species to their studies, an overwhelming majority responded affirmatively (86.7%, n = 85). The new species to be incorporated continued being non-human primates, birds, and invertebrates but also singular animals appeared (i.e., deer, beavers, Midshipman fish). Probably because of the singularity of some species, the research reasons to study these animals differed. The first position was tied between fascination about the animal (i.e., "I always dream as a child to study elephants and sometime in the future, I'll do it") and model species. This shows that, at least in this sample, researchers seem to combine both their scientific and loving interest at work, which may be a welcoming combination to expand our research to understudied taxa. Yet, the consideration of what a model species is should be revisited, since biases can happen (e.g., considering only male rodents as a model species in neuroethology has excluded reproductive behavior and phenotypic diversity to be analyzed [44]) as well as oversimplicity or funding biases (as a subject pointed out: "question-centered research means that researchers are encouraged to simply pick the easiest model organism for their question which encourages taxonomic bias").

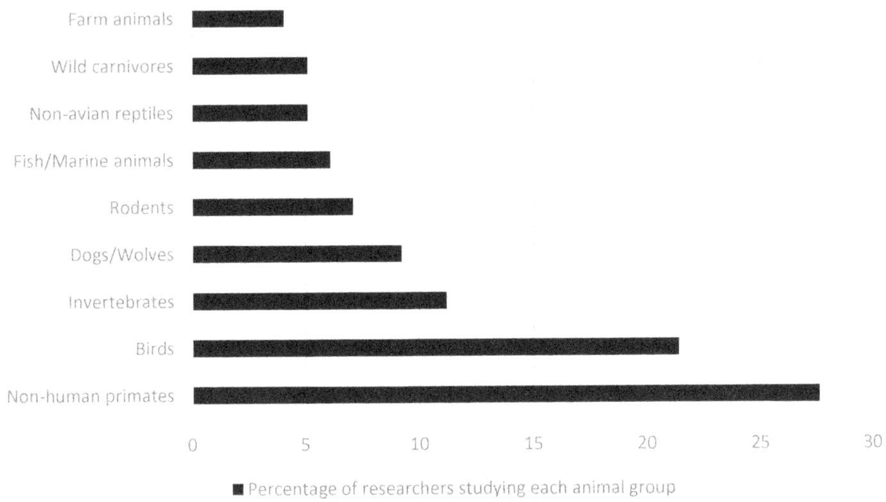

Figure 1. Distribution of the groups of animals studied in a sample of $N = 98$ Ethology researchers.

Some collective initiatives have emerged to fight species reductionism in Ethology. One example is studying the same cognitive process across different taxa with the same experimental protocol. As an example, in 2014, dozens of experimenters joined to study self-control in 36 species through two problem-solving tasks [45]. Thanks to this large-scale collaboration, the researchers could provide phylogenetic conclusions such as that absolute brain volume best predicted performance across species. Other initiatives have tried to explore the same taxa by conducting the same experiment across different species closely related, such as ManyPrimates. This is a project that congregates researchers worldwide aiming to investigate primate cognition and the ecological factors affecting its evolution. They do so through "large and diverse samples from a wide range of species" because "primate cognition research suffers from small sample sizes and is often limited to a handful of species, which constrains the evolutionary inferences we can draw" ([46], abstract). It is true that the enterprise is challenging and not all the factors influencing behavior might be controlled (i.e., upbringing, epigenetics, relation with the animal carers). However, so far, they have established an international network of collaboration both in the lab and in the wild studying very different primate species. Together, they select topics of study and design experimental protocols to be applied in all the working sites. With such a collaborative disposition, the diversity of species is guaranteed, as well as the power of their conclusions. Their success has permeated other animal taxa and other similar initiatives, such as ManyDogs and ManyBirds have emerged. It is true that including new species (whether or not from the same taxa) would not always be a panacea. One researcher in the questionnaire warned that "it may difficult the selection of proper reviewers to properly assess the quality of the research and [the] risk of becoming meaningless" is high. However, there is a solution for that, in specifying the Umwelt of whichever species studied in each manuscript by default.

4.3. *Umwelt*

> *"Our appreciation of what is important and what is accessory; what is big and what is small; lies on a false judgment, namely a truly anthropomorphic error."* (Santiago Ramón y Cajal)

"Umwelt" is a German term that von Uexküll [17] used in Ethology to refer to the perceptual and motor world of each animal: Not all species' senses and abilities to move are the same so "there is no real world but as many worlds as species" (pp. 92–93). Even when this is obvious, it is difficult to put oneself into another species' shoes. This difficulty has given path to a fruitful discussion about critical anthropomorphism and how we humans can easily fall into bias errors when studying animal behavior and cognition (e.g., [47,48]; see [49] vs. [50]).

However, this Umweltian effort should guide our experimental designs. Specifically, following Uexküll's classification [51], we should ensure in our tests that the specific way in which each species perceives stimuli is taken into account (i.e., *Merkwelt*) and that the actions we expect from the animals fall within their natural repertoire (i.e., *Wirkwelt*). Together with these two factors, it was proposed elsewhere to also consider in our tests how individuals of some species interact with conspecifics and the social factors that may influence its behavior and cognition (i.e., *Sozialwelt*) [18]. For instance, some authors have highlighted that "maintaining the social environment that is characteristic of a species' natural history as much as possible during cognitive testing improves the socio-ecological validity of the research. The increased validity should be especially pronounced for gregarious species and when social cognition is under investigation" ([41], p. 2). Failure to do so might wrongly lead us to misinterpret the species' capacity to solve problems and therefore bias any conclusion about their cognitive skills. To these three terms (*Merkwelt*, *Wirkwelt*, *Sozialwelt*) has been recently added the acronym STRANGE. This acronym is a parallel of the WEIRD characteristic of human subjects [22] that were raised to avoid generalizing conclusions with under-representative samples. In Ethology, STRANGE stands for Social background; Trappability and self-selection; Rearing history; Acclimation and habituation; Natural changes in responsiveness; Genetic make-up; and Experience [52]. Taking into account the three Umweltian factors plus the new acronym prior to conducting research could much improve the honesty, ecological validity, and cumulative knowledge in our discipline (indeed, the STRANGE framework has been recently adopted by some journals, such as Ethology or PLOSBiology). However, were participants aware of these terms?

Despite being so relevant for our discipline, the term "Umwelt" was unknown for a third of the sample (37.8%, $n = 37$) and its ignorance was not associated with working predominantly in the field or in the lab ($N = 98$, $U = 617.000$, $p = 0.410$). However, interestingly, not being familiar with the term might not imply that researchers do not regularly take the species' Umwelt into account in their experimental designs but just a by-product of the potential low frequency of use of the term in Ethology (i.e., "I will be transparent and say I did google Umwelt, because it's not a word that I usually use ... It isn't widely used. [However,] I try my best to take into account that the way my study subjects perceive the world is shaped by their evolutionary history and is different from mine").

Maybe it is time to bring the term back into fashion to consider whether we are truly taking all aspects into account. Indeed, considering whether the species' Umwelt was taken into consideration in the methods of some studies may help peer-review processes and replication. One idea is incorporating an Umwelt checklist into the canonical IMRaD paper structure. The checklist should ask, by default, for information about how the specific species' senses, motor, social, and STRANGE characteristics were taken into account across the study. Furthermore, the discussion should include a mandatory paragraph about how the object of study and the concrete tasks performed emulated the ecological challenges that the studied species usually face. Another idea would be producing Umweltian species-specific checklists, so that they become mandatorily included in each experimental or observational manuscript about that species, just as the ethical statement is also mandatory. Indeed, including standardized checklists in research reports is becoming popular in other disciplines (e.g., quantitative criminology [53]) and has been recently proposed for reviewing processes in general, with special attention to studies on ecology and evolution [54]. The benefits are evident: With this information, the scientific community could not only be convinced about the validity of the experiment, but they could also replicate the methods better, discuss the interpretation of the results better, and review the whole content better, even when it was about an understudied species or the reviewer was not familiar with the animal.

In an effort to start thinking about that potential checklist, a list of Umweltian factors was included in the questionnaire. Umweltian factors are defined as areas with species-specific differences that should be considered in experimental designs to avoid inappropriateness or low ecological validity (for example, an Umweltian factor is "psychological characteristics", since some species can be very neophobic whereas others can be very bold (i.e., parrots vs. hyenas), and therefore experimental methods should adapt the conditions and materials accordingly). The selected Umweltian factors were sensory modalities, ethogram, social factors, sex differences, psychological characteristics, and apparatus resembling natural challenges. The participants were asked to score (in a Likert scale from 0 = not at all to 3 = a lot) how much relevance they assigned to each one.

Interestingly, not all the Umweltian factors were equally assessed as relevant ($\chi^2 = 36.828$, df = 5, $p < 0.001$) by the sample ($N = 35$ in all factors except for the question about the ecological apparatus, $N = 98$). The least relevant Umweltian factor for this sample was the species' psychological characteristics ($M = 2.37$, $SD = 0.645$), although this may be species-dependent because some researchers specifically highlighted the need for considering them, particularly those working with hyenas (e.g.,: "We carefully applied knowledge of hyenas ... when developing test protocols to encourage/facilitate participation") and avian species (e.g., "Devised an aviary test that even a shy, conservative, non-problem solving species can pass"). The second least relevant factor for the sample was the use of ecological apparatuses ($M = 1.97$, $SD = 0.954$), with more than half of the participants (56.1%, $n = 55$) answering that they did not use apparatuses made only of natural (non-synthetic) material that reliably resembled the type of problems the species would find in its natural habitat. The justifications they gave were related to being unable to find a novel ecological method (e.g., "We attempt to get at ecologically valid set-ups, but sometimes, you have to do something very unnatural, like get dogs to look at screens") or the domestic nature of the species (e.g., "the "Umwelt" of dogs is the human environment. Even in free ranging dogs, they feed from human trash and they are familiar with human apparatus"). However, other authors particularly highlighted the relevance of this factor (e.g., "[We sould] Look at the problems elephants actually face and solve in their natural environments and explore those rather than applying artificial tests of cognition used in other species"). Indeed, using artificial (for the species) materials or situations is worrying in terms of ecological validity because we may arrive at unrealistic information since the animal is not used to that artificial environment.

In general, the species' perceptual system was considered as the most important factor to be taken into account when designing experiments ($M = 2.91$, $SD = 0.284$). The participants provided many illustrative examples about which sensory modality they considered in their studies (e.g., "Avian vision", "Dogs sense of their own size", "Goffins are very 'haptical' animals, they learn about the world by touch. We cannot rely on [visual stimuli] ... even knowing that their vision is good") and how they particularly adapted the experimental material to it (e.g., "Olfaction is very prevalent so [I] clean props") or how they needed to specifically code each trial (i.e., "Recording a song which couldn't be correctly analyzed with the use of only our ears because of it complexity"), thus acknowledging the human's sensory limitations compared to that of the studied species. The good news is that the rapid development of technology (sometimes coming from the artistic scene, see the "metaperceptual helmets" by Connolly and McKenzie) is allowing us to gain novel access to those specific-species alien worlds different from ours to improve our Umweltian empathy.

4.4. Technology

> *"We are on the brink of an age when technology will redefine birth, food, sex and death-the fundamental elements of our existence. (...) How much are we about to hand over technology?"* (Jenny Kleeman)

The evolution of technology, techniques, and new inventions can certainly revolutionize a discipline by allowing to explore something that was imperceptible to human senses (i.e., microscopes provided Biology, Medicine, or Physics the evidence and arguments for different elements such as virus or subatomic particles, e.g., [55]); by modifying rooted assumptions (i.e., genome sequencing contributed with evidence for the settlement of the evolution theory and forced taxonomy to be adjusted); by compiling multiple data, accelerating data processing, and providing opportunities to forecast and prevent events (i.e., deep learning is used to predict wind power, potential new crimes or diseases, e.g., [56]), or by providing new areas of interest that sometimes can collide with ethical issues (i.e., CRISPR method, artificial intelligence). Hence, no discipline can be alien to new discoveries and inventions and should be open to embrace those whose advantages surpassed the drawbacks. Which current available technologies can have an impact over Ethology? Although this question becomes rapidly obsolete, there are illustrative examples of how technology contributed to expand our knowledge in the area.

For instance, the question about whether non-human animals had theory of mind was controversial and produced numerous opposing correspondence between researchers for more than four decades (e.g., [48,57,58]). However, the arrival of eye tracker techniques, which allowed for following the gaze of subjects with a minimal margin error, allowed to incorporate new ways to explore the existence of theory of mind in apes with a simple experimental design based on the anticipatory look on others' movement [59,60]. Plus, the method is non-invasive and restraint-free [61]. Another milestone of technology in our discipline has been the use of drones in marine studies, because "behavioural observations are typically limited to records of animal surfacings obtained from a horizontal perspective" but "drones UAS provided three times more observational capacity than boat-based observations alone (300 vs. 103 min); provided more and longer observations plus enable documentation of multiple novel gray whale foraging tactics and social events not identified from boat-based observation" ([62], p. 359). To study songbirds in zebra finches, some authors have employed miniature, animal-wearable wireless microphones and brain recording devices [63], being able to conclude the neuronal activity of the birds' song system and their stack calls. Even scientific dissemination about animal behavior, such as broadcasted documentaries, have benefited from technology by using realistic "animatronic spy creatures" that infiltrated some species' environments to explore how they expressed complex emotions, such as grief [64].

Do ethologists regularly use technology in their experiments? In the sample, the majority of the participants reported using sophisticated technology (63.3%, n = 62) with no differences between working predominantly in the lab or in the field. Plus, the type of technology used did not vary a lot depending on their worksites either. The tech devices or methodologies included in the response options were extracted from the open responses in the first round (N = 63): Touchscreen, automatized feeders, artificial chambers, eye tracking, ECG/fMRI/Xray, sound analysis techniques, GPS/Tracking/Telemetry devices, OMICs techniques, drone, devices to overcome human sensory limitations, and computer simulations. The devices to overcome human sensory limitations and the devices to analyze sound were in the top three most frequently used technology both in the lab and in the field. The remaining position differed for obvious reasons (i.e., lab: Artificial chambers, field: GPS tracking and telemetry devices). Researchers were also provided with some blank space to indicate other technology they used, different from the options provided. The technology reported was very attached to the species' characteristics or to the type of study conducted (i.e., robotic raptors, RFID—PIT (Radio Frequency Identification through Passive Integrated Transponder) tag technology, Electrophysiology rig).

Researchers were also asked about the "dreaming device they would like to have". Interestingly, their responses did not seem that unfeasible in terms of the current technology possibilities but rather in terms of other factors such as budget, portability or size, power, accuracy, animal wellbeing, and all-proof material (see Table 3). It seems that Ethology would benefit from joining forces with engineers. To do so, some effort should be made to announce our needs within a public collaborative framework, such as some helping forum or specific website/app, this way promoting a new interdisciplinary network with many labor possibilities. Until then, researchers could find ways of collaboration with other colleagues sharing their tricks and affordable ways of doing comparative research (e.g., [19,65]).

Table 3. Main barriers to currently dispose more adequate tech devices in Ethology.

Barrier	Device Examples
Budget	Thermal imaging, cheaper micro imaging technology, top notch hydrophones
Portability/Size	Portable DNA sequencer, portable fMRI, GPS tag that could be deployed on <100 g birds, gaze tracking for fishes, wristband date loggers, walkie-talkies in the shape of smartwatches
Power	Higher-speed cameras, tracking multiple animals, tracking small organisms, micro imaging processing power, 3D simultaneously coding animal track and sounds recording, long battery life for microphones and for tags/telemetry—specially to study migrations
Accuracy	Better RFID, GPS satellite tags with high resolution (≤1 m accuracy), better nocturnal cameras
Animal wellbeing	Silent drones
All-proof material	Weather-proof cameras and microphones, GPS telemetry capable of being carried by a wild parrot without being chewed up within 10 min

According to our current knowledge and possibilities, the true dream technologies reported were artificial intelligence and automated devices that replace the human supervision or the human–computer interaction. This technology is desired because it may "reduce the current workload" and increase "objectivity" by reducing the "human observer" interpretation (see a revision of automated techniques in comparative psychology and affordable alternatives [65]). Some participants suggested to produce an "experienced ethologist robot", "programmes that could automatically record and identify different behaviours", and some "automated processing of data". It is true that robots are increasingly being used in Ethology [66]. However, one subject raised some concern about this potential technology: "the current trend in some areas to try and automate/quantify all behaviour without understanding either the behaviour or the ecology of the animal" is risky, since researchers' knowledge, adjustment to the research question, reasoning, ability to find errors, or to identify new, interesting interactions are (still) unreplaceable. It seems that we are facing a moment that is comparable to the 1960s' computer metaphor. Yet, as Powell and Rosenthal remind us [67], the use of artifices will not solve previous assessment difficulties alone, unless there is a rigorous experimental design.

Until then, researchers need to face much more worldly issues, such as the difficulties to afford sophisticated technology. That is why some free or cheap fee-subscription software initiatives specifically addressing ethological purposes have emerged. These programs allow one to conduct intricate experiments and observations (such as detecting and identifying maned wolves, e.g., [68]), but also, free software has evolved the way that researchers process, analyze, and report data.

4.5. Data

"'That is so,' replied Diagoras;
'it is because there are nowhere any pictures of those who have been shipwrecked and drowned at sea'" (Cicero)

Data are the main raw material upon which human knowledge is built. That is why it becomes relevant to discuss how that data are obtained and reported in each discipline. Recently, Science in general has been subject of the so-called "replication crisis" [69], an inability to reproduce the same data of previous studies, which challenges whether the original data could be considered real facts. Indeed, this crisis has extended to Psychology [70] and Ethology. The Animal Behavior and Cognition journal published a paper last year [71] to reveal some problems in the discipline with relation to data: Replication crisis, p-hacking, statistical analysis preferences, or small sample sizes, that other authors worldwide had also highlighted (e.g., [21,72–74]). Concretely, one pre-print entitled "The illusion of science in comparative cognition" became rapidly widespread [20]. The authors acknowledged some difficulties to achieve replicability: "(1) a lack of access to the species of interest; (2) real differences in animal behaviour across sites; and (3) sample size constraints producing very uncertain statistical estimates". Replication becoming a new daily practice seems not to be appealing, nor "part of our scientific culture" and frequently finds "general disdain by journal editors", ([75], abstract). However, the crisis, or at least the awareness of the crisis, has reached unprecedented levels, and the solution that authors more strongly endorsed is embracing Open Science: Data sharing; improving the visibility of negative results (so that Diagoras could be informed about the ships that drowned, see [76]); to improve statistical thinking, and to explore new infrastructures to create and combine the data necessary to understand how cognition evolves [20,71]. To see whether researchers in Ethology are inclined to adopt Open Science, the participants were asked about their sample sizes, the statistical program, and the type of analysis they most frequently used and the impact (in a Likert scale ranging from 0 = not much to 3 = a lot) they assigned to different issues related to reporting data: Small sample sizes, the reproducibility crisis, a publish or perish system, and the Open Science paradigm.

4.5.1. Sample Size

In general, the participants were divided into half who conducted studies with more than 20 subjects (49%, $n = 48$) and half whose studies ranged from 1 to 7 subjects (11.2%, $n = 11$) and from 8 to 20 subjects (39.8%, $n = 39$). Taking into account the worksite of the respondents, the studies done in the field (free range or wild) rarely had small samples while studies in the lab varied enormously ranging from a N = 1 sample to more than 40 subjects. The comparison between the group of animals and the number of subjects in the samples only approached significance ($\chi^2 = 48.883$, df = 36, $p = 0.074$). The studies with more than 20 subjects in their samples were mostly comprised by invertebrates (72.72%), birds (52.38%), and non-human primates (40.7%). It is therefore more likely that factors such as management, availability, and the research question itself might account better for the sample size.

4.5.2. Statistical Program

The most preferred software package was R (60.2%, $n = 59$) followed by SPSS (25.5%). The remaining percentage distributed similarly across other software (i.e., Matlab, Python). Being a junior or a senior researcher yielded significant differences in the software choice ($N = 98$, $\chi^2 = 15.719$, df = 2, $p < 0.001$) with juniors being more inclined to use R (77.7%, $n = 54$) than seniors (38.6%, $n = 44$). The arrival of free software also brought the need to learn programming, and it may have happened that senior researchers had more difficulties in incorporating this new skill.

4.5.3. Statistical Analysis

I provided the participants with different statistical analysis options, which entailed different levels of assumptions or control over the causation and prediction of any studied phenomenon to see which they usually perform. The participants could pick up several options from the list: Descriptive, non-parametric, parametric, correlation, logistic regression, GLMM, and Big Data analysis. The preferred type of analysis was GLMM (77.5%, $n = 76$). This shows that our discipline seems very interested in knowing which factors contribute, how much, and how they interact to explain some behavior. The problem is whether we are measuring and introducing all the potentially influential factors in the model. That is why some authors, rather than focusing on how to measure data and how to report the results, highlight pre-data actions, specifically, providing a good hypothesis [77]: "we don't just want science to be reproducible. Generating better hypotheses is at least as important (. . .). Who cares if you can replicate an experiment that found that people think the room is hotter after reading a story about nice people? Will this help us to develop better theories? You can craft a fun story about that result, but can you devise the next great scientific question?".

According to Table 4, Big Data was the least performed analysis (15.30%, n = 15). Due to its low frequency, it was interesting to explore where in the discipline it was being used. Concretely, it was performed similarly regardless of the worksite; it was mainly associated with sample sizes comprised of more than 40 subjects (60%, n = 9) of non-human primates, birds, rodents, and insect species and it was preferentially executed to compare two non-human species (66.6%, n = 10). Potentially, as fashion and technological possibilities influence which statistical analysis researchers usually perform [78], as it happened with GLMM and Bayesian models [78,79], the future may embrace Big Data and Deep Learning analysis in animal research just as they are already being embraced in many different disciplines [80].

Table 4. Most frequent type of statistical analysis conducted by Ethology researchers (N = 98).

Analysis	Percentage of Use (N = 98)
GLMM	77.55%
Descriptive	67.34%
Non-Parametric	61.2%
Parametric	57.14%
Correlations	55.10%
Logistic Regression	42.85%
Big Data	15.30%

4.5.4. Reporting Data

In this section, the participants were asked about the dissemination of the results, in particular, how much relevance they assigned to small sample sizes, the reproducibility crisis, the publish or perish system, and the Open Science paradigm. The results show that all these topics are deemed to have much or a lot of impact in our discipline by the majority of the sample (N = 35): publish or perish system (80%), small sample size (65.7%), reproducibility (63%) and Open Science paradigm (43%).

"Publish or perish" was the title of a letter disseminated by Nature in 1962 [81]. Already during the past century, the leading journal reported to receive six times more the number of studies they could publish. In this editorial, the author overtly blamed the academia about the publication system they had nurtured. In his opinion, the system was based more in quantity than in quality, and it had been established for lecturers and researchers to obtain academic positions: "It is also deplorable that a scientist should depend so much on his published works for his own professional advancement. This all tends towards rushing into print, writing of verbose articles and papers, claims for priority sometimes followed by voluminous and even acrimonious argument, little of which adds to the advancement, much less the dignity, of science itself" ([81], p. 709). Strikingly, this opinion seems to be valid more than fifty years later, since Nature published again an Editorial entitled the same [82], albeit this time to warn about the amount of falsified or fabricated data that researchers were producing (and even admitting to produce, see [83,84]) due to this pushing system. The low quality of Science that this system may produce was also highlighted by one of the participants in the sample: "researchers are pushed towards generating publishable data over scientific quality". That is why currently there are many movements fostering slow science, one example being the permanent manifesto exposed in the frontpage of the Slow Science Society website [85]. Maybe it is time, not only for Ethology, but for researchers in general, to stand for better academic policies that may reinforce not only publications but also fruitful discussions and thoughtful activities, which have always been the seed of Science.

The replication crisis also reached much concern in the sample whereas the Open Science paradigm did not. One constraint that replication finds is not only the high rate of rejection by editors [75], but also not complying with Open Science and not regularly publishing negative results. Indeed, the bias to positive results has been found in all disciplines in association with the so-called Hierarchy of Science, with social and animal sciences being at the top or intermediate levels of the bias [86]. Perhaps researchers, who despite being pursuing truth through Science have a human nature, feel embarrassed to report that their initial ideas were not achieved as they expected. One subject of the sample endorsed this idea adding that some research groups may be so strictly attached to some idea that this could preclude them to conduct certain experiments or disseminate opposite data: "I would say that people are "in love" with their hypothesis. Different labs defend different ideas and design studies explicitly to confirm them, usually over-selling their findings. I miss more objectivity." However, being frightened to be wrong, is being frightened of one of the main principles of science, falsifiability [87].

Therefore, embracing the Open Science paradigm may help to improve past barriers and shortcomings in the publication system. The Open Science scientific framework might be summarized as understanding that any piece of knowledge is part of humankind and therefore it should be available to everyone. The UNESCO Recommendation definition reads ([9], p. 4): "an umbrella concept that combines various movements and practices aiming to make scientific knowledge, methods, data and evidence freely available and accessible for everyone, increase scientific collaborations and sharing of information for the benefits of science and society, and open the process of scientific knowledge creation and circulation to societal actors beyond the institutionalized scientific community". To achieve this, some actions we could already take are being transparent in our methods, reporting both positive and negative results, and sharing datasheets with a common format so that systematic reviews and meta-analysis could be eased. To these practices we could also incorporate preregistering articles by default, since some authors demonstrated recently that standard articles reported more positive results than preregistered articles (96% vs. 44%) [88]. A new-born society, SORTEE (Society for Open, Reliable, and Transparent Ecology and Evolutionary biology) is trying to endorse these practices in our field. Perhaps the first step to face the changes to come for Ethology is establishing large-scale collaboration and networking.

4.6. Networking

"Are you nobody, too?/Then there's a pair of us . . . " (Emily Dickinson)

Discussing how social cognition could be better studied, at a Strüngmann Forum, the participants concluded: "In an ideal world, one useful approach would be to compare multiple species using the same or similar methods used by teams of researchers" ([42], p. 289). However, that is not an easy undertaking for one single person, nor even for one single research team. To conduct good science, we need many people, an authentic "team effort" ([40], p. 2). Indeed, von Frisch already acknowledge the existence of a team behind his Nobel nomination: "The effort of one individual is not sufficient for this. Helpers presented themselves, and I must express my appreciation to them at this time. If one is fortunate in finding capable students of whom many become permanent co-workers and friends, this is one of the most beautiful fruits of scientific work" ([89], p. 86).

This simple idea of connecting many people under the same task was the underlying motto for some networking projects, such as ManyPrimates: To start coordinated large-scale international collaborations. In their first publication together, the authors stated: "Inferring the evolutionary history of cognitive abilities requires large and diverse samples. However, such samples are often beyond the reach of individual researchers or institutions, and studies are often limited to small numbers of species. Consequently, methodological and site-specific-differences across studies can limit comparisons between species. Here we introduce the ManyPrimates project, which addresses these challenges by providing a large-scale collaborative framework for comparative studies in primate cognition" [46]. For similar projects to exist, and to achieve such level of coordination, researchers must be willing to network with peers. However, do researchers regularly interact between themselves?

Strikingly, a quarter of the participants (n = 26, 26.5%) reported not interacting at all during the year with other peers and, interestingly, being junior or senior did make a difference in this regard (N = 98, U = 918.000, $p < 0.05$), with juniors reporting an absence of interaction (33.33%, n = 54) more than seniors (18.18%, n = 44). Juniors are increasingly becoming digital natives and therefore familiarized with social networks; however, it seems that having access to the media does not imply using the media for academic or scientific purposes. Note that juniors in this sample were researchers with 10 or less years of experience, so it cannot be stated that they were at a very early stage of their career nor that they may be alien to most of the conference or discipline discussions. However, we could take this result as a message to improve collaboration between researchers when they initiate their careers.

Some examples of how to foster networking can be borrowed from other disciplines, such as the PsyResearchList for the study of moral psychology. Dr. Meltem Yucel led the dissemination of a public list of researchers interested in the topic who inscribed themselves on a voluntary basis (see https://www.psychresearchlist.com/moral-psychologists.html, accessed on 8 August 2021). This type of actions eases the act of locating peers with similar interests so further collaboration can arise. Indeed, it is demonstrated that the fresher a team is, the more original and multidisciplinary research they conduct [90] and the better they can face the future challenges to come.

4.7. Future

"I would like to be of help during this decisive but not easy period of your journey, in which you prepare for a direct confrontation with life." (Rita Levi-Montalcini)

The first questionnaire included an open question about what future events may shape the future of Ethology according to the researchers' opinion. From 53 valid and detailed responses, the issues that had a higher mode were extracted, yielding twelve: COVID-19 pandemic, funding, under-represented minorities, lack of dialogue, lack of interdisciplinarity, lack of collaboration, anthropocentrism, climate change, extinction, welfare ethics, absence of longitudinal data, robots and tech innovations. In the second round (N = 35), I asked the participants how much impact they attributed to these twelve issues in Ethology by using a Likert scale (ranging from 1 = low impact to 3 = lot of impact); see Table 5 for the results.

The main current challenges qualified as affecting Ethology "much" and "a lot" according to the participants (N = 35) were a lack of funding (85.7%), welfare ethics (77.1%), the COVID-19 pandemic (65.7%), and the absence of longitudinal data (63%). It is especially alarming that Ethology suffers from the exclusion of competitive funding sources, which is worsened by the abundant offer of precarious and seasonal positions. One subject in the sample stated: "It is extremely hard to continue after a master's or PhD, there are little to no position for early career researchers". Indeed, the discipline job websites are full of volunteer positions, in which the candidates are expected to pay for their travel costs and are offered scarce maintenance in compensation for their sometimes hard, long, and tiring observational field work. However, field work experience is much appreciated in ulterior job interviews as it is associated with successful training in research techniques. Therefore, if we do not stand against this situation, the precarious volunteer offers will be perpetuated.

Table 5. Level of impact that twelve events (obtained from qualitative responses in a previous survey, $N = 53$) may have in our discipline, according to researchers' opinion ($N = 35$).

Concept	Min.	Max.	Mean	Standard Deviation
Lack of Funding	1	3	2.46	0.741
Welfare Ethics	0	3	1.97	0.954
No Longitudinal Data	0	3	1.83	0.891
COVID-19	0	3	1.77	1.003
Climate Change	0	3	1.63	1.031
Extinction	0	3	1.63	1.114
Lack of Collaboration	0	3	1.60	0.914
Lack of Dialogue	0	3	1.57	0.948
Lack of Interdisciplinarity	0	3	1.54	0.980
Anthropocentrism	0	3	1.46	1.039
Underrepresented Minorities	0	3	1.29	0.926
Robots/Tech Innovations	0	3	1.23	0.910

Another concern shared by the sample is welfare ethics. Indeed, new knowledge can directly challenge our bioethics protocols, such as finding animal sentience in invertebrates [91]. Moreover, if we acknowledge the Anthropocene term for the geophysical epoch we live in, we may need to identify which human activities are affecting the welfare and the behavior of species (e.g., human noise affecting fish, [92]) to be ethically committed to modify certain habits and also to interpret observations accordingly. An evident consequence of human activities is climate change, which is threatening the survival of many species' groups to the extent of some alarming situations such as the so-called "insect apocalypse" [93]. Being aware of the potential disappearance of some taxa while being the responsible agents is unacceptable. Any living form deserves to be protected and its loss directly challenges the nature equilibrium plus the object of study of Ethology, which is species. Importantly, welfare ethics should not only be defended by researchers working in the field, but awareness should also be extended to those working in labs.

The factors that the participants considered to impact the discipline less were the use of robots or artificial intelligence (31.4%), low peer diversity of gender or ethnicity (35.1%), and lack of interdisciplinarity (42.9%). However, the absence of under-represented minorities in research teams is currently biasing and impoverishing Science. More diversity of gender and ethnicity in scientific teams has proven to yield more eclectic and integrative studies and even attracting more citations [94,95]. However, the access to the scientific pathway is not always easy for female foreign applicants, at least in higher-paying disciplines and private institutions, where departments usually endorse more Caucasian males in mentoring applications [96]. Indeed, even when the access barrier is overcome, a non-WEIRD ethnicity alone could be one key factor not to be eligible for an NIH (National Institutes of Health) award after controlling for factors such as educational background, country of origin, or previous awards [97]. A similar trend occurs with women's underrepresentation in Ethology, thus biasing the European narrative (see the Special Issue: A Historical Approach in Animal Behavior, 2020, edited by Zuleyma Tang-Martínez [98]). The underrepresentation of sexual diversity within researchers is also considered a current disadvantage. Their experience of negative emotions at science workplaces in general is high [99], but in Ethology in particular, their exclusion [100] can also imply disregarding potentially interesting perspectives about specific topics (i.e., sexual selection). Under this thought, the ABS (Animal Behavior Society) meeting held in August 2021 included a plenary talk entitled "Different People Ask Different Questions: A Queer Perspective on Studying Behavioural Diversity" given by Prof. Dr. Karen Warkentin. Therefore, Ethology should not only take into account all the scientific specific concerns commented on so far, but also the social, cultural, environmental, and political factors that also have an impact on the future of the discipline.

5. General Discussion

Ethology has had great historical momentum. However, "the early Giants are going or gone [and] new Giants are now being established and will lead the field" [101]. Now, it is time to think about the future of the field, and the aim of this paper was raising concern and discussing the most relevant challenges to come in seven areas (concepts, species, Umwelt, technology, data, networking, and future) through the responses of almost a hundred colleagues. First, it seems that there is a lack of consensus with regards to some core concepts in the discipline (and even with the name of the discipline itself), so World Consensus Conferences have been proposed to define a common scientific path and overcome misinterpretations and patched knowledge. Second, the underlying research motivations plus the species selection criteria we follow in our studies should be reconsidered to include less anthropocentric and more biocentric questions [12]. This has led to the third idea, the need to include mandatory checklists of Umweltian factors in published studies, as ethical statements are also mandatory, to ease peer-review processes when determining the ecological validity of the conclusions and to ease further replications. Fourth, the raising of new technology inventions will surely facilitate currently tiring tasks, but it should be accompanied by active human reasoning to provide full-sense knowledge. Plus, current and future tech needs could be easily addressed if collaboration with engineers was fostered through public and private initiatives. Fifth, the current untransparent, biased towards positivity and inaccessible publication system should be progressively replaced with an Open Science paradigm. To do so, preregistering studies, establishing standardized formats for sharing data sheets, losing the fear of contradiction and falsifiability, and substituting the publish or perish system for a more collaborative, fruitful, and slow science process would help. Sixth, networking should be integrated as another scientific principle from the very early stages of the researchers' careers, not only for the advancement of our discipline, since bigger samples and robust data can be better achieved through large-scale cooperation, but also for achieving a less competitive and more enjoyable scientific environment. Finally, the way climate change is affecting the diversity we study and the fact that some minorities are underrepresented in research teams could directly attack the survival of the subjects we study, the integration of the individuals that study them, and the new perspectives that our discipline needs.

6. Conclusions

To all these conclusions, one extra aspect should be included: The need to acknowledge and get knowledge from our predecessors prior to embarking on the future. This will help to ensure we do not take the findings and thoughts that have been already posed as novel ideas [28]. This does not mean to be stuck in the past: We also need to be more open-minded than our predecessors were and try to collaborate more than they could for the advancement of Ethology. Indeed, to make the insights of this essay useful, we should figure out how to educate the undergraduate students that will approach Ethology, so that they could already be prepared for the problems we are envisioning now. Thus, the postgraduate syllabus in Ethology should benefit from including modules about theoretical background on seminal contributions in Ethology, the principles of philosophy of science and Open Science paradigm, discussions on the definition of core theoretical concepts, the concept of Umwelt and its application to experimental designs, statistics programming, an interdisciplinary theoretical background (e.g., [102], p. 17; [103]), and welfare ethics.

Despite the limitations of this manuscript (i.e., sample size, potential underrepresentation of researchers in some animal taxa), it hopefully faithfully represented all the participants' opinion and may serve as a first boost to indagate more in the current status and future directions of Ethology. Moreover, despite all the shortcomings raised, no pessimistic approach should be established, because this discipline congregates three optimistic adjectives: It is varied, since we study biodiversity each day; it is enduring, since evolution never stops its action; and overall, it is eternal, because "a question answered usually raises new problems, and it would be presumptuous to assume that an end is ever achieved" ([89] p. 86).

Supplementary Materials: The following are available online at https://www.mdpi.com/article/10.3390/ani11092520/s1. The data sheet is included as a supplementary file. It can be downloaded from the Comillas Pontifical University's repository here: http://hdl.handle.net/11531/59820. The questionnaire is include in the annex.

Funding: This research received no external funding for its conception and writing. It received external funding from Comillas Pontifical University for its publication as a full-text access.

Institutional Review Board Statement: The study was conducted according to the guidelines of the Declaration of Helsinki, and approved by the Institutional Review Board of Ethics Committee of the Faculty of Human and Social Sciences of the Universidad Pontificia Comillas (approval code 2021/80).

Informed Consent Statement: Informed consent was obtained from all subjects involved in the study.

Data Availability Statement: The data sheet is included as a supplementary file.

Acknowledgments: The author would like to thank to two anonymous reviewers and to Gordon M. Burghardt because they provided insightful comments and provoked a fruitful discussion in earlier versions of this manuscript.

Conflicts of Interest: The author declares no conflict of interest.

References

1. Tinbergen, N. Nobel Lecture. NobelPrize.org. Available online: https://www.nobelprize.org/prizes/medicine/1973/tinbergen/lecture/ (accessed on 4 August 2021).
2. Lorenz, K. Banquet Speech. Available online: https://www.nobelprize.org/prizes/medicine/1973/lorenz/speech/ (accessed on 4 August 2021).
3. Kuhn, T. *The Structure of Scientific Revolutions*; Chicago University Press: Chicago, IL, USA, 1962.
4. Hinde, R. *Animal Behaviour: A Synthesis of Ethology and Comparative Psychology*; McGraw-Hill: New York, NY, USA, 1966.
5. Curd, M.; Cover, J.A.; Pincock, C. *Philosophy of Science. The Central Issues*, 2nd ed.; Norton & Company: New York, NY, USA, 2013.
6. Beach, F.A. The snark was a boojum. *Am. Psychol.* **1950**, *5*, 115–124. [CrossRef]
7. Shettleworth, S.J. The evolution of comparative cognition: Is the snark still a boojum? *Behav. Process.* **2009**, *80*, 210–217. [CrossRef]
8. Levitis, D.A.; Lidicker, W.Z.; Freund, G., Jr. Behavioural biologists do not agree on what constitutes behaviour. *Anim. Behav.* **2009**, *78*, 103–110. [CrossRef] [PubMed]
9. UNESCO. First Draft of the UNESCO Recommendation on Open Science. Available online: https://unesdoc.unesco.org/ark:/48223/pf0000374837 (accessed on 4 August 2021).
10. Stichweh, R. Differentiation of scientific disciplines: Causes and consequences. Encyclopedia of Life Support Systems. 2003. Available online: http://www.eolss.net/sample-chapters/c04/e6-49-01-02.pdf (accessed on 4 August 2021).
11. Call, J. *Handbook of Comparative Psychology*; American Psychological Association: Washington, DC, USA, 2017.
12. Bräuer, J.; Hanus, D.; Pika, S.; Gray, R.; Uomini, N. Old and new approaches to animal cognition: There is not "One Cognition". *J. Intell.* **2020**, *8*, 28. [CrossRef] [PubMed]
13. Alpi, A.; Amrhein, N.; Bertl, A.; Blatt, M.R.; Blumwald, E.; Cervone, F.; Dainty, J.; De Michelis, M.I.; Epstein, E.; Galston, A.W.; et al. Plant neurobiology: No brain, no gain? *Trends Plant Sci.* **2007**, *12*, 135–136. [CrossRef]
14. Nobel Prize. Press Release: The Nobel Prize in Chemistry 2020. Available online: https://www.nobelprize.org/prizes/chemistry/2020/press-release/ (accessed on 4 August 2021).
15. Beck, B.B. Chimpocentrism: Bias in cognitive ethology. *J. Hum. Evol.* **1982**, *11*, 3–17. [CrossRef]
16. Shettleworth, S.J. Clever animals and killjoy explanations in comparative psychology. *Trends Cogn. Sci.* **2010**, *14*, 477–481. [CrossRef]
17. von Uexküll, J. *Cartas Biológicas a Una Dama*; Cactus: Buenos Aires, Argentina, 1920/2014.
18. Bueno-Guerra, N. How to apply the concept of umwelt in the evolutionary study of cognition. *Front. Psychol.* **2018**, *9*. [CrossRef]
19. Bueno-Guerra, N.; Amici, F. *Field and Laboratory Methods in Animal Cognition: A Comparative Guide*; Cambridge University Press: Cambridge, UK, 2018.
20. Farrar, B.G.; Ostojic, L. The illusion of science in comparative cognition. *PsyArXiv* **2019**. [CrossRef]

21. Beran, M.J. Editorial: The value and status of replications in animal behavior and cognition research. *Anim. Behav. Cogn.* **2020**, *7*. [CrossRef]
22. Henrich, J.; Heine, S.J.; Norenzayan, A. The weirdest people in the world? *Behav. Brain Sci.* **2010**, *33*, 61–83. [CrossRef]
23. Poincare, H. *Ciencia y Método*; Espasa Calpe: Madrid, Spain, 1944/1963.
24. Powell, A.; O'Malley, M.A.; Müller-Wille, S.; Calvert, J.; Dupré, J. Disciplinary baptisms: A comparison of the naming stories of genetics, molecular biology, genomics and systems biology. *Hist. Philos. Life Sci.* **2007**, *29*, 5–32. [PubMed]
25. Martin, J.D. What's in a name change? Solid state physics, condensed matter physics, and materials science. *Phys. Perspect.* **2015**, *17*, 3–32. [CrossRef]
26. Moreno, C.B.; Muñoz-Delgado, J. An account on the history of ethology. *Suma Psicol.* **2007**, *14*, 213–224.
27. Egerton, F.N. History of ecological sciences, part 56: Ethology until 1973. *Bull. Ecol. Soc. Am.* **2016**, *97*, 31–88. [CrossRef]
28. Burghardt, G.M. Insights found in century-old writings on animal behaviour and some cautions for today. *Anim. Behav.* **2020**, *164*, 241–249. [CrossRef]
29. Jarvis, E.D.; Güntürkün, O.; Bruce, L.; Csillag, A.; Karten, H.; Kuenzel, W.; Medina, L.; Paxinos, G.; Perkel, D.J.; Shimizu, T.; et al. Avian Brain Nomenclature Consortium. Avian brains and a new understanding of vertebrate brain evolution. *Nat. Rev. Neurosci.* **2005**, *6*, 151–159. [CrossRef]
30. Kuhn, T. Objectivity, Value Judgment, and Theory Choice. In *The Essential Tension: Selected Studies in Scientific Tradition and Change*; Kuhn, T., Ed.; Chicago University Press: Chicago, IL, USA, 1977; pp. 320–339.
31. Wittgenstein, L. *Tracttatus Logico-Philosophacus*; Alianza Editorial: Madrid, Spain, 1921/2003.
32. Chalmers, D.J. Facing up to the problem of consciousness. *J. Conscious. Stud.* **1995**, *2*, 200–219.
33. Birch, J.; Schnell, A.K.; Clayton, N.S. Dimensions of animal consciousness. *Trends Cogn. Sci.* **2020**, *24*, 789–801. [CrossRef]
34. Levy, D. The ethical treatment of artificially conscious robots. *Int. J. Soc. Robot.* **2009**, *1*, 209–216. [CrossRef]
35. Uher, J. Conceiving "personality": Psychologist's challenges and basic fundamentals of the transdisciplinary philosophy-of-science paradigm for research on individuals. *Integr. Physiol. Behav. Sci.* **2014**, *49*, 398–458. [CrossRef] [PubMed]
36. Good, G.A. The assembly of geophysics: Scientific disciplines as frameworks of consensus. *Stud. Hist. Philos. Sci. Part B Stud. Hist. Philos. Mod. Phys.* **2000**, *31*, 259–292. [CrossRef]
37. Bateson, P.; Laland, K.N. Tinbergen's four questions: An appreciation and an update. *Trends Ecol. Evol.* **2013**, *28*, 712–718. [CrossRef] [PubMed]
38. Menzel, R.; Fischer, J. *Animal Thinking. Contemporary Issues in Comparative Cognition*; Massachusetts Institute of Technology: Cambridge, MA, USA; Frankfurt Institute for Advanced Studies: Frankfurt, Germany, 2011.
39. Rosenthal, M.F.; Gertler, M.; Hamilton, A.; Prasad, S.; Andrade, M.C. Taxonomic bias in animal behaviour publications. *Anim. Behav.* **2017**, *127*, 83–89. [CrossRef]
40. Fragaszy, D.M. Editorial. *J. Comp. Psychol.* **2018**, *132*, 1–3. [CrossRef] [PubMed]
41. Cronin, K.A.; Jacobson, S.L.; Bonnie, K.E.; Hopper, L.M. Studying primate cognition in a social setting to improve validity and welfare: A literature review highlighting successful approaches. *PeerJ* **2017**, *5*, e3649. [CrossRef]
42. Jensen, K.; Silk, J.B.; Andrews, K.; Bshary, R.; Cheney, D.L.; Emery, N.; Hemelrijk, C.K.; Holekamp, K.; Penn, D.C.; Perner, J.; et al. Social Knowledge. In *Animal Thinking. Contemporary Issues in Comparative Cognition*; Menzel, R., Fischer, J., Eds.; Massachusetts Institute of Technology: Cambridge, MA, USA; Frankfurt Institute for Advanced Studies: Frankfurt, Germany, 2011; p. 267291.
43. Call, J. Chimpocentrism: A continuing problem in ape cognition research. In *Advances in Primatology*; Casanova, C., Sousa, C., Eds.; University of Lisbon: Lisbon, Portugal, 2007.
44. Zilkha, N.; Sofer, Y.; Beny, Y.; Kimchi, T. From classic ethology to modern neuroethology: Overcoming the three biases in social behavior research. *Curr. Opin. Neurobiol.* **2016**, *38*, 96–108. [CrossRef]
45. MacLean, E.L.; Hare, B.; Nunn, C.L.; Addessi, E.; Amici, F.; Anderson, R.C.; Aureli, F.; Baker, J.M.; Bania, A.E.; Barnard, A.M.; et al. The evolution of self-control. *Proc. Natl. Acad. Sci. USA* **2014**, *111*, E2140–E2148. [CrossRef]
46. Altschul, D.M.; Beran, M.J.; Bohn, M.; Caspar, K.R.; Fichtel, C.; Försterling, M.; Grebe, N.M.; Hernandez-Aguilar, R.A.; Kwok, S.C.; Llorente, M.; et al. Collaborative open science as a way to reproducibility and new insights in primate cognition research. *Jpn. Psychol. Rev.* **2019**, *62*, 205–220.
47. Burghardt, G.M. Cognitive ethology and critical anthropomorphism: A snake with two heads and hognose snakes that play dead. In *Cognitive Ethology: The Minds of Other Animals*; Ristau, C.A., Ed.; Erlbaum: Mahwah, NJ, USA, 1991; pp. 53–90.
48. Povinelli, D.J.; Vonk, J. We don't need a microscope to explore the chimpanzee's mind. *Mind Lang.* **2004**, *19*, 1–28. [CrossRef]
49. Burghardt, G.M. Critical anthropomorphism, uncritical anthropocentrism, and naïve nominalism. *Comp. Cogn. Behav. Rev.* **2007**, *2*, 136–138. [CrossRef]
50. Wynne, C.D.L. What are animals? Why anthropomorphism is still not a scientific approach to behavior. *Comp. Cogn. Behav. Rev.* **2006**, *2*. [CrossRef]
51. von Uexküll, J. *A Foray into the Worlds of Animals and Humans with a Theory of Meaning*; University of Minnesota Press: Minnesota, MN, USA, 1934/2010.
52. Webster, M.; Rutz, C. How STRANGE are your study animals? *Nature* **2020**, *582*, 337–340. [CrossRef] [PubMed]
53. Sherman, L.W. An Introduction to Experimental Criminology. In *Handbook of Quantitative Criminology*; Piquero, A., Weisburd, D., Eds.; Springer: New York, NY, USA, 2010; pp. 399–436. [CrossRef]

54. Parker, T.H.; Griffith, S.C.; Bronstein, J.L.; Fidler, F.; Foster, S.; Fraser, H.; Forstmeier, W.; Gurevitch, J.; Koricheva, J.; Seppelt, R.; et al. Empowering peer reviewers with a checklist to improve transparency. *Nat. Ecol. Evol.* **2018**, *2*, 929–935. [CrossRef]
55. Van Helvoort, T.; Sankaran, N. How seeing became knowing: The role of the electron microscope in shaping the modern definition of viruses. *J. Hist. Biol.* **2018**, *52*, 125–160. [CrossRef]
56. Hinton, G. Deep learning—A technology with the potential to transform health care. *JAMA* **2018**, *320*, 1101–1102. [CrossRef] [PubMed]
57. Premack, D.; Woodruff, G. Does the chimpanzee have a theory of mind? *Behav. Brain Sci.* **1978**, *1*, 515–526. [CrossRef]
58. Call, J.; Tomasello, M. Does the chimpanzee have a theory of mind? 30 years later. *Trends Cogn. Sci.* **2008**, *12*, 187–192. [CrossRef] [PubMed]
59. Krupenye, C.; Kano, F.; Hirata, S.; Call, L.; Tomasello, M. Great apes anticipate that other individuals will act according to false beliefs. *Science* **2016**, *354*, 110–114. [CrossRef]
60. Kano, F.; Krupenye, C.; Hirata, S.; Call, J. Eye tracking uncovered great apes' ability to anticipate that other individuals will act according to false beliefs. *Commun. Integr. Biol.* **2017**, *10*, e1299836. [CrossRef]
61. Hopper, L.M.; Gulli, R.A.; Howard, L.H.; Kano, F.; Krupenye, C.; Ryan, A.M.; Paukner, A. The application of noninvasive, restraint-free eye-tracking methods for use with nonhuman primates. *Behav. Res. Methods* **2020**, *53*, 1003–1030. [CrossRef] [PubMed]
62. Torres, L.G.; Nieukirk, S.L.; Lemos, L.; Chandler, T.E. Drone up! Quantifying whale behavior from a new perspective improves observational capacity. *Front. Mar. Sci.* **2018**, *5*, 319. [CrossRef]
63. Ter Maat, A.; Trost, L.; Sagunsky, H.; Seltmann, S.; Gahr, M. Zebra finch mates use their forebrain song system in unlearned call communication. *PLoS ONE* **2014**, *9*, e109334. [CrossRef]
64. BBC. Spy in the Wild. Available online: https://www.bbc.co.uk/programmes/b088t67m (accessed on 4 August 2021).
65. Varnon, C.A.; Lang, H.; Abramson, C.I. Automated research in comparative psychology: Limitations and new directions. *Int. J. Comp. Psychol.* **2018**, *31*. Available online: https://escholarship.org/uc/item/8gf4d5tx (accessed on 4 August 2021). [CrossRef]
66. Klein, B.A.; Stein, J.; Taylor, R.C. Robots in the service of animal behavior. *Commun. Integr. Biol.* **2012**, *5*, 466–472. [CrossRef]
67. Powell, D.L.; Rosenthal, G.G. What artifice can and cannot tell us about animal behavior. *Curr. Zool.* **2017**, *63*, 21–26. [CrossRef]
68. Rocha, L.H.; Ferreira, L.; Paula, B.C.; Rodrigues, F.H.G.; Sousa-Lima, R.S. An evaluation of manual and automated methods for detecting sounds of maned wolves (Chrysocyon brachyurus Illiger 1815). *Bioacoustics* **2015**, *24*, 185–198. [CrossRef]
69. Baker, M. 1500 scientists lift the lid on reproducibility. *Nature* **2016**, *533*, 452–454. [CrossRef]
70. Open Science Collaboration. Estimating the reproducibility of psychological science. *Science* **2015**, *349*, aac4716. [CrossRef]
71. Farrar, B.; Boeckle, M.; Clayton, N.S. Replications in comparative cognition: What should we expect and how can we improve? *Anim. Behav. Cogn.* **2020**, *7*, 1–22. [CrossRef]
72. Nosek, B.A.; Lakens, D.D. Registered reports: A method to increase the credibility of published results. *Soc. Psychol.* **2014**, *45*, 137–141. [CrossRef]
73. Stevens, J.R. Replicability and reproducibility in comparative psychology. *Front. Psychol.* **2017**, *8*, 862. [CrossRef] [PubMed]
74. Tecwyn, E.C. Doing reliable research in comparative psychology: Challenges and proposals for improvement. *J. Comp. Psychol.* **2021**, in press.
75. Kelly, C.D. Replicating empirical research in behavioral ecology: How and why it should be done but rarely ever is. *Q. Rev. Biol.* **2006**, *81*, 221–236. [CrossRef] [PubMed]
76. Wagenmakers, E.-J.; Dutilh, G.; Sarafoglou, A. The creativity-verification cycle in psychological science: New methods to combat old idols. *Perspect. Psychol. Sci.* **2018**, *13*, 418–427. [CrossRef] [PubMed]
77. Smaldino, P. Better methods can't make up for mediocre theory. *Nat. Cell Biol.* **2019**, *575*, 9. [CrossRef] [PubMed]
78. McElreath, R. Bayesian Data Analysis. In *Field and Laboratory Methods in Animal Cognition. A Comparative Guide*; Bueno-Guerra, N., Amici, F., Eds.; Cambridge University Press: Cambridge, UK, 2018; pp. 422–423.
79. McElreath, R. *Statistical Rethinking: A Bayesian Course with Examples in R and Stan*; Chapman & Hall/CRC: London, UK, 2016.
80. LeCun, Y.; Bengio, Y.; Hinton, G. Deep learning. *Nature* **2015**, *521*, 436–444. [CrossRef]
81. Nature Editorial. Publish or Perish. *Nature* **1962**, *193*, 70924. Available online: https://www.nature.com/articles/193709a0#citeas (accessed on 4 August 2021).
82. Nature Editorial. Publish or perish. *Nature* **2015**, *521*, 159. Available online: https://www.nature.com/news/publish-or-perish-1.17574 (accessed on 4 August 2021). [CrossRef]
83. Fanelli, D. How many scientists fabricate and falsify research? A systematic review and meta-analysis of survey data. *PLoS ONE* **2009**, *4*, e5738. [CrossRef]
84. Kiai, A. To protect credibility in science, banish "publish or perish". *Nat. Hum. Behav.* **2019**, *3*, 1017–1018. [CrossRef] [PubMed]
85. SSO. The Slow Science Manifesto. Available online: http://slow-science.org/ (accessed on 4 August 2021).
86. Fanelli, D. "Positive" results increase down the hierarchy of the sciences. *PLoS ONE* **2010**, *5*, e10068. [CrossRef] [PubMed]
87. Popper, K. *The Logic of Scientific Discovery*; Routledge: London, UK, 1935/2002.
88. Scheel, A.M.; Schijen, M.R.M.J.; Lakens, D. An excess of positive results: Comparing the standard psychology literature with registered reports. *Adv. Methods Pract. Psychol. Sci.* **2021**, *4*. [CrossRef]

89. von Frisch, K. Nobel Lecture Decoding the Language of the Bee. Available online: https://www.nobelprize.org/uploads/2018/06/frisch-lecture.pdf (accessed on 4 August 2021).
90. Zeng, A.; Fan, Y.; Di, Z.; Wang, Y.; Havlin, S. Fresh teams are associated with original and multidisciplinary research. *Nat. Hum. Behav.* **2021**, 1–9. [CrossRef]
91. Baracchi, D.; Baciadonna, L. Insect sentience and the rise of a new inclusive ethics. Commentary on mikhalevich & powell on invertebrate minds. *Anim. Sentience* **2020**, *5*, 18. [CrossRef]
92. Popper, A.N. Effects of anthropogenic sounds on fishes. *Fisheries* **2003**, *28*, 24–31. [CrossRef]
93. Goulson, D. The insect apocalypse, and why it matters. *Curr. Biol.* **2019**, *29*, R967–R971. [CrossRef] [PubMed]
94. Freeman, R.B.; Huang, W. Collaboration: Strength in diversity. *Nature* **2014**, *513*, 305. [CrossRef]
95. Freeman, R.B.; Huang, W. Collaborating with people like me: Ethnic coauthorship within the United States. *J. Labor Econ.* **2015**, *33*, S289–S318. [CrossRef]
96. Milkman, K.L.; Akinola, M.; Chugh, D. What happens before? A field experiment exploring how pay and representation differentially shape bias on the pathway into organizations. *J. Appl. Psychol.* **2015**, *100*, 1678–1712. [CrossRef]
97. Ginther, D.K.; Schaffer, W.T.; Schnell, J.; Masimore, B.; Liu, F.; Haak, L.L.; Kington, R. Race, ethnicity, and NIH research awards. *Science* **2011**, *333*, 1015–1019. [CrossRef]
98. Tang-Martínez, Z. The history and impact of women in animal behaviour and the ABS: A North American perspective. *Anim. Behav.* **2020**, *164*, 251–260. [CrossRef]
99. Cech, E.A.; Pham, M.V. Queer in STEM organizations: Workplace disadvantages for LGBT employees in STEM related federal agencies. *Soc. Sci.* **2017**, *6*, 12. [CrossRef]
100. Lee, D.N. Diversity and inclusion activisms in animal behaviour and the ABS: A historical view from the U.S.A. *Anim. Behav.* **2020**, *164*, 273–280. [CrossRef]
101. Schein, M.W. A Guide to the Records of the Animal Behavior Society. Available online: https://www.animalbehaviorsociety.org/web/about-history.php (accessed on 4 August 2021).
102. Reznikova, Z.; Ryabko, B. Ants and bits. *IEEE Inf. Theory Soc.* **2012**, *65*, 17–20.
103. Krakauer, J.W.; Ghazanfar, A.A.; Marin, A.G.; MacIver, M.; Poeppel, D. Neuroscience needs behavior: Correcting a reductionist bias. *Neuron* **2017**, *93*, 480–490. [CrossRef] [PubMed]

animals

MDPI

Review

Context-Specific Habituation: A Review

Andrea Dissegna [1], Massimo Turatto [2] and Cinzia Chiandetti [1,*]

[1] Department of Life Sciences, University of Trieste, 34136 Trieste, Italy; andrea.dissegna@phd.units.it
[2] Center for Mind/Brain Sciences (CIMeC), University of Trento, 38068 Rovereto, Italy; massimo.turatto@unitn.it
* Correspondence: cchiandetti@units.it

Simple Summary: Habituation reflects the ability to learn to ignore irrelevant stimuli, which form the vast majority of the sensory input impinging on any organism's sensory systems at a given moment. However, although habituation is often described as one of the simplest forms of learning affected only by the stimulus features, such as frequency or intensity, in fact evidence exists showing that habituation can be specific for the context in which it takes place. This shows that habituation, in some cases, exhibits an associative nature, and that the underlying learning mechanism is more sophisticated than previously believed.

Abstract: Habituation consists of the progressive response decrement to a repeated stimulation, a response decline that is not accounted for by sensory or motor fatigue. Together with sensitization, habituation has been traditionally considered to be a prototypical example of non-associative learning, being affected only by the features of the stimulation, as for instance its intensity or frequency. However, despite this widespread belief, evidence exists showing that habituation can be specific to the context of the stimulation, thus suggesting that habituation can have an associative nature. Such an unexpected characteristic of habituation was in fact predicted by a theoretical model of associative learning proposed by Wagner in a series of works that appeared in the late 1970s. Here, we critically review the experimental data that since then have been accumulated in support of this hypothesis. What emerges from the literature is that context-specific habituation is common to several animal species and that the ability to form an association between the habituating stimulus and its context is independent of the complexity of the animal's nervous system. Finally, context-specific habituation is observed for a variety of organism's responses, ranging from visceral to motor and mental activities.

Keywords: associative learning; plasticity; evolution; habituation model

Citation: Dissegna, A.; Turatto, M.; Chiandetti, C. Context-Specific Habituation: A Review. *Animals* **2021**, *11*, 1767. https://doi.org/10.3390/ani11061767

Academic Editors: Lucia Regolin and Maria Loconsole

Received: 5 May 2021
Accepted: 10 June 2021
Published: 12 June 2021

1. Introduction

It is late at night and you feel sleepy. After putting down the book you were reading you turn the light off, and in a few moments your consciousness fades away. You have fallen asleep despite the distinct ticking of your bedside clock, which you did not notice at all. How is that possible? The answer is habituation, a widespread phenomenon showing that animals usually cease to respond to repetitive stimuli, especially if irrelevant, and that the nervous system is capable of learning to filter out certain sensory inputs [1]. The next day you have guests, and you kindly offer them your bedroom for the night. You then go sleeping on the sofa, and take your bedside clock with you. It is bedtime, your guests have retired and you feel like sleeping, but this time you cannot avoid noticing the ticking of the clock, which keeps you awake the whole night. What has happened? Isn't this the same clock you were able to ignore in your bedroom? Apparently yes, but sometimes the nervous system reacts in a different way to familiar stimuli when encountered in a different place. To put it differently, habituation can be specific for the context in which it has occurred. Thus, although habituation is usually considered a prototypical example of non-associative learning, in which the vanishing of the response is determined only by the characteristics of the stimulation, in fact habituation can be context specific, revealing that it can arise from an associative learning process that takes into account also the surrounding environment.

In agreement with this notion, we will review existing evidence of context-specific habituation, showing that this phenomenon is common to different animal species (see [2]). Furthermore, and quite surprisingly, this sophisticated learning ability is not dependent on the complexity of the animal's neural system, as context-specific habituation has been shown both in human beings, who possess a brain composed of about 80 billion neurons, and in different types of worms, whose behavior is controlled by a few hundred neurons (see Figure 1).

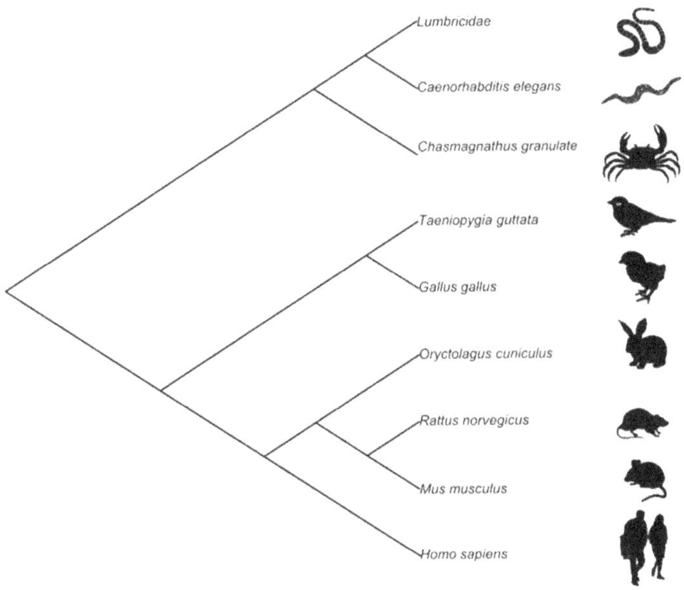

Figure 1. Cladogram of the species showing context-specific habituation.

However, before reviewing the experimental studies providing support to the associative nature of habituation, it may be useful to elucidate on which theoretical grounds habituation should be, under certain circumstances, context specific.

2. A Context-Specific Theory of Habituation: Wagner's Model

Historically, three main different models of habituation have been proposed: the stimulus-model comparator theory [3], the dual-process theory [4], and the gnostic-unit theory [5]. The models show both commonalities and differences [1], but only Wagner's model postulates that habituation can be context specific, and therefore it will be presented more in detail. Still, it may be worth also mentioning the key notions of both the stimulus-model comparator theory devised by Sokolov [3,6], and the dual-process theory elaborated by Groves and Thompson [7]. As for the former, it was mainly devised to account for habituation of the orienting response (OR), and its key mechanism consists of a comparison process between the current sensory input and a neural model of past stimulation. When the result of the comparison is a mismatch, an OR toward the new stimulation is triggered; by contrast, the more the stimulus matches the neural model, the more the OR is inhibited, and habituation arises [3,6]. The dual-process theory, instead, was developed on the basis of the acute-spinal cat model and grounded on the results of physiological recordings showing the existence of two main types of neurons in the spinal cord, those that reduce their response with stimulus repetition, causing habituation, and those that before habituating show an initial response increment called sensitization. The two classes of neurons form two independent systems, the S-R pathway leading to habituation, and the state-system controlling the arousal level and producing sensitization. Activity from both systems is then integrated at some synaptic level to determine the final response to the stimulation, which in general decreases with repetition [7]. As anticipated, neither of these two theories anticipate that habituation could be context specific.

Let us now turn to Wagner's model, which was originally elaborated and presented in three consecutive articles that appeared in 1976, 1978, and 1979, respectively [5,8,9], in the context of a more general conceptualization of how stimuli are differently processed in memory. The model was particularly inspired by the previous work of Konorski [10], who also proposed a similar habituation mechanism, and of Aktinson and Shiffrin [11] on the structure of memory, in which the functional distinction between short-term memory (STM) and long-term memory (LTM) was delineated. Although further developments of the original model have been proposed by Wanger and his collaborators [12,13], the basic principles governing habituation, schematically outlined below, remained those enounced in his seminal works, where more detailed explanations of the underlying mechanisms can be found [5,8,9].

Pivotal to Wagner's model is the idea that LTM can be conceptualized as a set of interconnected nodes forming an associative network, with each node (also known as a gnostic unit) coding for a given knowledge or stimulus representation. In the LTM the vast majority of gnostic units are inactive, whereas the STM corresponds to the subset of these units that are temporarily active [10,11]. The model also postulates that when a stimulus is initially presented, the corresponding representation becomes fully active in the STM, which strongly also activates the response generator. However, because of the transient nature of the STM, the activation of the stimulus representation will decay over time. A key notion of the model is that while the representation is active it prevents a subsequent instance of the same stimulus to activate its representation at full strength again, and consequently the response elicited by the repeated stimulus becomes weaker. In other words, " ... the likelihood of a response on trial *n* of a habituation sequence appears to be depressed by a short-term refractory-like effect generated by recent trial events ... " ([8], p. 109), an effect that Wagner termed "priming" of the STM. This simple refractory-like mechanism is capable of explaining one of the main features of habituation, namely that it is directly proportional to the frequency of stimulation. Indeed, each occurrence of the stimulus will elicit a self-generated priming, which however decays over time. Hence, short-term habituation, namely habituation observed within a sequence of stimulation, is usually stronger the shorter the inter-stimulus-interval (ISI) or, conversely, the higher the frequency of stimulation.

It is worth noting, however, that the ISI has another important consequence on habituation. Indeed, the longer the ISI the more the stimulus representation remains active in the STM without being disrupted or weakened by subsequent events (priming of the STM). This, in turn, also increases the probability that such representation will enter in associations with other contextual stimuli, and that these associations will be transferred and consolidated in the LTM [14,15]. Hence, long-term habituation, measured across different sequences of stimulation, should be stronger the longer the ISI [14,15]. Crucially, the contextual cues encountered again following the initial habituation phase can retrieve or prime representation of the associated habituating stimulus in the STM (associatively-generated priming). As a consequence, if the stimulus re-occurs in the same context the corresponding response retains a certain degree of habituation. Yet, if the stimulus appears in a different context, with which no previous associations were formed, contextual cues will not prime the stimulus representation in the STM, and the previously habituated response will recover. Hence, because of associative mechanisms analogous to those involved in Pavlovian conditioning, the Wagner model predicts that long-term habituation is context specific.

Although the model devised by Wanger is the only one that explicitly posits that long-term habituation is controlled by stimulus-context associations, in fact the model proposed by Sokolov [3,6] could also be compatible with context-specific habituation. To the extent that the neural model of the stimulus is defined not only by the specific stimulus features, as originally proposed, but incorporates also some characteristics of the surrounding environment, the stimulus appearance in a new context will cause a mismatch between the sensory input and the stored neural model of past stimulation, and consequently a recovery of the previously habituated response. On the contrary, contextual habituation is not easily accommodated in Groves and Thompson's dual-process theory [7], which does not postulate the existence of associative links between the stimulus and its context.

In fact, a fourth alternative account of habituation has been recently proposed by Hall and Rodriguez [16,17], which postulates that habituation for a given stimulus does not arise the more the stimulus is predicted (or primed), as assumed for example by Sokolov's or Wagner's models; rather, habituation is the consequence of a progressive reduction in the stimulus salience (which determines the ability to evoke its unconditioned response) due to the fact that across repetitions the stimulus is followed by no consequences. In other words, this account emphasizes the fact that habituation would depend on the predictiveness of the habituating stimulus rather than on its predictability. The model does not exclude the possibility that the context can modulate habituation but interprets this effect as an instance of occasion setting rather than evidence that associations have been formed between the context and the habituating stimulus.

3. Methods to Investigate Context-Specific Habituation

Three different approaches have been traditionally used to show that habituation can be specific for the context of stimulation. The most straightforward is the context-change method, whereby the context changes from the training to the test phase. Evidence of context-specific habituation is found when the response habituated during training recovers in a different context during the test. A second approach to reveal the associative nature of habituation relies on the extinction phenomenon. Here the logic is that during training the stimulus automatically forms associations, stored in the LTM, with the surrounding environmental cues [8]. When the context is further encountered in the future, these associations prime the stimulus (or anticipate its arrival) in the STM, thus maintaining previous habituation. However, if after habituation the organism is repeatedly exposed to the same context without the stimulus (the extinction condition), the previously formed associations will extinguish, and the habituated response will recover when the stimulus is reintroduced in the same context. By contrast, if the stimulus is omitted altogether with the context (the control condition), the stimulus-context associations will be retained in the LTM, so that habituation will remain effective when the stimulus occurs again in the same context. A third method is based on the latent inhibition phenomenon reported in conditioning studies [18]. In this phenomenon, the association between the conditioned and the unconditioned stimuli (CS and US) is delayed when the CS is pre-exposed in isolation before being paired with the US. One of the viable explanations is that, during the pre-exposure phase, the CS becomes associated with contextual cues, which then reduce its ability to enter in association with the US. Alternatively, during the pre-exposure phase the animal learns that the context predicts no consequences, namely that no US would follow. In the same vein, if habituation is context specific, then by presenting the context alone in a pre-exposure phase it will be less associated with the to-be habituated stimulus during training. Hence, at test the context would not activate, or would activate to a lesser extent, the stimulus representation in the STM, and the retention of habituation will be compromised as compared to when the pre-exposed context is different from that used during training.

To conclude this brief overview of the rationale behind the main methods used to address the associability of habituation, it should be worth mentioning that at least two alternative phenomena can explain the disruption of habituation when the context changes. One is that the presentation of a stimulus in a different environment may simply alter the stimulus representation. This would make the stimulus essentially new to an organism, leading to a disruption of the ongoing habituation. A second one is that a sudden change of the context of stimulation might lead to a neophobia reaction. This in turn increases the animal's arousal and propensity to respond to any stimulus, a result that could mimic a context-specificity effect [19–21]. To rule out this possible confound, in many studies animals have been given the possibility to familiarize with the new context before the test (see Table 1).

Table 1. Evidence of context specificity as a function of behavioral response in different species.

Behavioral Response	Species	Method	Evidence of Context Specificity	Study
Drug tolerance	*Rattus norvegicus*	CC	✓	[22]
Escape	*Lumbricidae*	CC *	✓	[23]
	Chasmagnathus granulate	CC *, LI, Ext	✓	[24–26]
	Caenorhabditis elegans	CC *, LI, Ext	✓	[27,28]
Gene expression	*Taeniopygia guttata*	CC	✓	[29]
Neophobia	*Rattus norvegicus*	CC *	×	[21]
Orienting response				
Attentional capture	*Homo sapiens*	CC, Ext	✓	[30,31]
Skin conductance	*Homo sapiens*	CC	×	[32,33]
Light approach	*Rattus norvegicus*	CC *, LI	×	[19]
Head orienting	*Rattus norvegicus*	Ext	✓	[34]
Startle	*Rattus norvegicus*	CC *	×	[34,35]
	Mus musculus	CC *	×	[36]
Suppression				
Bar press	*Rattus norvegicus*	CC *	✓/×	[20,34]
Licking	*Rattus norvegicus*	CC *, Ext	✓/×	[34,37,38]
Wheel running	*Gallus gallus*	CC *	✓	[39,40]
Vasomotor activity	*Oryctolagus cuniculus*	Ext	✓	[8]

CC = Context change; LI = Latent inhibition; Ext = Extinction. * indicates that, to attenuate neophobia, animals were exposed to the new context before testing, as suggested in [19,20]. ✓ evidence; × absence of evidence; ✓/× mixed results.

4. Context-Specific Habituation in Humans

Evidence of context-specific habituation in humans (*Homo sapiens*) is rather scant. In one of the few studies that have addressed this issue, Turatto, Bonetti, and Pascucci [30] investigated whether habituation of attentional capture, a covert component of the OR, is context specific or generalizes across different contexts. In three consecutive days, participants performed a visual discrimination task in focused attention, while a visual onset distractor appeared in the display, which also defined the context of stimulation. The results showed that on day 1 the capture of attention triggered by a repetitive visual onset distractor was subject to habituation. Then, on day 2, participants were divided into two groups: the extinction group performed the same visual discrimination task of day 1, but without the distractor, whereas the control group suspended the task. On day 3, both groups resumed the visual task with the distractor. While habituation of capture was still present in the control group, the attentional capture response recovered in the extinction group, a result consistent with a context-specific habituation view [5,8,9]. Evidence in favor of a context-specific habituation of attentional capture emerged also in a subsequent study [31]. Here the stimuli used to measure habituation of capture were presented over a background consisting, for example, of a countryside landscape. The next day, one group of participants continued the same task in the same context, whereas for another group the context changed, with the background image now depicting an industrial landscape. Habituation of capture was retained in the same-context group but was disrupted in the different-context group.

Studies investigating the context-specific habituation of different electrophysiological responses have provided less consistent results. An electrodermal study by Churchill, Remington, and Siddle [32] recorded the skin conductance response of participants repeatedly exposed to a geometrical shape projected on a monitor. The authors found the same level of long-term habituation when either local contextual cues—e.g., the background image of the monitor—or global contextual cues—e.g., the furniture in the experimental room—changed between the training and test sessions, thus showing generalization of habituation across different contexts. Similarly, an extinction session with the trained context did not produce a recovery of the habituated response. Schaafsma, Packer, and Siddle [33] studied the role of context in long-term habituation of the skin conductance response to stimuli with different motivational value. Specifically, all participants were exposed to tones and vibrations as habituation stimuli. The authors manipulated the motivational value of either tones or vibrations by instructing participants to press a microswitch at the offset of one of the two types of stimuli. The hypothesis was that instructing participants to perform a specific action in response to one of the two habituation stimuli would increase the amount of processing that one stimulus received from the STM relative to the other. Since stimuli that receive more processing in the STM are more likely to consolidate their association with contextual cues [5,8,9], the authors expected that the context change would produce less long-term habituation retention for the motivationally significant stimulus than for the other one. However, the authors found no effect of context for either of the stimuli.

5. Context-Specific Habituation in Non-Human Mammals

An initial evidence of context-specific habituation was reported by Wagner [8] in an unpublished work on rabbits. The author reported lower retention of long-term habituation of the vasoconstriction response to a repeated tone in rabbits that remained in the experimental apparatus between the training and test session compared to animals that returned to their home cage, a pattern of results indicating a disruption of habituation due to the extinction of the context-stimulus association. Several subsequent studies have used the rat as an animal model to study context-specific habituation. Of remarkable importance is the discovery made by Siegel [22] that the associative link between context and habituation is implicated in the regulation of drug tolerance. Tolerance is an instance of habituation because some of the drug effects decrease with its administration. For example, in rats the analgesic effects of morphine declines after repeated injections, leading rats to increase the amount of narcotic to resist pain. Siegel [22] demonstrated that rats injected with morphine in a given context developed a context-specific tolerance for that drug, but tolerance dropped significantly when rats were given the shot in a new context. He also demonstrated context-specific morphine tolerance using latent inhibition and extinction, thus attesting that the associative nature of habituation can interact with the biological processes underlying addiction.

The scenario, however, is a bit more intricate, as some responses appear to be more prone to show context-specific habituation than others. In particular, habituation has been shown to depend on context in the case of inhibitory responses (e.g., lick suppression or bar-press suppression [34,37]; see [20,38] for a null result), whereas there is no evidence for the context-specific habituation of the startle response [34,35] (see also [36] for a study on mice), and conflicting results for different aspects of the OR (null results have been reported in [19]; a positive result in [34]). Specifically, the study of Jordan, Strasser, and McHale [34] has found, in the same animal, evidence of the context-specific long-term habituation of lick and bar-press suppression, but not of the acoustic startle response. Moreover, they found that extinction of context disrupted the long-term habituation of lick-suppression and of the OR to a light. These results suggest that different responses supported by independent neural circuits can be differentially sensitive to the context, in agreement with the hypothesis that habituation does not represent a unitary phenomenon affecting all behaviors in the same fashion [41,42].

However, as briefly mentioned above, it is important to stress the fact that not all the response increments observed after a context change are instances of context-specific habituation. For example, Hall and Channell [19] showed that once rats stopped to orient toward a repetitive flashing light, orientation increases again when rats were moved into a new context—apparently supporting the context-specific hypothesis—but this effect disappeared if rats could familiarize with the new context before the test. Hall and Channell [19] hypothesized that when rats were tested in the new context, the OR increased as a consequence of the neophobia induced by the context change [43]. However, contrary to context-specific habituation, response increments due to neophobia or arousal should affect the overall responsivity of the animal [4]. Indeed, a similar response sensitization was measured in a new context also in rats tested for habituation of lick-suppression [20] and neophobia reaction induced by a new flavor [21].

A less tested hypothesis is that context-specific habituation might be modulated by the biological significance of the stimulation. Indeed, stimuli that are more relevant to an organism are more likely to form associative links with the surrounding environment [5,8,9]. In line with this hypothesis, Evans and Hammond [37] showed that long-term habituation of lick suppression elicited by the distress squeal of another rat was context-specific, whereas long-term habituation of the same response caused by an artificial sound with similar acoustical features was not.

6. Context-Specific Habituation in Birds

Studies with two types of avian species, zebra finches and chicks of domestic fowl, have provided converging evidence of context-specific habituation in birds. Kruse, Stripling, and Clayton [29] reported for the first time that habituation of a genetic response is context specific. They measured the expression of the *zenk* gene—a specific immediate early gene (IEG) synthetized in the auditory brain areas of zebra finches (*Taeniopygia guttata*)—in birds repeatedly exposed to a conspecific song. The results showed that the expression of this gene decreased when the same song was repeated in the same context. However, when finches listened the familiar song under different light conditions, the synthesis of the *zenk* gene spiked again, suggesting that habituation of its expression was specific for the context in which the song was experienced.

The research with domestic chicks has focused on the development of the associative mechanism underlying context-specific habituation. Chicks of domestic fowl (*Gallus gallus*) are precocial birds that develop almost completely in the egg. Thus, after hatching, they already have enough cognitive and motor skills to be independent of parental care. Chiandetti and Turatto [39] demonstrated that the associative learning process underlying context-specific habituation is also part of this early cognitive equipment. They measured the stop of the wheel-running behavior elicited by a loud sound in 4-day old chicks in four consecutive sessions of stimulation within the same context, comparing this performance with that of groups of chicks for which different aspects of the context were changed after the first two sessions. The degree of generalization vs. specificity of the habituated freezing response to the sound varied with the number of features that the training contexts shared with the test context. In addition, it should be also noted that the increased response observed when the context changed cannot be accounted for by a general increased arousal, as the response measured was the stop of an already ongoing activity (i.e., wheel-running behavior).

Furthermore, a recent study by Turatto, Dissegna, and Chiandetti [40] suggests that the ability to take into account the context of stimulation to filter unwanted sensory input is an innate cognitive ability in chicks. The authors exposed one group of animals in the last stage of their embryonic maturation to a repetitive sound, and then tested their freezing response to an analogous series of sounds in a running wheel two days after hatching. They compared habituation to the sounds in this group of animals with that of another group of chicks exposed to the sounds one day after birth in the running wheel or in a different context, namely in the chicks' home cage. The results showed that the prenatal group of chicks had a similar disruption of long-term habituation as the chicks trained in the home cage. Nevertheless, their degree of habituation to the sounds was higher than that of an untrained group, attesting that the prenatal stimulation had successfully induced habituation, and that the level of habituation was comparable to that of the group stimulated and tested in two different contexts.

7. Context-Specific Habituation in Invertebrates

Despite the relatively simple organization and modest dimension of invertebrate's nervous systems, the study of habituation in these species has confirmed that this form of learning can be context specific. Here we will briefly review the main studies conducted with the crab (*Chasmagnathus granulate*), the nematode (*Caenorhabditis elegans*) and, more recently, the earthworm of the *Lumbricidae* family [23].

There exists a long tradition of studies concerning the remarkable ability of crabs to form associations between stimuli and context and to flexibly adapt their escape responses. Tomsic and colleagues [24] demonstrated that changing the contextual cues between the training and test phase produced a recovery of habituation of the escape response elicited by a paddle moving above the animal. The authors found the same result when crabs were exposed to the context prior to or following habituation training, thus attesting that the escape response was also sensitive to latent inhibition and extinction [25]. Furthermore, the neurobiological mechanisms regulating context-specific habituation were also investigated by injecting crabs with an inhibitor of protein synthesis (*Cycloheximide).* Interestingly, injection of the drug before the training blocked the context-specificity of habituation in crabs that could fully retain long-term habituation even when the context changed. By contrast, the injection of the inhibitor after the training impaired the formation of long-term habituation in crabs tested either in the same or different contexts [44,45]. This pattern of results suggests the presence of two distinct cellular processes that lead to context-specific habituation: one responsible for the formation of contextual memories, that was immediately activated as crabs were placed in the training context; the second deputed to the consolidation of long-term habituation triggered by repeated exposure to the stimulus.

The research on crabs has also revealed that the frequency of stimulation is also critical for the emergence of context-specific habituation. For example, it has been suggested that the longer a stimulus representation remains active in the STM before being displaced by the next stimulus occurrence, the more likely it will consolidate its association with the representations of contextual cues in the LTM [15]. In line with this, Hermitte et al. [26] demonstrated that context change produced a recovery of the habituated response in a group of crabs trained with a 171 s inter-stimulus-interval (ISI), but not in a group of crabs trained with a 0 s ISI. Moreover, the injection of the protein inhibitor *Cycloheximide* impaired the retention of long-term memory only in crabs trained at a 171 s ISI, attesting that training with long and short ISIs recruited separate cellular processes (for a replication of this result, see [46]).

However, it remains unclear whether only training with long ISIs results in context-specific habituation. Of considerable importance to this topic is a study in which Rankin [27] investigated context-specific habituation of the tap-withdrawal response in the *C. Elegans*, with stimulations delivered with either a 10 or a 60 s ISI. The context was defined by the presence of a given chemical substance (sodium acetate, $NaCh_3COO$) in the petri-dish housing the nematode. The results showed greater retention of habituation at both 10 and 60 s ISI in animals trained and tested in the presence of $NaCh_3COO$, as compared with a group trained in the same context but tested in a different one (i.e., plain agar). Rankin also found that context-specific habituation was abolished by latent inhibition and extinction if the animals were exposed to the chemosensory context in the absence of the tap. The fact that context-specific habituation emerged also with the shorter ISI is surprising because with a 10 s ISI long-term habituation is usually not observed in *C. Elegans* [47]. This paradoxical result was found also by Lau, Timbers, Mahmoud, and Rankin [28] in which another chemical substance (diacetyl, $C_4H_6O_2$) was used as the context. Lau et al. also compared several mutant strains of *C. Elegans* to identify genes involved in context-specific habituation. They found that worms with a mutation in the *nmr-1* gene (NMDA-type glutamate receptor subunit) showed comparable long-term habituation to the non-mutant group when trained and tested in a plain petri-dish, but they showed a lack of context-specific effect in the presence of the olfactory cue, indicating a deficit in their capacity to associate the tap with the context. This suggests that context-specific habituation and long-term habituation involve different biological mechanisms, that long-term habituation does not necessarily require context-stimulus associations, and that the mechanisms underlying context-specific habituation are activated by both short and long ISIs, whereas those responsible for long-term habituation are activated only by long ISIs.

8. Conclusions

Textbooks often describe habituation as a prototypical example of non-associative learning consisting of the simple vanishing of basic reflexes for repetitive stimulation (e.g., [48]). Despite this simplistic description, habituation is observed for a multitude of pivotal and complex animal behaviors, such as sexual partner choice [49] and food intake [50]. Furthermore, the underlying cognitive mechanisms of habituation have been the focus of a great bulk of research for almost a century [1]. However, the predominant belief that habituation is an instance of non-associative learning [1,51] may have considerably limited the research efforts devoted to study whether habituation may, instead, have an associative nature. Here we have briefly summarized evidence, from phylogenetically distant species, showing that habituation can be context specific for a large set of responses (see Table 1), and irrespective of the complexity of the nervous systems considered (see Figure 1). Therefore, the ability to take into account the context of stimulation to regulate the organism response must have had a high adaptive value, such that natural selection has favored the spreading of this sophisticated form of learning across different animal and plant species.

Finally, the reviewed results also suggest that context specificity should be included among the features of habituation [1], and that, in our view, the model proposed by Wagner [5,8,9] is perhaps the most exhaustive in accounting for habituation (at least in non-human animals), including its feature of being, for certain types of responses, context specific.

Author Contributions: A.D., M.T. and C.C.: conceptualized, drafted and edited the manuscript. All authors have read and agreed to the published version of the manuscript.

Funding: This research received no external funding.

Institutional Review Board Statement: Not applicable.

Conflicts of Interest: The authors declare no conflict of interest.

References

1. Thompson, R.F. Habituation: A history. *Neurobiol. Learn. Mem.* **2009**, *92*, 127–134. [CrossRef] [PubMed]
2. Hall, G. *Perceptual and Associative Learning*; Clarendon Press: Oxford, UK, 1991. [CrossRef]
3. Sokolov, E.N. Higher Nervous Functions: The Orienting Reflex. *Annu. Rev. Physiol.* **1963**, *25*, 545–580. [CrossRef] [PubMed]
4. Thompson, R.F.; Spencer, W.A. Habituation: A model phenomenon for the study of neuronal substrates of behavior. *Psychol. Rev.* **1966**, *73*, 16–43. [CrossRef]
5. Wagner, A.R. Habituation and memory. In *Mechanisms of Learning and Motivation: A Memorial Volume for Jerzy Konorski (1973)*; Dickinson, A., Boakes, R.A., Eds.; Erlbaum: Hillsdale, NJ, USA, 1979; pp. 53–82.
6. Sokolov, E.N. Neural model and the orienting influence. In *The Central Nervous System and Behavior: III*; Brazier, M.A.B., Ed.; Macy Foundation: New York, NY, USA, 1960.
7. Groves, P.M.; Thompson, R.F. Habituation: A dual-process theory. *Psychol. Rev.* **1970**, *77*, 419–450. [CrossRef] [PubMed]
8. Wagner, A.R. Priming in STM: An information-processing mechanism for self-generated or retrieval-generated depression in performance. In *Habituation: Perspectives from Child Development, Animal Behavior, and Neurophysiology*; Tighe, T.J., Leaton, R.N., Eds.; Erlabaum: Hillsdale, NJ, USA, 1976; pp. 95–128.

9. Wagner, A.R. Expectancies and the Priming of STM. In *Cognitive Processes in Animal Behavior*; Hulse, S.H., Honig, K.H., Fowler, H., Eds.; Erlbaum: Hillsdale, NJ, USA, 1978; pp. 177–209.
10. Konorski, J. *Integrative Activity of the Brain*; University of Chicago Press: Chicago, IL, USA, 1967.
11. Aktinson, R.C.; Shiffrin, R.M. *Human Memory: A Proposed System and its Control Processes*; Academic Press: Cambridge, MA, USA, 1968; Volume 2, pp. 89–195.
12. Wagner, A.R. SOP: A Model of Automatic Memory Processing in Animal Behavior. In *Information Processing in Animals: Memory Mechanisms*; Spear, N.E., Miller, R.R., Eds.; Erlbaum: Hillsdale, NJ, USA, 1981; pp. 5–47.
13. Vogel, E.H.; Ponce, F.P.; Wagner, A.R. The development and present status of the SOP model of associative learning. *Q. J. Exp. Psychol.* **2019**, *72*, 346–374. [CrossRef]
14. Uribe-Bahamonde, Y.E.; Becerra, S.A.; Ponce, F.P.; Vogel, E.H. A quantitative account of the behavioral characteristics of habituation: The sometimes opponent processes model of stimulus processing. *Front. Psychol.* **2019**, *10*, 504. [CrossRef] [PubMed]
15. Davis, M. Effects of interstimulus interval length and variability on startle-response habituation in the rat. *J. Comp. Physiol. Psychol.* **1970**, *72*, 177–192. [CrossRef] [PubMed]
16. Hall, G.; Rodríguez, G. Habituation and conditioning: Salience change in associative learning. *J. Exp. Psychol. Anim. Learn. Cogn.* **2017**, *43*, 48–61. [CrossRef]
17. Hall, G.; Rodríguez, G. When the stimulus is predicted and what the stimulus predicts: Alternative accounts of habituation. *J. Exp. Psychol. Anim. Learn. Cogn.* **2020**, *46*, 327. [CrossRef] [PubMed]
18. Lubow, R.E.; Moore, A.U. Latent inhibition: The effect of nonreinforced pre-exposure to the conditional stimulus. *J. Comp. Physiol. Psychol.* **1959**, *52*, 415. [CrossRef]
19. Hall, G.; Channell, S. Differential effects of contextual change on latent inhibition and on the habituation of an orienting response. *J. Exp. Psychol. Anim. Behav. Process.* **1985**, *11*, 470–481. [CrossRef]
20. Hall, G.; Honey, R.C. Contextual Effects in Conditioning, Latent Inhibition, and Habituation—Associative and Retrieval Functions of Contextual Cues. *J. Exp. Psychol. Anim. Behav. Process.* **1989**, *15*, 232–241. [CrossRef]
21. Honey, R.C.; Pye, C.; Lightbown, Y.; Rey, V.; Hall, G. Contextual factors in neophobia and its habituation: The role of absolute and relative novelty. *Q. J. Exp. Psychol.* **1992**, *45*, 327–347. [CrossRef]
22. Siegel, S. Morphine tolerance acquisition as an associative process. *J. Exp. Psychol. Anim. Behav. Process.* **1977**, *3*, 1–13. [CrossRef] [PubMed]
23. Reyes-Jimnez, D.; Iglesias-Parro, S.; Paredes-Olay, C. Contextual specificity of habituation in earthworms. *J. Exp. Psychol. Anim. Learn. Cogn.* **2020**, *46*, 341–353. [CrossRef] [PubMed]
24. Tomsic, D.; Massoni, V.; Maldonado, H. Habituation to a danger stimulus in two semiterrestrial crabs: Ontogenic, ecological and opioid modulation correlates. *J. Comp. Physiol. A* **1993**, *173*, 621–633. [CrossRef]
25. Tomsic, D.; Pedreira, M.E.; Romano, A.; Hermitte, G.; Maldonado, H. Context-US association as a determinant of long-term habituation in the crab chasmagnathus. *Anim. Learn. Behav.* **1998**, *26*, 196–209. [CrossRef]
26. Hermitte, G.; Pedreira, M.i.a.E.; Tomsic, D.; Maldonado, H. Context Shift and Protein Synthesis Inhibition Disrupt Long-Term Habituation after Spaced, but Not Massed, Training in the CrabChasmagnathus. *Neurobiol. Learn. Mem.* **1999**, *71*, 34–49. [CrossRef]
27. Rankin, C.H. Context conditioning in habituation in the nematode Caenorhabditis elegans. *Behav. Neurosci.* **2000**, *114*, 496–505. [CrossRef]
28. Lau, H.L.; Timbers, T.A.; Mahmoud, R.; Rankin, C.H. Genetic dissection of memory for associative and non-associative learning inCaenorhabditis elegans. *Genes Brain Behav.* **2012**, *12*, 210–223. [CrossRef]
29. Kruse, A.A.; Stripling, R.; Clayton, D.F. Context-specific habituation of the zenk gene response to song in adult zebra finches. *Neurobiol. Learn. Mem.* **2004**, *82*, 99–108. [CrossRef] [PubMed]
30. Turatto, M.; Bonetti, F.; Pascucci, D. Filtering visual onsets via habituation: A context-specific long-term memory of irrelevant stimuli. *Psychon. Bull. Rev.* **2017**, 1–7. [CrossRef] [PubMed]
31. Turatto, M.; Bonetti, F.; Chiandetti, C.; Pascucci, D. Context-specific distractors rejection: Contextual cues control long-term habituation of attentional capture by abrupt onsets. *Vis. Cogn.* **2019**, *27*, 291–304. [CrossRef]
32. Churchill, M.; Remington, B.; Siddle, D. Is the electrodermal orienting response sensitive to background context? *Biol. Psychol.* **1985**, *21*, 253–254. [CrossRef]
33. Schaafsma, M.F.; Packer, J.S.; Siddle, D.A.T. The effect of context change on long-term habituation of the skin conductance response to signal and non-signal stimuli in humans. *Biol. Psychol.* **1989**, *29*, 181–191. [CrossRef]
34. Jordan, W.P.; Strasser, H.C.; McHale, L. Contextual control of long-term habituation in rats. *J. Exp. Psychol. Anim. Behav. Process.* **2000**, *26*, 323–339. [CrossRef] [PubMed]
35. Marlin, N.A.; Miller, R.R. Associations to contextual stimuli as a determinant of long-term habituation. *J. Exp. Psychol. Anim. Behav. Process.* **1981**, *7*, 313–333. [CrossRef]
36. Pilz, P.K.D.; Arnold, S.W.; Rischawy, A.T.; Plappert, C.F. Longterm-habituation of the startle response in mice is stimulus modality, but not context specific. *Front. Integr. Neurosci.* **2014**, *7*, 1–10. [CrossRef]
37. Evans, J.G.M.; Hammond, G.R. Differential generalization of habituation across contexts as a function of stimulus significance. *Anim. Learn. Behav.* **1983**, *11*, 431–434. [CrossRef]

38. Leaton, R.N. Long-term retention of the habituation of lick suppression in rats. *J. Comp. Physiol. Psychol.* **1974**, *87*, 1157–1164. [CrossRef]
39. Chiandetti, C.; Turatto, M. Context-specific habituation of the freezing response in newborn chicks. *Behav. Neurosci.* **2017**, *131*, 437. [CrossRef] [PubMed]
40. Turatto, M.; Dissegna, A.; Chiandetti, C. Context learning before birth: Evidence from the chick embryo. *Biol. Lett.* **2019**, *15*, 20190104. [CrossRef] [PubMed]
41. McDiarmid, T.A.; Yu, A.J.; Rankin, C.H. Habituation Is More Than Learning to Ignore: Multiple Mechanisms Serve to Facilitate Shifts in Behavioral Strategy. *Bioessays* **2019**, *41*, 1900077. [CrossRef] [PubMed]
42. Randlett, O.; Haesemeyer, M.; Forkin, G.; Shoenhard, H.; Schier, A.F.; Engert, F.; Granato, M. Distributed plasticity drives visual habituation learning in larval zebrafish. *Curr. Biol.* **2018**. [CrossRef] [PubMed]
43. Mackintosh, N.J. Neurobiology, psychology and habituation. *Behav. Res. Ther.* **1987**, *25*, 81–97. [CrossRef]
44. Pedreira, M.E.; Dimant, B.; Maldonado, H. Inhibitors of protein and RNA synthesis block context memory and long-term habituation in the crab Chasmagnathus. *Pharmacol. Biochem. Behav.* **1996**, *54*, 611–617. [CrossRef]
45. Pedreira, M.E.; Dimant, B.; Tomsic, D.; Quesada-Allue, L.A.; Maldonado, H. Cycloheximide inhibits context memory and long-term habituation in the crab Chasmagnathus. *Pharmacol. Biochem. Behav.* **1995**, *52*, 385–395. [CrossRef]
46. Pereyra, P.; Portino, E.G.; Maldonado, H. Long-Lasting and Context-Specific Freezing Preference Is Acquired after Spaced Repeated Presentations of a Danger Stimulus in the Crab Chasmagnathus. *Neurobiol. Learn. Mem.* **2000**, *74*, 119–134. [CrossRef]
47. Beck, C.D.O.; Rankin, C.H. Long-term habituation is produced by distributed training at long ISIs and not by massed training or short ISIs inCaenorhabditis elegans. *Anim. Learn. Behav.* **1997**, *25*, 446–457. [CrossRef]
48. Carew, T.J.; Pinsker, H.M.; Kandel, E.R. Long-Term habituation od Defensive Withdrawal Reflex in Aplysia. *Science* **1972**, *175*, 451–454. [CrossRef]
49. Daniel, M.J.; Koffinas, L.; Hughes, K.A. Habituation underpins preference for mates with novel phenotypes in the guppy. *Proc. R. Soc. Lond.* **2019**, *286*, 20190435. [CrossRef] [PubMed]
50. Epstein, L.H.; Temple, J.L.; Roemmich, J.N.; Bouton, M.E. Habituation as a determinant of human food intake. *Psychol. Rev.* **2009**, *116*, 384–407. [CrossRef] [PubMed]
51. Rankin, C.; Abrams, T.; Barry, R.J.; Bhatnagar, S.; Clayton, D.; Colombo, J.; Coppola, G.; Geyer, M.A.; Glanzman, D.; Marsland, S.; et al. Habituation Revisited: An Updated and Revised Description of the Behavioral Characteristics of Habituation Catharine. *Neurobiol. Learn. Mem.* **2009**, *92*, 135–138. [CrossRef] [PubMed]

animals

MDPI

Article

Experimental Tests for Measuring Individual Attentional Characteristics in Songbirds

Loïc Pougnault [1], Hugo Cousillas [1], Christine Heyraud [1], Ludwig Huber [2], Martine Hausberger [1] and Laurence Henry [1,*]

1 CNRS, EthoS (Éthologie Animale et Humaine)-UMR 6552, University Rennes, Normandie University, F-35000 Rennes, France; loic.pougnault@univ-rennes1.fr (L.P.); hugo.cousillas@univ-rennes1.fr (H.C.); christine.heyraud@univ-rennes1.fr (C.H.); martine.hausberger@univ-rennes1.fr (M.H.)

2 Comparative Cognition, Messerli Research Institute, University of Veterinary Medicine Vienna, 1210 Vienna, Austria; Ludwig.Huber@vetmeduni.ac.at

* Correspondence: laurence.henry@univ-rennes1.fr

Simple Summary: Attention is at the core of all cognitive processes such as learning, memorization, and categorization. However, quantifying animals' attention is challenging, and there is still a need to have standardized and easy-to-use tests. This article describes experimental tests aimed to evaluate spontaneous attention of a songbird, the European starling, in response to the presentation of different types of stimuli: visual non-social or social stimuli and auditory hetero or species-specific stimuli. Our results reveal intra-individual variations but also consistency of individual attentional characteristics in the presence of a species-specific stimulus or different auditory stimuli. They demonstrate further that attention is not a uniform concept and may depend upon the modality and the type of stimulus perceived.

Citation: Pougnault, L.; Cousillas, H.; Heyraud, C.; Huber, L.; Hausberger, M.; Henry, L. Experimental Tests for Measuring Individual Attentional Characteristics in Songbirds. *Animals* **2021**, *11*, 2233. https://doi.org/10.3390/ani11082233

Academic Editors: Lucia Regolin and Maria Loconsole

Received: 26 May 2021
Accepted: 21 July 2021
Published: 29 July 2021

Abstract: Attention is defined as the ability to process selectively one aspect of the environment over others and is at the core of all cognitive processes such as learning, memorization, and categorization. Thus, evaluating and comparing attentional characteristics between individuals and according to situations is an important aspect of cognitive studies. Recent studies showed the interest of analyzing spontaneous attention in standardized situations, but data are still scarce, especially for songbirds. The present study adapted three tests of attention (towards visual non-social, visual social, and auditory stimuli) as tools for future comparative research in the European starling (*Sturnus vulgaris*), a species that is well known to present individual variations in social learning or engagement. Our results reveal that attentional characteristics (glances versus gazes) vary according to the stimulus broadcasted: more gazes towards unusual visual stimuli and species-specific auditory stimuli and more glances towards species-specific visual stimuli and hetero-specific auditory stimuli. This study revealing individual variations shows that these tests constitute a very useful and easy-to-use tool for evaluating spontaneous individual attentional characteristics and their modulation by a variety of factors. Our results also indicate that attentional skills are not a uniform concept and depend upon the modality and the stimulus type.

Keywords: spontaneous attention; social and non-social visual attention; auditory attention; experimental tests; starlings; *Sturnus vulgaris*

1. Introduction

Attention is defined as the ability to process selectively one aspect of the environment over others [1]. It involves several processes by which the nervous system apprehends and organizes sensory inputs and generates coordinated behavior [2]. Attentional mechanisms are essential to most cognitive processes, such as learning [3], memorization [4,5], and categorization [6]. Attentional processes allow individuals to adapt to their physical and social environments [7–9].

Most studies of animals' attentional characteristics aimed to promote valid animal models of human attention disorders [10] and to explore the attentional mechanisms, the cerebral substrates which underlie these functions [7,11], as well as the phylogeny of these complex behaviors [8]. Tests used vary according to the attentional process being investigated, but they are generally based on operant conditioning [2]. The test evaluating attention used the most frequently is probably the "Five Choice Serial Reaction Time Task" (5-CSRTT) developed by Carli et al. [12] for rodents to investigate the main types of attention. During this test, a rat is placed facing five openings set in a horizontal array along a curved wall of the test chamber. A discriminative visual stimulus is presented, consisting of a brief illumination of one of the five openings (0.5 s). When the subject reaches the opening with the stimulus, it receives a positive reinforcement such as a food pellet. This task tests the ability of rats to divide attention among a number of locations, to sustain attention over a large number of trials (about 100), and to express selective attention by ignoring brief bursts of white noise while detecting the visual targets [10]. Recently, Rochais et al. [13] successfully adapted the 5-CSRTT for horses. However, this test requires extensive training, taking weeks or even months to learn the task according to the species [14], and behavior may be modulated by inter-individual differences related to individuals' motivation to search for food. On the contrary, by analyzing unconditioned behavior, animals' attention can be recorded rapidly and in an economical way [2]. Analyses of the spontaneous expression of attention reveal variations related to intrinsic (e.g., age, sex, [15]) or extrinsic (e.g., human–animal relationship, [16,17] factors). Few studies, however, measured the structure of spontaneous attention. Blois-Heulin [18] showed that the structure of social attention varies according to social organization and distinguished short glances from more durable gazes (i.e., longer attention, lasting more than 1 s). Various studies based on social or human-animal contexts suggest that visual attentional structures present intrinsic differences between species [8,19,20].

Characterizing and evaluating attention and its structure in order to assess individual variations and their causes in a predictable way is a challenge. Most studies of spontaneous attention were performed either in naturalistic settings or in tests based on social visual stimuli (e.g., [21]).

Songbirds are relatively little studied in this respect. Attention is, however, one of the mechanisms proposed to explain selective song copying [22,23]. This is especially the case of European starlings *(Sturnus vulgaris)* for which social selective attention was proposed as a major factor involved in song copying between conspecifics [24–27]. The aim of the present study was to test the feasibility and the interest of developing attention tests that could be used in further studies to examine the impact of different factors, experience in particular, on individual attentional characteristics. We adapted the social visual attention tests (attention to conspecifics) developed by Range et al. [19] for keas (*Nestor notabilis*), dogs (*Canis familiaris*), and humans (*Homo sapiens*; pre-school children) and the non-social (visual or auditory) attention tests recently developed by Rochais et al. [13,28] for horses. Range et al. [19] demonstrated that the attentional behavior of the three species they studied depended on age/sex/identity of the model but that birds (keas) looked longer at a feeding than a non-feeding model, while children paid more attention to a model manipulating objects. Dalmaso et al. (2020) [9] indicated that gaze-mediated orienting of attention in humans can be shaped by social variables. Rochais et al. [28] found that individual variations of visual attentional characteristics remained relatively stable over time and across situations, as individuals ranked similarly after 6 months and the results of this test were predictive of attention in other contexts (e.g., working situation). They found opposite results between the visual and the auditory attention tests, the former predicting attention and the latter distractibility in working situations. In humans, it is well known that one can be more easily distracted from an ongoing task by auditory stimuli that do not require one to look than by visual stimuli [29].

The aim of the present study was to develop tests of attention as a tool for future comparative research on the impact of intrinsic or extrinsic factors on songbirds' attentional processes. European starlings' vocal repertoire and social integration present important individual variations, and song learning is highly dependent upon attention towards social models (e.g., [30,31]). Therefore, we expected attentional characteristics to present important individual variations and, in these song learners, visual and auditory attention to converge when submitted to species-specific stimuli. Three tests were used to characterize the structure of spontaneous attention of adult male starlings that did not have any specific training. Overall, starlings, as with most species, react to auditory stimuli by increasing their visual attention towards the sound source so that the same types of measures (e.g., number and duration of gazes towards the source) can be used to quantify attention towards visual or auditory stimuli (e.g., [32]). In accordance with Cohen [33], we differentiated attention capture/getting that triggers an orienting response [34] (and attention holding which is generally associated with a decrease of movement and more focused attention [24]. Passerine birds often scan the environment by repeated head movements associated with short looks ("glances" of less than one second) but may focus and remain immobile when gazing (e.g., [32]). Therefore, we measured attention characteristics on the basis of these two types of visual attention. A previous study showed that gazes towards a novel heterospecific stimulus are more lateralized than glances in starlings, suggesting further that they reflect individual characteristics and perception of different types of stimuli [35].

2. Material and Method

2.1. Subjects

The tests ran between 1 April 2017 and 14 April 2017 included 10 male starlings that were wild-caught as adults (they were more than 2 years old according to their throat feathers [36,37]) in Normandy, 4 individuals in November 2012 and 6 in November 2014. They were then kept with other starlings in mixed groups in outdoor aviaries with open shelters, grassy grounds, water and food ad libitum, and numerous perches. The choice of sample size was based on experimental logistic constraints but also and mainly on a wish to keep a balance between statistical validity and the "reduction" principle of the 3R ethical principles that aim to reduce as much as possible the number of experimental animals involved in research programs, especially as we used wild-caught animals.

Before the attention tests in April 2017, the individuals were transferred to the EthoS laboratory (UMR6552) on the Rennes1 University campus and placed in individual wire cages (60 × 39 × 65 cm), where they had an unlimited access to food and water throughout the course of the study. The cages had one grid side in the front and three opaque sides.

After 4 days of habituation without any stimulation, they were submitted individually to 3 experimental conditions with the goal to characterize their spontaneous attention towards 3 different types of stimuli.

2.2. Experimental TESTS

Three tests were conducted: (a) the non-social visual attention test (VAT) inspired by Rochais et al. [13], (b) the social visual attention (SVAT) inspired by Range et al. [19], and (c) the auditory test (AAT) inspired by Rochais et al. [28]. Both VAT and AAT were performed during the first week of experimentation, and the visual social test was performed during the second week. We chose to perform the non-social tests first, as we supposed that the social tests may have a more excitatory or motivational effect on the birds, which could then bias the results of the non-social tests (e.g., seeking more conspecific behind the screen or the loudspeaker). The order of the non-social tasks was changed between two subgroups of birds: five birds were tested first in the non-social visual task and then in the auditory task, and the other five birds were tested in the opposite order. During the second week, the 10 birds were tested in the social visual test.

2.3. The Non-Social Visual Attention Test (VAT)

Each cage was located in an anechoic chamber (35 dB attenuation). An LCD screen 15″ (1024 × 768 pixels, AccuSync LCD52VM©) was attached to one lateral opaque cage side. Two perches were positioned respectively 15 cm and 35 cm from the screen. A USB camera (Logitec CarlZeiss Tessar2.0/3.7©) was used to record the whole cage screen, and the cameras were controlled by an external computer (HP elitbook1470P©).

The stimulus was a green spot (2 cm in diameter) selected in accordance with the perceptual capacities of the birds [38]. Previous studies showed that European starlings can react to 2D images (e.g., [39–41]). Moreover, this novel moving visual stimulus was chosen to avoid biases due to familiarity and to take into account starlings' abilities to detect movement [12]. It was projected on the black screen for 2 min when the bird was positioned on the perch nearest the screen and facing it. The trajectory of the spot followed the same pattern of movement as in Rochais et al.'s [13] VAT by using the Processing software©. The object was continuously moving, therefore, it slowly transformed into a successive pattern following a circular movement (Figure 1). As suggested by Davidson et al. (2014) [42], we adjusted the test duration to species-specific characteristics. Songbirds are known for presenting rapid shifts of attention [43], much more rapid than horses. Each bird was tested four times, once each half-day for two consecutive days. Recording began 2 min before the stimulation was switched on and continued for 2 min during the stimulation. Thus, in all, we obtained 16 min of recording for each bird.

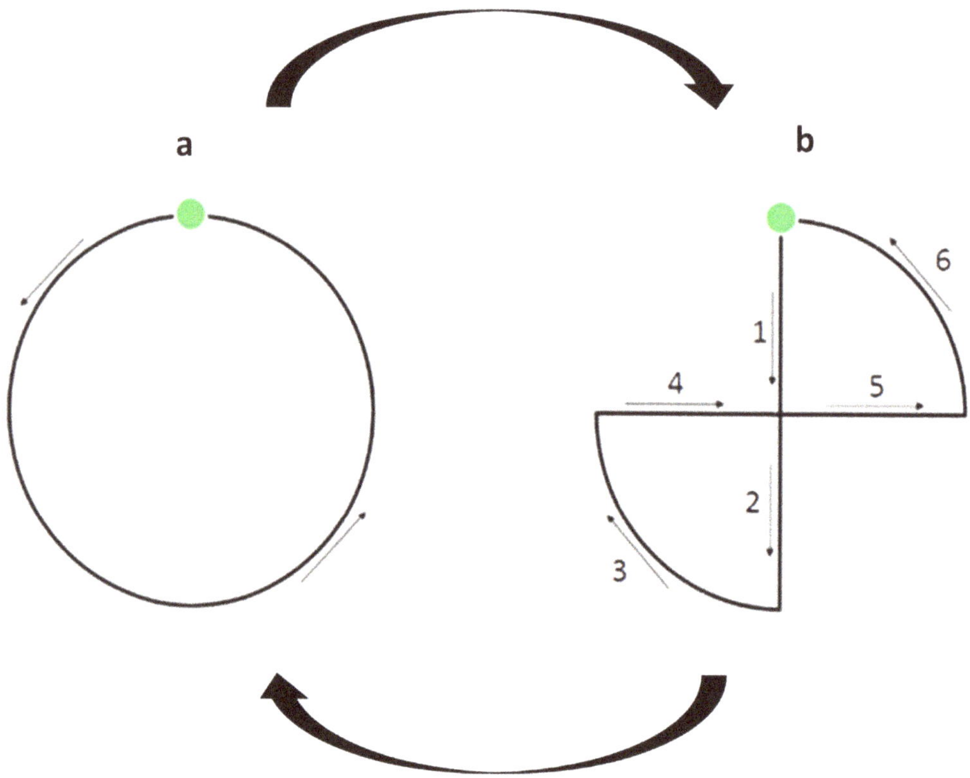

Figure 1. Visual stimulus used during the non-social visual attention test, adapted from [26]. The light was moved repeatedly, regularly, and continuously over 20 s using a computer program, starting anticlockwise (**a**) to continue over a "8" shape (**b**). Six cycles were broadcasted during each 2 min test.

2.4. The Social Visual Attention Test (SVAT)

For this test, five pairs of starlings were randomly constituted from the initial group of 10 subjects. Observations in their aviary revealed no clear hierarchical ranking, therefore, we could not take this parameter into account, although some authors found a relationship between hierarchical status and cognitive abilities in starlings [44] or between dominance and attention in other species [45].

Four days before a test, each dyad was placed in an experimental wire cage (121 × 50 × 39 cm) separated into two parts by a white opaque plexiglass board with one bird on each side of the cage. A horizontal slot (6 × 2 cm) at bird's eye level was drilled in the board. The length of this horizontal slot could be modified using a white sliding plexiglass tongue. The behaviors of the test bird (named "A") were recorded by a camera (Sony Handycam, DCR-SR75®) located 1 m in front of the experimental cage. To reduce the probability of interactions between the two subjects, they were deprived of food one hour before the experiment and the observed bird (hereafter, "B") had access to palatable food (insectivorous mix for birds).

The test bird ("A") had visual access to his conspecific ("B") twice a day for 5 min: in the morning, with the slot fully opened (6 cm) so that bird "A" could learn the presence of his conspecific "B"; and in the afternoon, when the slot was only opened 1 cm to promote potential monocular observation of "B" by "A" (hence potentially laterality). We decided to systematically test our birds first in the morning with a large opening (6 cm) for giving the possibility to the tested bird (A) to be aware of its congener presence (B) and to see him binocularly. In the afternoon, the limited opening (1 cm) allowed only monocular attentional behaviors. Pilot experiments showed that the birds did not always see that there was a conspecific when the slot was small (see Range et al. [19], who allowed a familiarization period to birds before tests). Since we wanted to make sure the animals saw that they had access to a visible neighbor, we were obliged to have a fixed order, even though, of course, random allocation would have been ideal.

2.5. The Auditory Attention Test (AAT)

For this test, subjects were placed individually in the experimental cage without any visual contact with conspecifics. The 10 cages were placed against a wall, allowing simultaneous video recording of all birds. The auditory stimuli were broadcasted using a loudspeaker (Philips SPA5300 2.1®). A camera (Sony Handycam, DCR-SR75®) was placed above the loudspeaker (see [35]). The two devices were placed 2 m in front of the experimental cages, and the behaviors of all the subjects were recorded at the same time to avoid habituation due to repetition of the stimulus. Two types of stimuli were broadcasted: (a) 4 different whistles from unfamiliar individuals recorded in New Zealand to ensure that the subjects could not have heard them before and (b) 4 different vocalizations of humpback whales (*Megaptera novaeangliae*) as unknown heterospecific stimuli. Durations (0.50 s) and intensities (~70 dB at two meters) were standardized for all stimuli using ANA software [46]. Four different vocalizations (with different acoustic structures) of each type, one of each per half day, were broadcasted in random order and various time intervals (between 6 to 21 min) between stimulations to avoid biases due to habituation. Therefore, birds heard each stimulus only once. Video recording began 1 min before the stimulation and finished 1 min after the stimulation; in all, we obtained 8 min of recording for each type of stimulus.

2.6. Data and Statistical Analyses

Video recordings were analyzed using continuous focal sampling [47]. All analyses were made by the same single observer (LP), but interobserver reliability was tested with a second observer on a sample of recordings (Christiane Rössler), leading to 84% agreement, a reasonable assessment [48], especially on very short items such as glances.

The parameters measured were:

- Glances: monocular or binocular looks towards the stimulus lasting less than one second;
- Gazes: a monocular or binocular look towards the stimulus lasting more than one second;
- Total gaze duration: total time during which a subject gazed at the stimulus during a test session.

The distinction between gazes and glances was based on previous studies [20,49]. Furthermore, the first reaction after the stimulus presentation was also recorded (see also [41])

Normality and homogeneity of variance were assessed by inspection of residuals with Shapiro-Wilk W tests. Given the non-normality distribution of data, we ran non-parametric statistical tests [50] using R 3.3.2 software® (R Core team, Vienna, Austria) [51] with a significance threshold set at 0.05. The aim of this work was to assess individual variations in attention structure for each given test separately using two aspects (number of glances and gazes, duration of gazes). Since the question and the data were quite simple and the sample size was reasonable, we stayed on "conventional" simple non-parametric statistical analyses. In the same line, we compared temporal changes in behavior within tests using Wilcoxon signed-rank test. Spearman's correlations tests were used to assess possible individual consistency between tests (i.e., VAT; SVAT and AAT), since it compares the relative ranking of given individuals independently of the actual precise measures (Tables S1 and S2). In the figures, we did, however, use the rate per minute in order to facilitate an overview of the results at the study's level.

3. Results

3.1. Visual Attention Tests

Non-SOCIAL (VAT)

All starlings paid attention to the stimulus, and none expressed any withdrawal or any other fear behavior. The first reaction of all of them when the stimulus appeared was to glance at the screen. However, frequencies of glances did not vary significantly between before and during the stimulation (Wilcoxon signed-rank test, $T = 10$, $N = 15$ $p = 0.2$). On the contrary, projection of the visual stimuli was associated with a significant increase of the frequency and the total duration of gazes ($T_{Gaze} = 5$, $N = 10$, $p = 0.02$; $T_{Total\ gaze\ duration} = 1$, $p = 0.007$), but not of glances ($T = 15$, $N = 10$, $p = 0.2$) towards the screen, although glances were overall more frequent than gazes at all times ($T = 0$, $N = 10$, $p = 0.006$) (Figure 2a). Starlings looked more at the stimulus with monocular than binocular gazes ($T = 0$, $N = 9$, $p = 0.006$) and glances ($T = 1$, $N = 10$, $p = 0.007$), but no laterality bias could be observed ($p > 0.5$ in all cases).

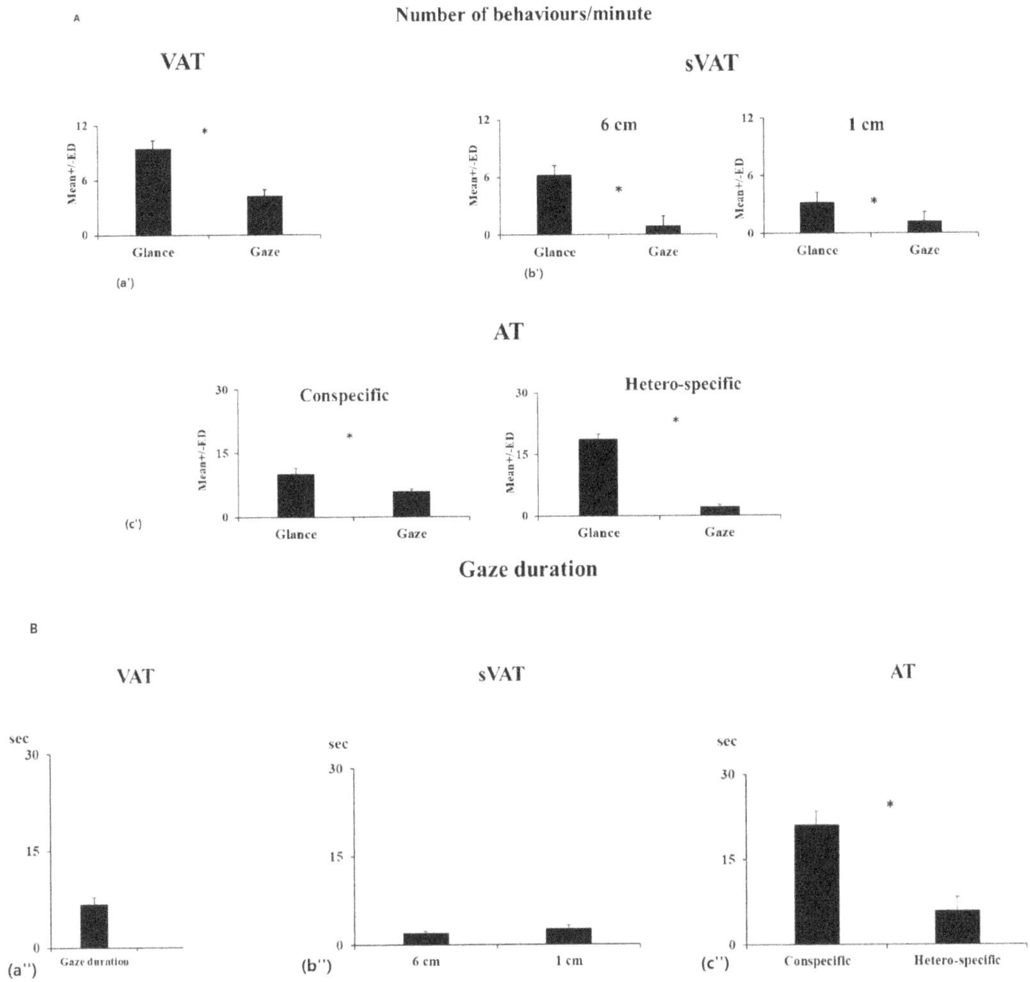

Figure 2. (**A**). Attention characteristics during tests: occurrences of glances and gazes, (**B**): duration of gazes during stimulus presentation, (data: mean number per minute ± ED [52], *: Wilcoxon test: $p \leq 0.05$): (**a′**,**a″**) Non-social visual test (VAT). (**b′**,**b″**) Social visual test (sVAT) with maximum opening (6 cm) and minimum opening (1 cm). (**c′**,**c″**) Auditory test (AT): attention characteristics after the broadcast of conspecific vocalizations and hetero-specific (Whale).

3.2. Social Visual Attention (sVAT)

The birds expressed more glances than gazes towards their conspecifics no matter the size of the slot opening (Wilcoxon signed-rank test, $T_{Max} = 0$, N = 10, $p = 0.006$; $T_{Min} = 3.5$, $p = 0.01$; 1, Figure 2b). Total gaze durations and frequencies of gazes did not differ significantly between minimal or maximal slot opening tests ($T_{gaze\ duration} = 12$, N = 9, $p = 0.2$, $T_{gazes\ frequencies} = 3$, N = 7, $p = 0.06$), but the subjects glanced more when the slot was large than when it was small ($T_{glances\ frequencies}$ T = 2.5, N = 10, $p = 0.01$). Interestingly, when the slot was fully open, although they had the possibility to use either their binocular or monocular visual fields to look at their conspecifics, starlings performed more monocular than binocular glances and gazes ($T_{Glance} = 8$, N = 10, $p = 0.04$; $T_{Gaze} = 8$, p0.04). When using monocular vision, they still performed more glance than gazes (T = 0, N = 10, $p = 0.005$). However, we could not evidence a laterality bias (preferred side) ($T_{Glance} = 17.5$, N = 10, $p = 0.3$; $T_{Gaze} = 9.5$, N = 9 $p = 0.1$; $T_{Total\ gaze\ duration} = 4$, N = 7, $p = 0.09$).

3.3. Auditory Test (AT)

When the auditory stimuli were broadcasted, all individuals displayed attentional behaviors (glances and gazes) in the direction of the loudspeaker. Although birds expressed more glances than gazes (Wilcoxon signed-rank test, T = 0, N = 10 $p = 0.006$ in both cases) (Figure 2c), their attentional structure differed according to the type of stimulus. Following the broadcast of heterospecific sounds, all subjects increased glances towards the loudspeaker (Wilcoxon signed-rank test, N = 10, T = 1, $p = 0.006$). Frequency or total duration of gazes did not change significantly ($T_{Gaze} = 24$, $p = 0.7$; $T_{Total\ gaze\ duration} = 23$, $p = 0.6$). On the contrary, in response to the conspecific stimuli, all subjects increased numbers of gazes and total duration of gazes ($T_{Gaze} = 0$, N = 10, $p = 0.006$; $T_{Total\ gaze\ duration} = 0$, N = 10, $p = 0.006$), but numbers of glances did not change significantly (T = 11, N = 10, $p = 0.09$). Overall, our subjects emitted more glances (total number) towards the loudspeaker in response to the heterospecific than to the conspecific stimuli (Wilcoxon signed-rank test, N = 10, T = 3, $p = 0.01$) and more gazes and longer total gaze durations in response to the conspecific than the heterospecific stimulus (N = 10, $T_{Gaze} = 0$, $p = 0.006$; $T_{Total\ gaze\ duration} = 0$, $p = 0.006$) (ratios), Figure 2c).

3.4. Individual Consistency between Tests

The coefficients of variation observed for the different traits in the different tests showed that individuals differed most in gaze frequency and/or duration in the presence of the visual social stimulus and the heterospecific auditory stimulus. Interestingly, the numbers of glances towards their conspecific during the visual social test (minimal opening) and the numbers of glances after the broadcast of a conspecific vocalization were positively correlated (Spearman correlation test, rs = 0.70, N = 10, $p = 0.05$, Figure 3a).

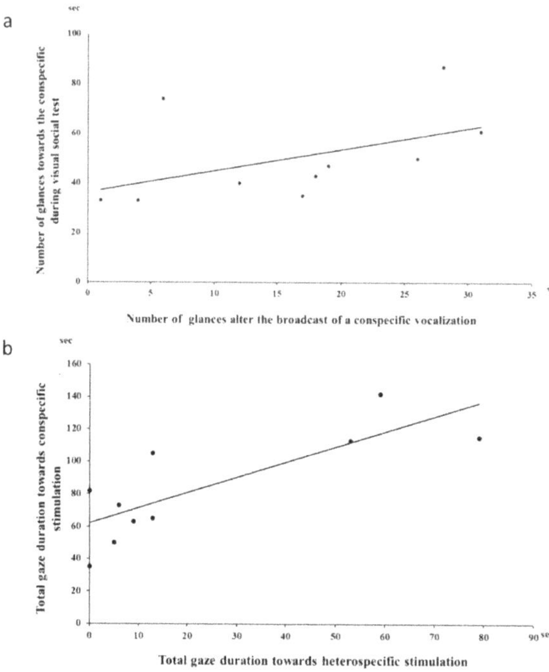

Figure 3. (**a**) Numbers of glances towards a conspecific during the visual social test (minimum opening) in relation to numbers of glances after the broadcast of a conspecific vocalization (correlation rs = 0.70, N = 10, $p = 0.05$). (**b**) Total gaze durations after the broadcast of the conspecific vocalization in relation to total gaze durations following the hetero-specific vocalization during auditory tests (correlation r = 0.79, N = 10, $p = 0.01$). Each dot represents an individual.

Thus, the individuals the most attentive to the view of a conspecific were also the most attentive to conspecific sounds, but only in terms of repeated glances. There was no evidence for any other correlation of attentional data between the different tests, indicating no consistent attentional individual stability of their reactions towards visual versus auditory non-social stimuli or towards social versus non-social visual stimuli (Spearman's correlation, for all $p > 0.05$). However, in the auditory test, total gaze durations towards the loudspeaker after the broadcast of a conspecific song and after a hetero-specific song were positively correlated (Spearman's correlation, rho = 0.79, $p = 0.01$; Figure 3b).

4. Discussion

The results of the present study demonstrate the feasibility and the importance of attentional tests adapted for European starlings for the investigation of spontaneous attentional structures. These structures show individual variations and depend on the type of stimuli presented. Overall, the starlings' attentional structure differed according to the target of their visual attention. They showed more durable attention when looking at an unusual non-social object than when they had access to the view of a familiar live conspecific. In all cases, they preferred to use a monocular visual field although without any laterality bias. Results reveal that starlings pay more attention towards unusual visual stimuli and conspecific auditory stimuli, whereas they glanced more at social visual stimuli and heterospecific auditory stimuli. These results suggest opposite patterns between visual and auditory social attention comparable to Rochais et al.'s [28] observations showing opposite individual processes between these two modalities. Nevertheless, the numbers of glances towards visual and auditory species-specific were correlated, suggesting that social motivation could trigger bimodal attention. Although we found no correlations between social and non-social visual tests, individual responses to the two types of auditory stimuli (conspecific and heterospecific sounds) were correlated, suggesting that individual auditory attention was more consistent. Individual variations were the highest for gazes, i.e., for more durable attention.

Classical experimental attention tasks require individuals to monitor the location of a stimulus to obtain food and are based on operant conditioning. Thus, differences between individuals then are related to an individual's motivation to work to obtain food, and reward delivery can alter the processing of specific stimuli directly by increasing their attentional priority [10,12,53]. Moreover, training in operant conditioning can take time (e.g., 5-CSRTT: 2–4 months for rats [54]; 3–5 months for mice [14]. In addition, long-term experimental procedures can negatively affect a subject's attentional function, as it may experience repeated stress such as restriction of food or water [55]. Therefore, these procedures are not adapted to study the "spontaneous attention" of a large number of animals. The three attentional tests we applied (i.e., VAT, SVAT, AT) for starlings and adapted from pre-existing attentional tests [13,19] proved again useful for assessing individual differences in spontaneous attention and variations in attentional structure according to the type of stimulus and the modality involved.

All our subjects performed many glances, and this could be related to starlings' status of "prey-species", which requires paying permanent attention to the environment while performing other behaviors such as feeding and to change rapidly the direction of visual attention rather than to fix their attention during a long period [7].

The attentional structure of our starlings appeared to depend both on the targets and the modality involved. As Davidson et al. (2014) [42] pointed out, gaze behaviors must be described in term of "observable behavior". For example, a shorter latency and a longer time of looking may reflect gaze preference. Our birds paid more sustained attention, through an increase of gaze occurrences and their total duration, to unknown auditory or visual stimuli. However, glances were expressed more towards visual or auditory stimuli that they already knew, such as a conspecific or a "neutral" stimulus. These results suggest further that longer and frequent gazes reflect a higher level of attention than repeated glances (see also [21]).

Studies showed that, when facing a usual stimulus, carrying no information of interest, horses quickly scanned their environment by glances and then resumed their previous activities [56]. Conversely, Rochais et al. [13] described fixed attentional behaviors towards an unusual visual stimulus, revealing selective attention promoted by endogenous attentional processes. In the same context, our birds demonstrated a similar visual attention pattern. This same kind of behavior can be noted in a natural environment during foraging. Indeed, detection of food can be difficult (i.e., cryptic items such as seeds or insects) so that an individual must base its choice on particular aspects of the visual target [57–59]. Goto et al. [60] demonstrated a similar type of spontaneous visual attention in blue jays (*Cyanocitta cristata*) using the "search image" paradigm proposed by Tinbergen [59]. An important point is that, contrarily to Rochais et al. [13], we were able to remove the bias potentially induced by the presence of a human experimenter during the attentional test by using an electronic apparatus. This new method would be easy to use and to adapt the VAT test for other species in this same way to estimate spontaneous non-social visual attention.

Playback of species-specific vocalizations (i.e., AAT) induced starlings' sustained attention. Poirier et al. [24] observed a global reduction of the activity of young starlings raised without adults when adult songs were broadcasted, suggesting an increase of auditory attention. These observations were confirmed by Ten Cate [23]: young zebra finches (*Taeniopygia guttata*) similarly approached and looked at an adult that was vocalizing. Because they lack external auditory organs, we cannot conclude which modality was privileged in this context. However, the head movements observed (see also [32]) suggest that the attentional process of young birds towards an adult was bimodal, i.e., auditory and visual attention. George et al.' studies [39,61] revealed that the responses of auditory neurons in the primary auditory area of starlings was influenced by the broadcast of social visual stimuli. This bimodal treatment of information allows individuals to perceive information concerning the emission source of the stimulus [62]. Thus, during playback of conspecific auditory stimuli, starlings probably tried to complete the auditory information with visual information, orienting their gazes towards the emission source. On the contrary, our subjects expressed a majority of glances after hearing the hetero-specific auditory vocalization, which, according to San Miguel et al. [34], may have captured attention and triggered an orienting response but also induced "curiosity", i.e., visual exploration [35]. Yet, total gaze durations appeared to remain consistent for each individual between these two auditory attentional tests, indicating interest differences between conspecific versus hetero-specific stimuli rather than differences between preferred modalities.

In addition, visual social attention was expressed mainly by glances in our experimental situations. Social life requires one to be attentive towards others and their actions [21]. Attention—particularly social attention—is modulated by social factors in most social species, including humans (Dalmaso et al. [9,45]. One important aspect is the quality of social relations between individuals such as hierarchical rank [63] or affinities ([30] Blois-Heulin [18] and Lemasson et al. [64] showed opposite trends of visual attention between red-capped mangabeys (*Cercocebus torquatus torquatus*) characterized by a "despotic" type of society, where group members focus their attention towards the dominant male, and Campbell monkeys, characterized by a tolerant type of society where members' visual attention network reflects affinities. In rhesus macaque (*Macaca Rhesus*), the lower ranking individuals reacted quicker than high ranking animals to visual cues from conspecifics [63]. Studies performed in large naturalistic aviaries showed that most social relationships were expressed through tolerance and spatial proximities between preferred partners, while the hierarchy was more rarely expressed and only in some particular contexts [44,65,66]. This was also the case in the observations performed on our experimental birds when in groups (Pougnault et al. in prep). In canaries, observational learning was enhanced when watching a preferred social partner [67]. We did not investigate the role of social status nor of being or not being social partners on the attentional pattern. Further studies will have to be developed in order to test specifically this question. At least, these standardized tests can help testing further such questions. Finally, there are differences between species in social attention due to other reasons than just social. Starlings express attention towards conspecifics while feeding in the form of "scans" (i.e., glances) or gazes if obstacles are present (e.g., tall grass), which allow them to anticipate a potential predation risk but also to locate a food source by looking at the body position of their conspecifics [68–70]. This is different from what was found in keas, which paid quite long attention to a conspecific that was feeding (up to 50% of the total time of demonstration) [19]. Jackdaws are much less inclined to watch conspecifics than ravens [21]. Finally, in our tests, birds emitted a majority of monocular visual behaviors, even when they had the possibility to look at their conspecific with binocular vision. Given the width of the opening (i.e., 2 cm width), their beak may have constituted a handicap because it could obstruct their field of view, and this could explain this result [71].

Individual differences clearly appeared, showing consistency mostly for auditory stimuli but also a transmodal consistency for social stimuli. Limited vision through a narrow opening or lack of vision of the conspecific seemed to induce similar individual differences in visual and in auditory attention, and this probably reflects individual differences in the motivation to seek social contact [40]. Future studies of larger samples with various social experiences should enlighten the processes underlying such individual differences and show how social attention and social motivation are interrelated (Perret et al. in rev., Henry et al. sub.). Horses' responses to the auditory test were influenced by their welfare state [72] and by the visual test by their working conditions (Rochais et al. sub.).

5. Conclusions

This study aimed to validate new methodological tools, inspired from those used for other species, to evaluate individual starlings' attentional characteristics. Given our conclusive results, these methodologies offer interesting ways to assess birds' spontaneous attention whilst avoiding extensive training that can imply weeks or even months of training.

Our three tests appear to be promising and simple tools to investigate and predict the attention characteristics of several species. This kind of investigations seems able to highlight intrinsic (e.g., age, welfare state) and/or extrinsic (e.g., living conditions) factors of variations in attentional structure.

Supplementary Materials: The following are available online at https://www.mdpi.com/article/10.3390/ani11082233/s1, Table S1: Occurrences of glances and gazes during all attentional tests. Table S2: Total gaze duration (s) during the attention tests.

Author Contributions: L.P., H.C., L.H. (Ludwig Huber), M.H. and L.H. (Laurence Henry), designed the procedure; L.P. performed the experiment and analyzed the data; C.H. took care of the birds and helped with preparing the experiment; L.P., M.H., L.H. (Laurence Henry), L.H. (Ludwig Huber) and H.C. wrote the paper. All authors have read and agreed to the published version of the manuscript.

Funding: This research was funded by the CNRS and University of Rennes (France) as well as by an Amadeus PHC program.

Institutional Review Board Statement: All our experiments complied with the current French laws (Centre National de la Recherche Scientifique) related to animal experimentation. Only behavioral observations were performed, and laboratory experiments were approved by the ethic council for animal experimentation (APAFIS#19211-2019021511353031V2). Animal husbandry and care were under management of the animal caregivers of the UMR6552.

Data Availability Statement: Data are available on reasonable request to the corresponding author.

Acknowledgments: We are grateful to Christiane Rössler for her contribution to inter-observer reliability agreement and Stephane Louazon for his help with the technical aspects of the study. We are also grateful to Ann Cloarec for improving the English.

Conflicts of Interest: The authors declare no conflict of interest.

References

1. Posner, M.I.; Snyder, C.R.; Davidson, B.J. Attention and the detection of signals. *J. Exp. Psychol. Gen.* **1980**, *109*, 160–174. [CrossRef]
2. Bushnell, P. Behavioral approaches to the asessment of attention in animals. *Psychopharmacology* **1998**, *138*, 231–259. [CrossRef] [PubMed]
3. Kruschke, J.K. Attention in learning. *Curr. Dir. Psychol. Sci.* **2003**, *12*, 171. [CrossRef]
4. Cowan, N. Activation, attention, and short-term memory. *Mem. Cogn.* **1993**, *21*, 162–167. [CrossRef] [PubMed]
5. Hanggi, E.B.; Ingersoll, J.F. Long-term memory for categories and concepts in horses (*Equus caballus*). *Anim. Cog.* **2009**, *12*, 451–462. [CrossRef] [PubMed]
6. Cowan, N. *Attention and Memory*; University Press: Oxford, UK, 1997.
7. Emery, N. The eye have it: The neuroethology, function and evolution of social gaze. *Neurosci. Biobehav. Rev.* **2000**, *24*, 581–604. [CrossRef]
8. Shepherd, S.V. Following Gaze: Gaze-Following Behavior as a Window into Social Cognition. *Front. Integr. Neurosci.* **2010**, *4*, 5. [CrossRef] [PubMed]
9. Dalmaso, M.; Castelli, L.; Galfano, G. Social Modulators of Gaze-Mediated Orienting of Attention: A Review. *Psychon. Bull. Rev.* **2020**, *27*, 833–855. [CrossRef] [PubMed]
10. Robbins, T.W. The 5-choice serial reaction time task: Behavioural pharmacology and functional neurochemistry. *Psychopharmacology* **2002**, *163*, 362–380. [CrossRef]
11. Posner, M.I. *Cognitive Neuroscience of Attention*; Guilford Press: New York, NY, USA, 2012.
12. Carli, M.; Robbins, T.W.; Evenden, J.L.; Everitt, B.J. Effects of lesions to ascending noradrenergic neurones on performance of a 5-choice serial reaction task in rats; implications for theories of dorsal noradrenergic bundle function based on selective attention and arousal. *Behav. Bran Res.* **1983**, *9*, 361–380. [CrossRef]
13. Rochais, C.; Sébilleau, M.; Houdebine, M.; Bec, P.; Hausberger, M.; Henry, S. A novel test for evaluating horses' spontaneous visual attention is predictive of attention in operant learning tasks. *Naturwissenschaften* **2017**, *104*, 1–16. [CrossRef] [PubMed]
14. Steckler, T.; Sauvage, M.; Holsboer, F. Glucocorticoid receptor impairment enhances impulsive responding in transgenic mice performing on a simultaneous visual discrimination task. *Eur. J. Neurosci.* **2000**, *12*, 2559–2569. [CrossRef] [PubMed]
15. Range, F.; Huber, L. Attention in common marmosets: Implications for social-learning experiments. *Anim. Behav.* **2007**, *73*, 1033–1041. [CrossRef]
16. Horn, L.; Range, F.; Huber, L. Dogs' attention towards humans depends on their relationship, not only on social familiarity. *Anim. Cogn.* **2013**, *16*, 435–443. [CrossRef] [PubMed]
17. Rochais, C.; Henry, S.; Sankey, C.; Nassur, F.; Góracka-Bruzda, A.; Hausberger, M. Visual attention, an indicator of human-animal relationships? A study of domestic horses (*Equus caballus*). *Front. Psychol.* **2014**, *5*, 108. [CrossRef]
18. Blois-Heulin, C. Visual gaze in Red-Capped Mangabey (*Cercocebus torquatus torquatus*) and the Grey-Cheeked Mangabey (*Cercocebus albigena albigena*). *Folia Primatol.* **1999**, *70*, 264–268. [CrossRef] [PubMed]
19. Range, F.; Horn, L.; Bugnyar, T.; Gajdon, G.K.; Huber, L. Social attention in Keas, dogs, and human children. *Anim. Cogn.* **2009**, *12*, 181–192. [CrossRef] [PubMed]
20. Grandgeorge, M.; Gautier, Y.; Bourreau, Y.; Mossu, H.; Hausberger, M. Visual Attention Patterns Differ in Dog vs. Cat Interactions with Children with Typical Development or Autism Spectrum Disorders. *Front. Psychol.* **2020**, *11*, 2047. [CrossRef]
21. Scheid, C.; Range, F.; Bugnyar, T. When, What, and Whom to Watch? Quantifying Attention in Ravens (*Corvus corax*) and Jackdaws (*Corvus monedula*). *J. Comp. Psychol.* **2007**, *121*, 380–386. [CrossRef]
22. Kuhl, P.K. Human speech and birdsong: Communication and the social brain. *Proc. Natl. Acad. Sci. USA* **2003**, *100*, 9645–9646. [CrossRef]
23. Ten Cate, C. Listening behaviour and song learning in zebra finches. *Anim. Behav.* **1986**, *34*, 1267–1268. [CrossRef]
24. Poirier, C.; Henry, L.; Mathelier, M.; Lumineau, S.; Cousillas, H.; Hausberger, M. Direct Social Contacts Override Auditory Information in the Song-Learning Process in Starlings (*Sturnus vulgaris*). *J. Comp. Psychol.* **2004**, *118*, 179–193. [CrossRef] [PubMed]
25. Bertin, A.; Hausberger, M.; Henry, L.; Richard-Yris, M.-A. Adult and peer influences on starling song development. *Dev. Psychobiol.* **2007**, *49*, 362–374. [CrossRef] [PubMed]
26. Bertin, A.; Hausberger, M.; Henry, L.; Richard-Yris, M.-A. Adult:young ratio influences song acquisition in female European starlings (*Sturnus vulgaris*). *J. Comp. Psychol.* **2009**, *123*, 195–203. [CrossRef] [PubMed]
27. Cousillas, H.; George, I.; Henry, L.; Richard, J.-P.; Hausberger, M. Linking Social and Vocal Brains: Could Social Segregation Prevent a Proper Development of a Central Auditory Area in a Female Songbird? *PLoS ONE* **2008**, *3*, e2194. [CrossRef] [PubMed]
28. Rochais, C.; Henry, S.; Hausberger, M. Spontaneous attention-capture by auditory distractors as predictor of distractibility: A study of domestic horses (*Equus caballus*). *Sci. Rep.* **2017**, *7*, 15283. [CrossRef] [PubMed]
29. Murphy, S.; Dalton, P. Ear-catching? Real-world distractibility scores predict susceptibility to auditory attentional capture. *Psychol. Bull. Rev.* **2014**, *21*, 1209–1213. [CrossRef]
30. Hausberger, M. Social influences on song acquisition and sharing in the European starling (*Sturnus vulgaris*). In *Social Influences on Vocal Development*; Snowdon, C.T., Hausberger, M., Eds.; Cambridge University Press: Cambridge, MA, USA, 1997; pp. 128–156.
31. Cousillas, H.; Henry, L.; George, I.; Marchesseau, S.; Hausberger, M. Lateralization of social signal brain processing correlates with the degree of social integration in a songbird. *Sci. Rep.* **2020**, *10*, 14093. [CrossRef]

32. Hausberger, M.; Forasté-Mathelier, M.; Richard-Yris, M.A.; Nygren, C. Differential response of female starlings to shared and nonshared song types. *Etología* **1997**, *5*, 31–38.
33. Cohen, L.B. Attention-Getting and Attention-Holding Processes of Infant Visual Preferences. *Child Dev.* **1972**, *43*, 869–879. [CrossRef] [PubMed]
34. SanMiguel, I.; Linden, D.; Escera, C. Attention Capture by Novel Sounds: Distraction versus Facilitation. *Eur. J. Cogn. Psychol.* **2010**, *22*, 481–515. [CrossRef]
35. Hausberger, M.; Henry, L.; Rethoré, B.; Pougnault, L.; Kremers, D.; Rössler, C.; Aubry, C.; Cousillas, H.; Boye, M.; Lemasson, A. When Perceptual Laterality Vanishes with Curiosity: A Study in Dolphins and Starlings. *Laterality* **2021**, *26*, 238–259. [CrossRef] [PubMed]
36. Suthers, H.B. Analysis of a flock of starlings. *Bird Band.* **1978**, *49*, 35–46. [CrossRef]
37. Feare, C. *The Starling*; Oxford University Press: Oxford, UK, 1984.
38. Hart, N.S.; Partridge, J.C.; Cuthill, I.C. Visual pigments, oil droplets and cone photoreceptor dictribution in the European starling. *J. Exp. Biol.* **1998**, *201*, 1433–1446. [CrossRef] [PubMed]
39. George, I.; Richard, J.P.; Cousillas, H.; Hausberger, M. No need to talk, I know you: Familiarity influences early multisensory integration in a songbird's brain. *Front. Behav. Neurosci.* **2011**, *4*, 19. [CrossRef] [PubMed]
40. Perret, A.; Henry, L.; Coulon, M.; Caudal, J.-P.; Richard, J.-P.; Cousillas, H.; Hausberger, M.; George, I. Social visual contact, a primary "drive" for social animals? *Anim. Cogn.* **2015**, *18*, 657–666. [CrossRef] [PubMed]
41. Belin, L.; Formanek, L.; Heyraud, C.; Hausberger, M.; Henry, L. Influence of early experience on processing 2D threatening pictures by European starlings (*Sturnus vulgaris*). *Anim. Cog.* **2018**, *21*, 749–758. [CrossRef]
42. Davidson, G.L.; Butler, S.; Fernández-Juricic, E.; Thornton, A.; Clayton, N.S. Gaze Sensitivity: Function and Mechanisms from Sensory and Cognitive Perspectives. *Anim. Behav.* **2014**, *87*, 3–15. [CrossRef]
43. Sridharan, D.; Schwarz, J.S.; Knudsen, E.I. Selective attention in birds. *Curr. Biol.* **2014**, *24*, 510–513. [CrossRef]
44. Boogert, N.J.; Reader, S.M.; Hoppitt, W.; Laland, K.N. The Origin and Spread of Innovations in Starlings. *Anim. Behav.* **2008**, *75*, 1509–1518. [CrossRef]
45. Dalmaso, M.; Pavan, G.; Castelli, L.; Galfano, G. Social Status Gates Social Attention in Humans. *Biol. Lett.* **2012**, *8*, 450–452. [CrossRef] [PubMed]
46. Richard, J.-P.; Leppelsack, H.-J.; Hausberger, M. A Rapid Correlation Method for the Analysis of Spectrotemporal Receptive-Fields of Auditory Neurons. *J. Neurosci. Methods* **1995**, *61*, 99–103. [CrossRef]
47. Altmann, J. Observational Study of Behavior: Sampling Methods. *Behaviour* **1974**, *49*, 227–267. [CrossRef]
48. McHugh, M.L. Interrater Reliability: The Kappa Statistic. *Biochem. Med.* **2012**, 276–282. [CrossRef]
49. Blois-Heulin, C.; Girona, B. Patterns of Social Visual Attention in the Red-Capped Mangabey (Cercocebus Torquatus Torquatus) in the Context of Food Competition. *Folia Primatol.* **1999**, *70*, 180–184. [CrossRef]
50. Siegel, S.; Castellan, N.J. *Nonparametric Statistics for the Behavioral Sciences*, 2nd ed.; McGraw-Hill Humanities: New York, NY, USA, 1988.
51. R Core Team. *R: A Language and Environment for Statistical Computing*; R Foundation for Statistical Computing: Vienna, Austria, 2021; Available online: https://www.R-project.org/ (accessed on 22 July 2021).
52. Barlow, G.W. Ethological units of behavior. In *Central Nervous System and Fish Behavior*; Ingle, D., Ed.; University of Chicago Press: Chicago, IL, USA, 1968; pp. 217–232.
53. Chelazzi, L.; Perlato, A.; Santandrea, E.; Della Libera, C. Rewards teach visual selective attention. *Vis. Res.* **2013**, *85*, 58–72. [CrossRef]
54. Bari, A.; Dalley, J.W.; Robbins, T.W. The application of the 5-choice serial reaction time task for the assessment of visual attentional processes and impulse control in rats. *Nat. Protocol.* **2008**, *3*, 759–767. [CrossRef]
55. Alkam, T.; Hiramatsu, M.; Mamiya, T.; Aoyama, Y.; Nitta, A.; Yamada, K.; Kim, H.C.; Nabeshima, T. Evaluation of object-based attention in mice. *Behav. Brain Res.* **2011**, *220*, 185–193. [CrossRef]
56. Waring, G.H. *Horse Behavior (Park Ridge)*; Noyes Publications/William Andrew Pub: New York, NY, USA, 2003.
57. Dawkins, M. Shifts of "attention" in chicks during feeding. *Anim. Behav.* **1971**, *19*, 575–582. [CrossRef]
58. Pietrewicz, A.T.; Kamil, A.C. Search image formation in the blue jay (*Cyanocitta cristata*). *Science* **1979**, *204*, 1332–1333. [CrossRef] [PubMed]
59. Tinbergen, L. The Natural Control of Insects in Pinewoods. *Arch. Nerlandaises Zool.* **1960**, *13*, 265–343. [CrossRef]
60. Goto, K.; Bond, A.B.; Burks, M.; Kamil, A.C. Visual search and attention in blue jays (*Cyanocitta cristata*): Associative cuing and sequential priming. *J. Exp. Psychol. Anim. Learn. Cogn.* **2014**, *40*, 185–194. [CrossRef] [PubMed]
61. George, I.; Cousillas, H.; Richard, J.-P.; Hausberger, M. Experience with Adults Shapes Multisensory Representation of Social Familiarity in the Brain of a Songbird. *PLoS ONE* **2012**, *7*, e38764. [CrossRef] [PubMed]
62. Lemasson, A.; Boutin, A.; Boivin, S.; Blois-Heulin, C.; Hausberger, M. Horse (Equus Caballus) Whinnies: A Source of Social Information. *Anim. Cogn.* **2009**, *12*, 693–704. [CrossRef]
63. Shepherd, S.V.; Deaner, R.O.; Platt, M.L. Social status gates social attention in monkeys. *Curr. Biol.* **2006**, *16*, R119–R120. [CrossRef]
64. Lemasson, A.; Blois-Heulin, C.; Jubin, R.; Hausberger, M. Female Social Relationships in a Captive Group of Campbell's Monkeys (*Cercopithecus campbelli campbelli*). *Am. J. Primatol.* **2006**, *68*, 1161–1170. [CrossRef]

65. Hausberger, M.; Richard-Yris, M.-A.; Henry, L.; Lepage, L.; Schmidt, I. Song Sharing Reflects the Social-Organization in a Captive Group of European Starlings (*Sturnus vulgaris*). *J. Comp. Psychol.* **1995**, *109*, 222–241. [CrossRef]
66. Henry, L.; Bourguet, C.; Coulon, M.; Aubry, C.; Hausberger, M. Sharing Mates and Nest Boxes Is Associated with Female "Friendship" in European Starlings, *Sturnus vulgaris*. *J. Comp. Psychol.* **2013**, *127*, 1–13. [CrossRef]
67. Cadieu, J.C.; Cadieu, N. Influence of Some Interactions between Fledglings and Adults on the Food Choice in Young Canaries (Serinus Canarius). *J. Ethol.* **1996**, *14*, 99–109. [CrossRef]
68. Clergeau, P. Attractivité et utilisation du milieu chez les étourneaux en alimentation. *Acta Oecologica* **1982**, *3*, 307–320.
69. Fernández-Juricic, E.; Smith, R.; Kacelnik, A. Increasing the costs of conspecific scanning in socially foraging starlings affects vigilance and foraging behaviour. *Anim. Behav.* **2005**, *69*, 73–81. [CrossRef]
70. Rodriguez, A.; Hausberger, M.; Clergeau, P. Flexibility in European starlings' use of social information: Experiments with decoys in different populations. *Anim. Behav.* **2010**, *80*, 965–973. [CrossRef]
71. Martin, G.R. The eye of a passeriform bird, the European starling (*Sturnus vulgaris*): Eye movement amplitude, visual fields and schematic optics. *J. Comp. Physiol. A* **1986**, *159*, 545–557. [CrossRef]
72. Rochais, C.; Henry, S.; Fureix, C.; Hausberger, M. Investigating Attentional Processes in Depressive-like Domestic Horses (Equus Caballus). *Behav. Process.* **2016**, *124*, 93–96. [CrossRef] [PubMed]

animals

MDPI

Article

Learning by Doing: The Use of Distance, Corners and Length in Rewarded Geometric Tasks by Zebrafish (*Danio rerio*)

Greta Baratti [1,*], Angelo Rizzo [2], Maria Elena Miletto Petrazzini [3] and Valeria Anna Sovrano [1,4,*]

1 CIMeC, Center for Mind/Brain Sciences, University of Trento, 38068 Rovereto, Italy
2 School of Natural Sciences, University of Torino, 10124 Torino, Italy; angelo.rizzo691@edu.unito.it
3 Department of Biomedical Sciences, University of Padova, 35121 Padova, Italy; mariaelena.milettopetrazzini@unipd.it
4 Department of Psychology and Cognitive Science, University of Trento, 38068 Rovereto, Italy
* Correspondence: greta.baratti@unitn.it (G.B.); valeriaanna.sovrano@unitn.it (V.A.S.); Tel.: +39-0464-808703 (V.A.S.)

Simple Summary: Geometric navigation allows animals to efficiently move towards essential life-spaces by taking advantage of macrostructural information such as distance, angular magnitude, and length, in relation to left-right positional sense. In natural contexts, these cues can be referred to extensive three-dimensional surfaces such as a slope or a riverbed, thus becoming crucial to orient and find useful supplies. In controlled contexts, it is possible to set apart these components by handling the global shape of the experimental space (rectangular or square) as well, with the aim to specially probe the impact of each of them on navigation behavior of animals, including fishes. The present study aimed at investigating whether a well-known vertebrate, the zebrafish, could learn to encode and retain in memory such metric information (in terms of distances, corners, and lengths) in association with left–right directions, to gain rewards. Our results showed that zebrafish learned to use all these geometric attributes when repeatedly exposed to them, over a period of training, thereby giving strength to the ecological relevance of environmental geometry as a source of spatial knowledge. Generally, the engagement of zebrafish may consent to assess computations underlying large-scale-based navigation, also by drawing targeted comparisons, due to its behavioral, cognitive, and even emotional similarities with mammals.

Abstract: Zebrafish spontaneously use distance and directional relationships among three-dimensional extended surfaces to reorient within a rectangular arena. However, they fail to take advantage of either an array of freestanding corners or an array of unequal-length surfaces to search for a no-longer-present goal under a spontaneous cued memory procedure, being unable to use the information supplied by corners and length without some kind of rewarded training. The present study aimed to tease apart the geometric components characterizing a rectangular enclosure under a procedure recruiting the reference memory, thus training zebrafish in fragmented layouts that provided differences in surface distance, corners, and length. Results showed that fish, besides the distance, easily learned to use both corners and length if subjected to a rewarded exit task over time, suggesting that they can represent all the geometrically informative parts of a rectangular arena when consistently exposed to them. Altogether, these findings highlight crucially important issues apropos the employment of different behavioral protocols (spontaneous choice versus training over time) to assess spatial abilities of zebrafish, further paving the way to deepen the role of visual and nonvisual encodings of isolated geometric components in relation to macrostructural boundaries.

Keywords: navigation; spatial learning; environmental geometry; geometric components; zebrafish

Citation: Baratti, G.; Rizzo, A.; Miletto Petrazzini, M.E.; Sovrano, V.A. Learning by Doing: The Use of Distance, Corners and Length in Rewarded Geometric Tasks by Zebrafish (*Danio rerio*). *Animals* **2021**, *11*, 2001. https://doi.org/10.3390/ani11072001

Academic Editor: Rafael Freire

Received: 2 June 2021
Accepted: 30 June 2021
Published: 5 July 2021

1. Introduction

In recent years, the growing interest in the engagement of fish species for neuroscientific purposes has allowed comparative psychologists to deeply investigate cognitive skills and related underlying neural mechanisms of this heterogeneous group of organisms [1,2]. Since there is high variability among underwater ecosystems and habitats, fish naturally face spatial orientation challenges, further showing noteworthy navigation capacities such as the use of compasses [3,4], landmarks [5,6], and cognitive maps [7,8]. Besides these skills, fish have been experimentally observed also in relation to the use of environmental geometric layouts to solve place-finding demands after induced disorientation [9–23].

The term "environmental geometry" traditionally refers to macrostructural terrain-like characteristics of life-spaces where animals have to pinpoint worthwhile resources such as food, companions, and safe shelters, for survival. As a consequence, geometric navigation embraces the differential encoding of Euclidean concepts (for instance, "point", "surface", "boundary") that can characterize an enclosed space by which animals have to orient. Such a capacity has been demonstrated to be spread across species, both vertebrates [24–28] and invertebrates [29–32], thus highlighting the adaptive value of cognitive map and geometric representations over the phylogenetic history of animals.

Cheng's original investigations with *Rattus norvegicus* [33,34] have brought to light the use of spatial geometry in terms of metric (short/long) and directional sense (left/right) information within rectangular arenas, as well as the effect of behavioral protocols on the integrated use of geometric and nongeometric (i.e., local landmarks) cues. A differential encoding of these two types of information has been observed, especially in relation to the memory system involved: while both in working and reference memory tasks, rats made systematic rotational errors (by confusing the right position with the position at 180° on the arena's diagonal), only in reference memory tasks could rats combine geometric and nongeometric cues to pick out the place where the goal-object was present, thus solving the two-way spatial symmetry. In other words, the encoding of metric-plus-sense properties of surfaces versus features seems to depend on two dissociable cognitive systems, the former being liable to experience (in the presence of learning) while the latter mainly being driven by spontaneous computations (in the absence of learning).

Differences in disoriented navigation under spontaneous or rewarded experimental conditions have been found also in fishes [22,23], in a similar way to mammals [33–35]. The behavioral protocols with fish species provide for two well-validated spatial problems: the "social cued memory task" [16–18,23] and the "rewarded exit task" [9,10,12,13,15,20–23]. The social cued memory task consists in assessing spontaneous choices, where fish are required to approach the location of a social object (i.e., a companion) no-longer present, after a phase of passive disorientation. On the other hand, the rewarded exit task consists in assessing reorientation performances over time, where fish are required to learn the location of one or two exit-corridors to gain appealing rewards (i.e., companions, food, a cozy zone). In both cases, fish have to navigate by taking advantage of quantitative and/or qualitative properties of three-dimensional layouts that are marked out by their physical (visible or not) extended boundaries. Currently, however, we do not fully understand the role of isolated quantitative attributes such as distances, angular magnitudes, and lengths, together with the sensitivity to them, when they are not enclosed within a single, interconnected polygon. Most of all, it is still unknown whether and how learning mechanisms (for instance, repetitive trial-and-error training procedures) could aid to bypass the limits of spontaneous reorientation.

The opportunity to test apart these geometric components within fragmented surface arrays arose proper with the aim to specify what kind of metric representations are crucially involved in navigation by geometry. Lee and colleagues [36] found that young children were able to reorient by computing distances, but not lengths, among wall-like surfaces. Again, children could not use neither freestanding objects nor fragmented corners to face and solve disoriented navigation demands [37–40]. Focusing on nonhuman species, an effect of behavioral protocols has been found in *Gallus gallus*: while trained chicks learned to reorient by means of freestanding objects arrays [41–43], untrained chicks did not [44]. Likewise in zebrafish (*Danio rerio*), Lee and colleagues [17] observed that naïve fish spontaneously use distance and directional relationships among three-dimensional extended surfaces ("the goal is left/right of the closer/farther wall) to reorient within a rectangular arena but failed to take advantage of either an array of freestanding corners or an array of unequal-length surfaces to search for a no-longer-present goal, thus being unable to use the information supplied by corners and length without some kind of differential reinforcement. Starting from such an evidence, would zebrafish be able to compute both angular information and length, besides distances, for place-finding after disorientation? Additionally, the use of zebrafish as a model may provide a powerful tool to address navigation issues by applying a combinational approach [45] targeted to link research from multiple neuroscientific fields at both behavioral and neural levels of investigation [46,47]. Nonetheless, the role of zebrafish in multilevel scientific research is currently accepted and established, due to its characteristics and versatility. From pharmacology [48–51], to genomics [52–54], to behavioral studies as well [47], this teleost is becoming potentially useful, even with respect of its emotional and cognitive phenotype.

The purpose of the present work was to investigate whether zebrafish could learn to use the geometric components characterizing a rectangular enclosure, in terms of distance, corners, and length, to gain a reward. Therefore, we adapted the experimental conditions by Lee and colleagues [17] within an apparatus previously used to test both spontaneous and rewarded extra-visual encodings of pure geometry in three-eyed fishes [23]. We performed four experiments, two of them in a rectangular transparent arena, whereas the other two in a square transparent arena. More in details, the four experiments were scheduled as follows: Experiment 1 ("Use of distance as a geometric cue within a rectangular transparent arena"); Experiment 2 ("Use of corners as a geometric cue within a rectangular transparent arena"); Experiment 3 ("Use of length as a geometric cue within a square transparent arena"); Experiment 4 ("Control condition within a square transparent arena").

All the experiments employed a behavioral protocol widely used in the last 20 years to investigate navigation by geometry in fishes [9,10,12,13,15,20–23]. Briefly, it consists in training animals over time to locate one or two goal-exits inside a geometrically characterized arena to get a reward, a procedure named "rewarded exit task". Thus, in order to assess the learning performance of zebrafish, we compared, together with the percentages of choice towards the correct-geometry diagonal versus the incorrect-geometry diagonal. If the opportunity to experience spatial geometric contingencies is a factor to be taken into account for reorientation, then we expected that fish could learn to encode isolated metric components when consistently exposed to them.

2. Materials and Methods

2.1. Subjects and Housing

Subjects were 30 wild-type mature male zebrafish (*D. rerio*), ranging from 4 to 5 cm in body-length and coming from breeding stocks in our laboratory. Eight fish were engaged in Experiment 1, Experiment 2, and Experiment 3, while six fish were engaged in Experiment 4. All subjects were naïve, and each of them was observed in only one condition. To excite and catch the attention of the experimental fish, we used two female companions as sexual and social stimuli [55]. Outside of training, fish were maintained under a 10:14-h light/dark cycle and raised within glass home tanks (25 L capacity). Each tank was subdivided in four compartments with the aim to individually breed the experimental subjects. These tanks were further enriched with gravel and plants, thus cleaned with suitable hang-on filters (Niagara 250, Wave) to ensure comfortable habitats. The water temperature was maintained at 26 $\pm$ 1 °C by means of 25-watt heaters (NEWA Therm®, NEWA, Padua, Italy). Fish were fed with dry food (GVG-Mix Nature, sera® GmbH, Munich, Germany) exclusively within the experimental apparatus from Monday to Friday (the 5-day weekly period of training), to keep the motivation as high as possible.

2.2. Experimental Apparatus

The apparatus was the same used by Sovrano and colleagues [23] to test both spontaneous and rewarded extra-visual encodings of pure geometry in three-eyed fishes, where we introduced some adaptations with the aim to replicate the experimental conditions by Lee and colleagues [17], but by employing a different behavioral protocol (see below for details).

In general, the apparatus was placed in a darkened room and consisted of a circular amaranth tank (diameter × height: 175 × 27 cm) laying on a turntable that allowed the experimenter to rotate the whole structure at the beginning of each trial. This tank was surrounded by a circular black curtain fixed on a wood and metal frame, and centrally lit from above (height: 100 cm) through a 24-watt fluorescent white light tube (Lumilux, Osram GmbH, D, Munich, Germany). The water temperature was maintained at 26 ± 1 °C by means of a 50-watt heater (NEWA Therm®, NEWA, Padua, Italy), while a filter (NEWA Duetto®, NEWA, Padua, Italy) made sure a good quality. Both the heater and filter were not present during the training sessions.

In the center of such a tank, one of two experimental arenas could be placed: a rectangular transparent arena or a square transparent arena, the former for Experiment 1 and 2, whereas the latter for Experiment 3 and 4. As regards the rectangular transparent arena, it was the same used by Sovrano and colleagues [23], and consisted of an enclosure made by glass (length × width × height: 30 × 20 × 8 cm), thus composed of two long walls (length × height: 26 × 11 cm) and two short walls (length × height: 16 × 11 cm). In order to create a totally nonvisible environment, we did not glue the four walls together, but instead they were inserted in "single-track" supports made by polyvinyl chloride (PVC) and fixed on a Plexiglas base (length × width: 50 × 50 cm) at the bottom of the circular tank, then covered with a thick layer of homogeneous dark gravel (depth: 5 cm). As regards the square transparent arena, it was built ex novo for the purpose of this study and consisted of an enclosure made of glass (length × width: 30 × 30 cm; height: 8 cm), thus composed of four walls of equal size (length × height: 26 × 11 cm). As the rectangular enclosure, the square one stood on a basement (length × width: 50 × 50 cm), while its walls were installed within single-track supports then covered with gravel.

Both arenas were equipped with four "tunnels of choice" (conventionally, "corridors") placed at the meeting point of each pair of tangent walls. These corridors were positioned towards the outside of the transparent enclosures at precisely 2.5 cm from corners. More in detail, each corridor was composed of three parts: a rectangular transparent glass (length × height: 3 × 11 cm) and two rectangular transparent acetate sheets (length × height: 2.5 × 9 cm). These flexible sheets were perpendicularly glued on the glass in order to create a C-shape design, and they were characterized by a specific pattern of three-series vertical fissures depending on the scope. "Traversable" corridors had to allow fish to exit the arena to gain rewards (i.e., a cozier outer zone provided with food and companions); thus, they had a thick central fissure (length × height: 1 × 7.43 cm) and two thin lateral fissures (length × height: 0.2 × 7.43 cm). Conversely, "nontraversable" corridors had to not allow fish to exit the arena; thus, they had a 3 × 3 matrix of thin fissures (superior and inferior series: length × height: 0.3 × 2.5 cm; central series: length × height: 0.3 × 2 cm). Detail of corridors are depicted in Figure 1.

These slight differences in size were needed to balance the overall perimeter among traversable and nontraversable corridors (47.4 cm) by equalizing hydrodynamic effects due to movements of fish against the physical walls and potentially detectable through the lateral line [23]. For both the arenas, two traversable corridors were placed on the rewarding diagonal (correct-geometry: $C_{1,2}$), whereas two nontraversable corridors were placed on the opposite nonrewarding diagonal (incorrect-geometry: $X_{1,2}$). To be specific, the correct corners were labelled "C_1" and "C_2", while the incorrect ones "X_1" (close to C_1) and "X_2" (close to C_2). The arenas were entirely inside water, but not submerged (with a gap of 0.5 cm in respect of their height), in order to create visual continuity without light reflections.

Figure 1. Detail of corridors. At level of the arena's corners, four corridors (two per side) made of glass and acetate were placed: two traversable corridors having a large central hole were arranged on the correct-geometry diagonal, whereas two nontraversable corridors having smaller holes were arranged on the incorrect-geometry one. While the former allowed fish to exit the arena, thus achieving food and companions, the latter did not.

With the aim to investigate if fish could learn to use visual geometric cues in terms of distance, length, and corners, for the first three experiments, we added different features made of white polypropylene (Poliplak®, Röhm Italia SRL, Milan, Italy): four panels equal in length (length × height: 15 × 10 cm) for Experiment 1; four fragmented corners composed of three panels (central panel: length × height: 3 × 10 cm; lateral panels: length × height: 6 × 10 cm) for Experiment 2; four panels of 2:1 ratio in length (long panels: length × height: 20.4 × 10 cm; short panels: 10.6 × 10 cm) for Experiment 3. Experiment 4 was run in the absence of any additional visual geometric cues, in order to validate the square transparent arena that we built ex novo to replicate the experimental conditions by Lee and colleagues [17], but by applying a different behavioral protocol. The two arenas equipped for all the experiments are depicted in Figure 2.

2.3. *Experimental Procedure*

The training procedure was the same used by Sovrano and colleagues for the nonvisual geometric task with prolonged experience [23], and by Baratti and colleagues to test the encoding of visual geometry in zebrafish [21]. It consisted of a training subdivided in a maximum of 10 daily sessions (one per day), from Monday to Friday, for two successive weeks of work. Each daily session of training consisted in eight trials, until fish reached a learning criterion $\geq 70\%$ of correctness for diagonal $C_{1,2}$ in two consecutive sessions (one per day), the former as a measure of learning achieved and the latter as a validation of performance (accuracy). In case fish were able to reach the criterion within the first two sessions, the training was protracted for two additional sessions (as confirmation of learning). One group of fish was locally rewarded on diagonal 1, while one other group on diagonal 2, aiming to balance potential biases associated to intra/extra-arena cues.

At the beginning of each trial, fish were gently transferred from their home tank into the experimental apparatus, more precisely within a glass cylinder (diameter × height: 6 × 8) that was placed in the center of the transparent arena. At the upper end of such a cylinder, there was a transparent nylon wire that, by means of a rigid and nonvisible metal pulling mechanism, allowed the experimenter to vertical lift upward the cylinder without creating biases during this tricky operation. After 30 s of acclimation and visual exploration of the environment, the cylinder was slowly lifted up, leaving fish free to swim around and familiarize with the experimental space. Within each trial, fish had a 10-min limit to make choices targeted to the corridors: these attempts were sequentially noted down until fish correctly exited the arena. A correction method was used [56]: fish were allowed to change one or more wrong choices (i.e., towards corridors X_1 and X_2), up until they were able to choose one of the two right corridors (i.e., C_1 or C_2) or up until the 10-min limit was elapsed. Intervals among trials, where fish could stay within the cozier outer zone to enjoy the rewards, were managed as follows: six minutes (completely rewarding: small amount of food and companions) if fish identified the correct-geometry diagonal $C_{1,2}$ as single attempt, which allow it to get out of the arena, or two minutes (nonrewarding: neither food nor companions) if fish made two or more wrong attempts. Multiple choices for the correct corridors could occur, particularly when fish approached them without exiting of the arena. An attempt was considered as valid if fish entered the corridor with their whole body length; moreover, actual exit attempts were clearly visible in video recordings on the basis of characteristic tail-and-body movements made by fish against the acetate sheets. Whenever a fish did not make a choice within the 10-min limit, it was given a 5-min break inside the outer zone. After two null trials without valid choices, the training session could be interrupted and rescheduled for the day next. Before starting the following trial, the apparatus was rotated 90° right in order to remove any potential uncontrolled cues and reduce the use of compass and inertial information.

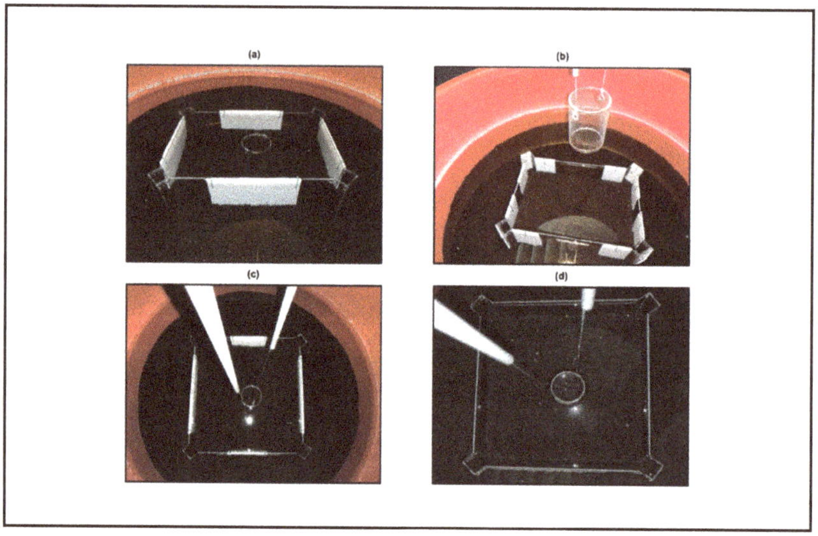

Figure 2. Photographs of the experimental arenas. (**a**) The rectangular transparent arena equipped with four white equal-length panels to assess the distance component. (**b**) The same rectangular arena but provided with four white angular panels to assess the corners component. (**c**) The square transparent arena equipped with four white unequal-length panels of 2:1 ratio to assess the length component. (**d**) The same square arena devoid of any supplementary visual cues to bear out the absence of potential biases associated to intra/extra-arena cues. All the variations were supplied with a transparent cylinder, placed in the center, where experimental fish were hosted before starting each training trials. Such a cylinder was lifted by the experimenter through a pulley system from the outside in order to not be seen by animals.

2.4. Statistical Analysis

With respect to Experiments 1, 2, and 3, we measured the following dependent variables: the mean number of trials to reach a criterion $\geq 70\%$ of correctness in two consecutive learning sessions (learning and validation), together with the percentages of choice towards the correct-geometry diagonal $C_{1,2}$ (thus, by cumulating the percentages of choice towards corners $C_{1,2}$) versus the incorrect-geometry diagonal $X_{1,2}$ (thus, by cumulating the percentages of choice towards corners $X_{1,2}$). With respect to Experiment 4, we inspected the percentages of choice towards $C_{1,2}$ versus $X_{1,2}$ over time, that is, within the 10 sessions scheduled to finalize the training. In the absence of geometry, in terms of metric (short/long) and sense (left/right), the correctness was referred to the traversable corridors placed on the rewarding diagonal of the square arena.

For the first three experiments, Student's *t*-test for unpaired samples was applied to bear out the lack of cues locally associated to the arenas' diagonals by comparing the group of fish that was trained on diagonal 1 versus 2 as the correct-geometry $C_{1,2}$. Student's t-test for paired samples was further applied to compare the percentages of choice towards the correct-geometry diagonal $C_{1,2}$ versus the incorrect-geometry diagonal $X_{1,2}$, with the aim to corroborate the idea that fish had actually learned how to solve the reorientation task. Moreover, Student's *t*-test for paired samples was applied to compare the two corners laying on the same diagonal (correct-geometry: C_1 versus C_2; incorrect-geometry: X_1 versus X_2). On the other hand, for the last experiment, a repeated measures ANOVA was applied to compare the percentages of choice towards $C_{1,2}$ versus $X_{1,2}$ over time (1–10 training sessions).

The Shapiro–Wilk test was performed to verify the normality, whereas Levene's test of equality of error variances and Mauchly's sphericity test were performed to assess the homoscedasticity. To estimate the effect size of significant data analysis, we reported 95% confidence intervals as an index for Student's *t*-test. Raw data were analyzed by means of IBM® SPSS Statistic 27 software package, and they are available in a submitted supplementary Excel file (see Supplementary Data S1).

3. Results

3.1. Experiment 1: Use of Distance as a Geometric Cue within a Rectangular Transparent Arena

Experiment 1 aimed at investigating if zebrafish could learn to use distance as a geometric cue to reorient in association with directional sense (distance-plus-sense relationships).

The mean number of trials needed to reach the 70% criterion for the correct-geometry diagonal $C_{1,2}$ was 39.625 ± 6.276 (mean ± SEM), which was ≈ 5 consecutive training sessions. Since one group of fish was locally rewarded on diagonal 1 and the other on diagonal 2 to balance any potential uncontrolled biases related to intra/extra-arena cues, the mean number of trials to 70% of correctness for diagonal 1 versus 2 was compared by performing a *t*-test for unpaired samples, which did not reveal significant differences ($t(6) = 0.927$, $p = 0.390$). For this reason, the following analyses were applied by collapsing the two samples.

Results of the learning session, where fish obtained a spatial performance ≥ 70%, and results of the validation session, where fish kept an above-threshold behavior, are shown in Figure 3.

A *t*-test for paired samples applied on both the percentages of first and the total choices in the learning session revealed a significant effect of Geometry ($C_{1,2}$ versus $X_{1,2}$) (first choices: $t(7) = 14.150$, $p \leq 0.001$, 95% IC [51.362, 71.973]; total choices: $t(7) = 11.471$, $p \leq 0.001$, 95% IC [41.485, 63.029]). There were no differences between the two corners laying on the same diagonal, by independently comparing the correct-geometry $C_{1,2}$ (first choices: $t(7) = 0.241$, $p = 0.817$; total choices: $t(7) = -0.254$, $p = 0.807$), and the incorrect-geometry $X_{1,2}$ (first choices: $t(7) = 2.165$, $p = 0.067$; total choices: $t(7) = 2.064$, $p = 0.078$).

In a similar way, a *t*-test for paired samples applied on both the percentages of first and total choices in the validation session revealed a significant effect of Geometry ($C_{1,2}$ versus $X_{1,2}$) (first choices: $t(7) = 6.769$, $p \leq 0.001$, 95% IC [39.039, 80.960]; total choices: $t(7) = 8.365$, $p \leq 0.001$, 95% IC [43.323, 77.467]). There were no differences between the two corners laying on the same diagonal, by independently comparing the correct-geometry $C_{1,2}$ (first choices: $t(7) = 0.346$, $p = 0.740$; total choices: $t(7) = 0.056$, $p = 0.957$), and the incorrect-geometry $X_{1,2}$ (first choices: $t(7) = -0.500$, $p = 0.632$; total choices: $t(7) = 0.228$, $p = 0.826$).

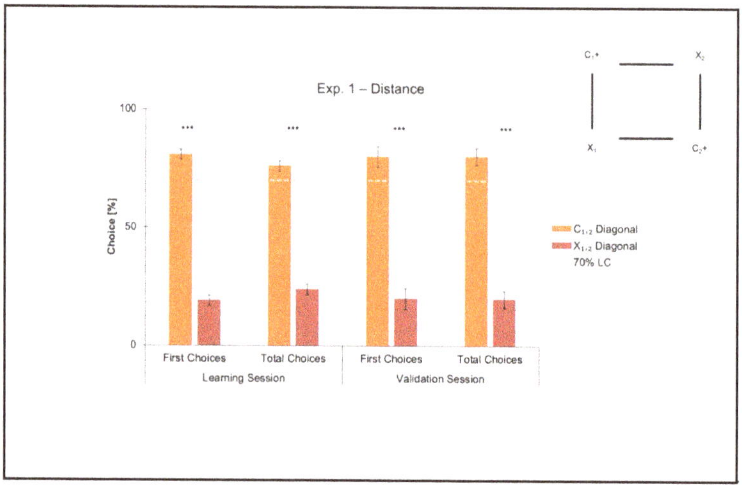

Figure 3. Results of Experiment 1. The bar chart shows the percentages of first and total choices (mean, SEM) obtained by fish towards the two diagonals ($C_{1,2}$ versus $X_{1,2}$) in both learning and validation sessions. The gray dashed line represents the 70% learning criterion (LC), while the asterisks (***) indicate a p-value ≤ 0.001.

Results showed that in the presence of nonvisual geometric surfaces (i.e., the rectangular transparent arena) that were visually defined by distance (i.e., the four white panels equal in length), zebrafish easily learned to reorient on the basis of such a visual cue, there perceived in terms of the physical space between them and the four walls, within the 10-session limit allowed to finalize the training. In fact, the percentages of choice towards the correct-geometry diagonal $C_{1,2}$ versus the incorrect-geometry one $X_{1,2}$ statistically departed from chance level (=50%), both in the learning and validation sessions, thus highlighting that fish were able to efficiently navigate to pick out the rewarded corridors by taking advantage of distance-plus-sense spatial relationships.

3.2. *Experiment 2: Use of Corners as a Geometric Cue within a Rectangular Transparent Arena*

Experiment 2 aimed at investigating if zebrafish could learn to use corners as a geometric cue to reorient in association with directional sense (corners-plus-sense relationships).

The mean number of trials needed to reach the 70% criterion for the correct-geometry diagonal $C_{1,2}$ was 60.375 ± 5.568 (mean $\pm$ SEM), which was $\approx$8 consecutive training sessions. Since one group of fish was locally rewarded on diagonal 1 and the other on diagonal 2 to balance any potential uncontrolled biases related to intra/extra-arena cues, the mean number of trials to 70% of correctness for diagonal 1 versus 2 was compared by performing a t-test for unpaired samples, which did not reveal significant differences ($t(6) = -1.693$, $p = 0.141$). For this reason, the following analyses were applied by collapsing the two samples.

Results of the learning session, where fish obtained a spatial performance $\geq 70\%$, and results of the validation session, where fish kept an above-threshold behavior, are shown in Figure 4.

A t-test for paired samples applied on both the percentages of first and the total choices in the learning session revealed a significant effect of Geometry ($C_{1,2}$ versus $X_{1,2}$) (first choices: $t(7) = 14.705$, $p \leq 0.001$, 95% IC [43.533, 60.217]; total choices: $t(7) = 18.772$, $p \leq 0.001$, 95% IC [45.813, 59.017]). There were no differences between the two corners laying on the same diagonal, by independently comparing the correct-geometry $C_{1,2}$ (first choices: $t(7) = -0.473$, $p = 0.651$; total choices: $t(7) = 0.130$, $p = 0.900$), and the incorrect-geometry $X_{1,2}$ (first choices: $t(7) = -0.060$, $p = 0.954$; total choices: $t(7) = -0.305$, $p = 0.769$).

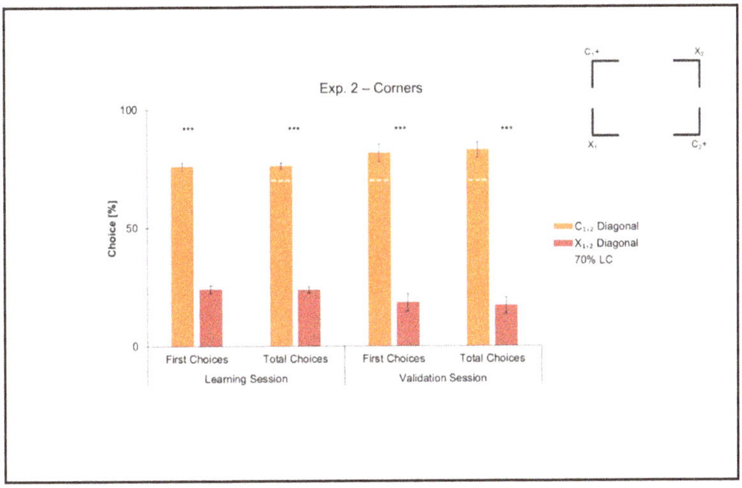

Figure 4. Results of Experiment 2. The bar chart shows the percentages of first and total choices (mean, SEM) obtained by fish towards the two diagonals ($C_{1,2}$ versus $X_{1,2}$) in both learning and validation sessions. The gray dashed line represents the 70% learning criterion (LC), while the asterisks (***) indicate a p-value ≤ 0.001.

In a similar way, a t-test for paired samples applied on both the percentages of first and total choices in the validation session revealed a significant effect of Geometry ($C_{1,2}$ versus $X_{1,2}$) (first choices: $t(7) = 8.355$, $p \leq 0.001$, 95% IC [45.813, 80.991]; total choices: $t(7) = 10.013$, $p \leq 0.001$, 95% IC [50.213, 81.262]). There were no differences between the two corners laying on the same diagonal, by independently comparing the correct-geometry $C_{1,2}$ (first choices: $t(7) = 1.391$, $p = 0.207$; total choices: $t(7) = 2.130$, $p = 0.071$), and the incorrect-geometry $X_{1,2}$ (first choices: $t(7) = 0.135$; $p = 0.897$; total choices: $t(7) = 0.751$, $p = 0.477$).

Results showed that in the presence of nonvisual geometric surfaces (i.e., the rectangular transparent arena) that were visually defined by their corners (i.e., the four white angular panels), zebrafish easily learned to reorient on the basis of these cues within the 10-session limit, allowing the training to be finalized. In fact, the percentages of choice towards the correct-geometry diagonal $C_{1,2}$ versus the incorrect-geometry one $X_{1,2}$ statistically departed from chance level (=50%), both in the learning and validation sessions, thus highlighting that fish were able to efficiently navigate to pick out the rewarded corridors by taking advantage of corners-plus-sense spatial relationships.

3.3. *Experiment 3: Use of Length as a Geometric Cue within a Square Transparent Arena*

Experiment 3 aimed at investigating if zebrafish could learn to use length as a geometric cue to reorient in association with directional sense (length-plus-sense relationships).

The mean number of trials needed to reach the 70% criterion for the correct-geometry diagonal $C_{1,2}$ was 48.750 ± 10.007 (mean $\pm$ SEM), which was $\approx$6 consecutive training sessions. Since one group of fish was locally rewarded on diagonal 1 and the other on diagonal 2 to balance any potential uncontrolled biases related to intra/extra-arena cues, the mean number of trials to 70% of correctness for diagonal 1 versus 2 was compared by performing a t-test for unpaired samples, which did not reveal significant differences ($t(6) = 1.088$, $p = 0.318$). For this reason, the following analyses were applied by collapsing the two samples.

Results of the learning session, where fish obtained a spatial performance $\geq 70\%$, and results of the validation session, where fish kept an above-threshold behavior, are shown in Figure 5.

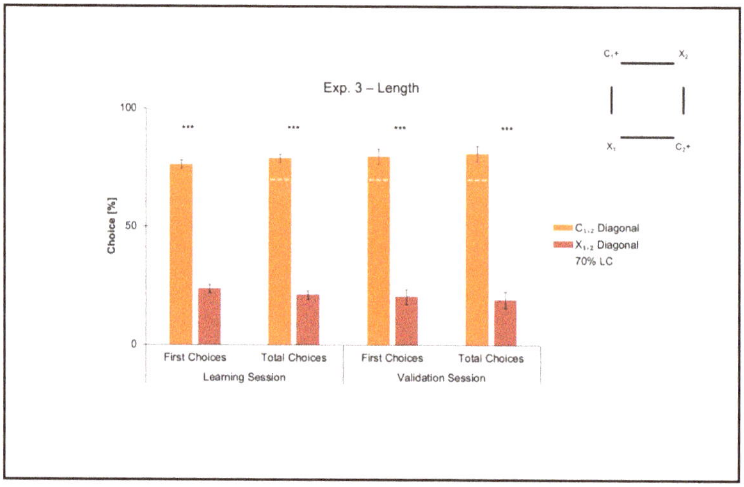

Figure 5. Results of Experiment 3. The bar chart shows the percentages of first and total choices (mean, SEM) obtained by fish towards the two diagonals ($C_{1,2}$ versus $X_{1,2}$) in both learning and validation sessions. The gray dashed line represents the 70% learning criterion (LC), while the asterisks (***) indicate a p-value ≤ 0.001.

A t-test for paired samples applied on both the percentages of first and the total choices in the learning session revealed a significant effect of Geometry ($C_{1,2}$ versus $X_{1,2}$) (first choices: $t(7) = 13.899$, $p \leq 0.001$, 95% IC [44.605, 62.895]; total choices: $t(7) = 15.918$, $p \leq 0.001$, 95% IC [49.229, 66.406]). There were no differences between the two corners laying on the same diagonal, by independently comparing the correct-geometry $C_{1,2}$ (first choices: $t(7) = -1.193$, $p = 0.272$; total choices: $t(7) = -0.698$, $p = 0.508$), and the incorrect-geometry $X_{1,2}$ (first choices: $t(7) = 0.174$, $p = 0.867$; total choices: $t(7) = -0.390$, $p = 0.708$).

In a similar way, a t-test for paired samples applied on both the percentages of first and total choices in the validation session revealed a significant effect of Geometry ($C_{1,2}$ versus $X_{1,2}$) (first choices: $t(7) = 9.029$, $p \leq 0.001$, 95% IC [43.825, 74.925]; total choices: $t(7) = 9.092$, $p \leq 0.001$, 95% IC [45.833, 78.052]). There were no differences between the two corners laying on the same diagonal, by independently comparing the correct-geometry $C_{1,2}$ (first choices: $t(7) = 0.341$, $p = 0.743$; total choices: $t(7) = 0.080$, $p = 0.939$), and the incorrect-geometry $X_{1,2}$ (first choices: $t(7) = 0.814$; $p = 0.442$; total choices: $t(7) = -1.014$, $p = 0.344$).

Results showed that in the presence of nonvisual nongeometric surfaces (i.e., the square transparent arena) that were geometrically and visually defined by their length (i.e., the four white panels of 2:1 ratio in length), zebrafish easily learned to reorient on the basis of these cues within the 10-session limit allowed to finalize the training. In fact, the percentages of choice towards the correct-geometry diagonal $C_{1,2}$ versus the incorrect-geometry one $X_{1,2}$ statistically departed from chance level (=50%), both in the learning and validation sessions, thus highlighting that fish were able to efficiently navigate to pick out the rewarded corridors by taking advantage of distance-plus-sense spatial relationships.

3.4. Control Condition within a Square Transparent Arena

Experiment 4 was a control condition aimed at validating the square arena that we had built ex novo for the purpose of the present study.

Fish did not reach the 70% criterion for the correct-geometry diagonal $C_{1,2}$ within the 10-session limit allowed to finalize the training. Results of training over time (1–10 sessions) are shown in Figure 6.

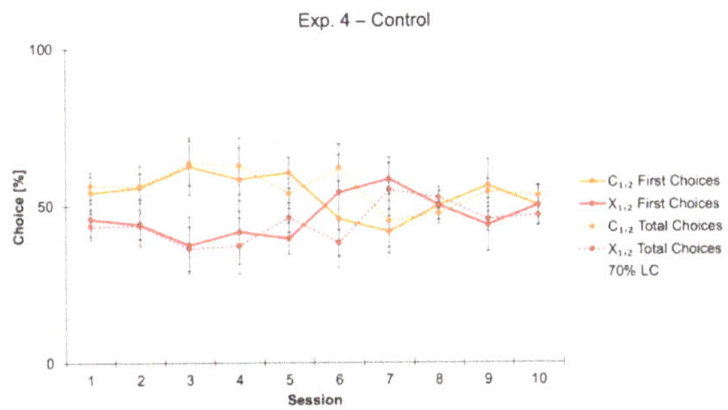

Figure 6. Results of Experiment 4. The line chart shows the percentages of choice (mean, SEM) obtained by fish towards the two diagonals ($C_{1,2}$ versus $X_{1,2}$) over time, that is, for each of the 10 training sessions. Solid lines represent first choices, and dotted lines represent total choices, while the gray dashed line represents the 70% learning criterion (LC).

A repeated measures ANOVA was performed by considering all data collected per fish towards the two diagonals ($C_{1,2}$ versus $X_{1,2}$) throughout the summed 10 sessions of training. The ANOVA with Geometry ($C_{1,2}$ versus $X_{1,2}$) and Session (1–10) as within-subject factors, and Diagonal (1 versus 2) as between-subject factor, then applied on both the percentages of first and total choices, did not reveal any significant differences (first choices: Geometry: $F(1, 4) = 1.232$, $p = 0.329$; Geometry × Diagonal: $F(1, 4) = 1.962$, $p = 0.234$; Session: $F(9, 36) = 0.949$, $p = 0.496$; Session × Diagonal: $F(9, 36) = 0.952$, $p = 0.494$; Geometry × Session: $F(9, 36) = 0.692$, $p = 0.712$; Geometry × Session × Diagonal: $F(9, 36) = 0.365$, $p = 0.944$; Diagonal: $F(1, 4) = 1.001$, $p = 374$) (total choices: Geometry: $F(1, 4) = 5.590$, $p = 0.077$; Geometry × Diagonal: $F(1, 4) = 2.442$, $p = 0.193$; Session: $F(9, 36) = 0.953$, $p = 0.493$; Session × Diagonal: $F(9, 36) = 0.952$, $p = 0.494$; Geometry × Session: $F(9, 36) = 1.098$, $p = 0.389$; Geometry × Session × Diagonal: $F(9, 36) = 0.587$, $p = 0.799$; Diagonal: $F(1, 4) = 1.000$, $p = 0.374$).

Results showed that in the presence of nonvisual nongeometric surfaces (i.e., the square transparent arena) that were neither geometrically nor visually defined by some kind of additional cues such as distance (closer/farther), corners (the meeting point of a pair of tangent surfaces), and length (shorter/longer), together with directional sense (left/right), zebrafish did not learn to reorient within the 10-session limit allowed to finalize the training. In fact, the percentages of choice towards the correct-geometry diagonal $C_{1,2}$ versus the incorrect-geometry one $X_{1,2}$ statistically did not depart from chance level (=50%), thus highlighting that fish were totally disoriented in the course of their navigation inside the square transparent arena.

4. Discussion

The present study aimed at studying what kind of spatial relationships could be learned by zebrafish during disoriented navigation within a space characterized by three-dimensional extended surfaces. To be specific, we wanted to investigate whether zebrafish could learn to use the geometric components typically characterizing a rectangular-shaped enclosure, in terms of distance (between the center and the walls), corners (as the meeting point of a pair of tangent walls), and length (as the pure metric information short/long), in association with directional sense (left/right). In order to set apart these three components, we required two transparent arenas (i.e., rectangular and square) equipped with distinctive opaque feature (i.e., white polypropylene panels). Starting from evidence by Lee and colleagues [17], where they found that zebrafish were able to spontaneously use only distance-plus-sense relationships to reorient at the expense of corners and length, we wondered if the behavioral protocol they adopted (social cued memory task) might have somehow affected the spatial performance. In this view, two studies by Sovrano and colleagues [22,23] have clearly showed a differential effect of spontaneous and rewarded protocols on geometric navigation of fish.

Therefore, we adapted the experimental conditions by Lee and colleagues [17] within an experimental apparatus previously used to test both spontaneous and rewarded extra-visual encodings of pure geometry in three-eyed fishes [23]. Four experiments were scheduled and then carried out by employing a behavioral protocol widely spread in the last 20 years to investigate navigation by geometry in fishes [9,10,12,13,15,20–23]. Briefly, it consists in training animals over time to locate one or two goal-exits inside a geometrically characterized arena to get something enticing (food and companions), a procedure named rewarded exit task [22,23]. Fish were required to reach a learning criterion $\geq$ 70% of correctness in two successive training sessions towards the correct-geometry diagonal of the arena, which they had to identify by encoding the geometric information specially provided by the environment.

Results of Experiment 1 showed that zebrafish easily learned to use the distance component, in terms of the physical space between their starting position in the center of the arena and the four extended surfaces, in order to reorient within a rectangular enclosure that was geometrically defined by visible, white-colored panels equal in length (15 cm). Such evidence strictly replicates findings by Lee and colleagues [17], where they observed that untrained zebrafish spontaneously used distance-plus-sense spatial relationships when subjected to a social cued memory task. In brief, experimental fish were placed inside a glass cylinder in the center of a completely bounded arena that was equipped with four white panels with the same length. At each corner, there was a glass jar (in total, four) where one of them hosted a rewarding sexual companion. Under these conditions, fish were first required to observe and memorize the goal-position, and then to approach it in the absence of the attractor on the basis of the distance component provided by the environment. Zebrafish exhibited a spontaneous tendency for the goal-position, thus making rotational errors targeted to the position at 180° on the diagonal as well.

Another study by Lee and colleagues with children [36] gave strength to the "inborn" predisposition to use the oneself-surface distance, since human babies (partially supplied with speech, abstract symbols, and cognitive maps) were also able to encode such a component during disoriented navigation within different geometric layouts. In this way, it is plausible that the encoding of distance-plus-sense spatial relationships (i.e., the goal is on the left/right of the closer/farther wall) could be both spontaneous and acquired over experience when animals have to face with place-finding demands. With the purpose to clarify the role of either instinct or learning, neurobiological investigations would be needed, as it has been observed by O'Keefe and Burgess with rats [57], or by comparing the behavior of fish larvae versus adults.

Results of Experiment 2 showed that zebrafish easily learned to use the corners component, in terms of the meeting point of a pair of tangent extended surfaces, to reorient within a rectangular enclosure that was geometrically defined by visible, white-colored panels resembling corners. Such evidence does not replicate findings by Lee and colleagues [17], wherein they observed that untrained zebrafish did not spontaneously use corners-plus-sense spatial relationships if subjected to a social cued memory task (working memory).

In the case of the rewarded exit task (reference memory), the behavioral protocol exerted a relevant part on spatial performances of fish: by consistently exposing them to a fragmented array of corners, *de facto*, zebrafish were able to infer the macrostructural morphology of the rectangular arena to efficiently reorient. Despite the absence of an "occludent", such a situation may be compared to the "amodal completion" phenomenon, since there is a kind of reconstruction of surfaces without direct stimulation of vision (i.e., without object-to-retina physiological correspondence). The amodal completion effect has been observed in fish, particularly with respect to bidimensional shape discrimination, both in controlled settings [58] and in natural contexts [59]. In this vein, such a perception could play a role as for goal-oriented navigation by zebrafish, which typically inhabit muddy environments where the exposure to partially covered objects is recurrent.

Results of Experiment 3 showed that zebrafish easily learned to use the length component, in terms of the pure metric information refereed to the extended surfaces (short/long), to reorient within a square enclosure that was geometrically defined by visible, white colored panels of 2:1 ratio in length. Such evidence does not replicate findings by Lee and colleagues with humans [37–40] and zebrafish [17], wherein they observed that untrained people and fish did not spontaneously use length-plus-sense spatial relationships if subjected to working memory conditions. On the contrary, we demonstrated that zebrafish are able to take advantage of length-plus-sense spatial relationships (i.e., the goal is on the left/right of the shorter/longer wall) if subjected to reference memory conditions. Analogously to Experiment 2, it seems that the chance to be exposed over time to specific geometric contingencies may have aided fish to retain the only attribute available around them, this being the different ratio in length provided by the four opaque panels. Hence, these findings highlight how motivational factors driven by some kind of recompense can affect the ability to represent in mind such a spatial relationship.

Although both untrained two-year-old babies and zebrafish ignore metric information as a cue to reorient within three-dimensional layouts, recent studies have shown that they were susceptible to the length of pictures and small objects [60]. In more detail, by applying a paradigm called "deviant detection paradigm" where five out of six types of "L" shared a geometric feature, children aged 4–10 were able to recognize the "geometric deviant" that diverged with respect to the control stimuli on the basis of their axis' length. Therefore, a predisposition to encode and make use of this characteristic may exist in certain circumstances [61]. Similarly, it has been observed that guppies (*Poecilia reticulata*) showed a tendency to overestimate the length of vertical stimuli such as, for example, the bisector of a reverse "T", thereby providing additional support to the capacity to represent metric attributes in humans as well as in aquatic species [62].

Experiment 4 was a control aimed at evaluating whether the square transparent arena we built ex novo to replicate, under training, the experimental condition under spontaneous navigation by Lee and colleagues [17] did not have spurious nongeometric cues thereby affecting spatial performances achieved by fish in Experiment 3. In line with our hypothesis, zebrafish were totally disoriented inside the square transparent arena devoid of any opaque panels, thus further highlighting the actual role of length on reorientation. Nevertheless, the validation of the square transparent arena is relevant also in light of the following experiments targeted to investigate if zebrafish can learn to use salient pattern-specific landmarks at corners in the absence of metric. The idea behind it is to evaluate more in depth if discrimination learning abilities could be eased or hampered by spatial geometric contingencies.

Results of Experiment 4 are at odds with previous findings by Sovrano and colleagues [23], wherein they found that three species of fish (*D. rerio*, *X. eiseni*, and *C. auratus*) learned to reorient within a rectangular transparent arena, thereby in the absence of visibility, by means of extra-visual sensory modalities (for instance, the lateral line or the sense of touch). In fact, fish were able to encode the environment's shape that was characterized by two short and two long walls only in cases of consistent trial-and-error experience, but not during spontaneous navigation, further supporting the observations by Lee and colleagues with the social cued memory task [17,18]. Because of all this evidence, the key role of behavioral protocols (spontaneous versus rewarded) and, as a consequence, of the memory system mainly recruited (working versus reference), become strongly evident in the solution of spatial reorientation problems.

Altogether, the results of our study allow us to better understand what kind of geometric relationships zebrafish are susceptible to, thereby providing a threefold scientific contribution. First, we confirmed the tendency of this species to use distance-plus-sense relationships, and, in addition, we demonstrated that fish could learn to use corners-plus-sense (through a kind of amodal reconstruction of the nonvisible surfaces) and length-plus-sense as well, in order to pinpoint the two geometrically equivalent positions. Distance, corners, and length are easily available in rectangular visible three-dimensional contexts; in fact, it has been shown that fishes are highly responsive to them [9–11,21]. Furthermore, the spatial performance of fish in the four experiments is ascribable to a context where one only geometric component was present: for such a reason, a higher number of training trials to learn may be due to interindividual differences [63]. Second, the present work focuses on the role of protocols (operant conditioning versus spontaneous choices), somehow highlighting that both can determine what sort of abilities animals display when solving high levels cognitive demands. In more depth, we showed the remarkable impact of rewards on goal-oriented behavior of zebrafish, a species that it has been observed to be sensitive to food [64,65] and companions [48,65] as motivational factors. As regards geometric navigation, Sovrano and colleagues [22] observed in *X. eiseni* improvements in landmark-based reorientation under rewarded training over time. In general, the two protocols showed different results because they clearly implement different behavioral techniques to test geometric reorientation capacities of fish. While the social cued memory task focuses on unconstrained approaches made by animals in the absence of a reward that follows a correct choice for the time being, the rewarded exit task takes proper advantage of forced motivation levels (on the basis of operant conditioning rules), which probably undergo an increment over time (e.g., the more days fish are "hungry and alone", the more they will search for food and companions). Indeed, for future similar studies with zebrafish, it would be interesting to deepen the impact of training on disoriented navigation by replicating another study by Lee and colleagues in working memory [18]. The broad idea could be to adapt in reference memory all the experimental conditions where fish failed to reorient by means of nonvisual quantitative and visual qualitative cues (i.e., the use of 2D form, 3D proximal landmark, proximal light source, distal landmark) in order to better clarify the effect of persistent incentives on such a spatial skill. Third, our results underline the role of memory as independent systems (short versus long term) on the encoding of isolated metric attributes. Specifically, how the opportunity to consistently experience macrostructural cues is essential to consolidate and fix distances, corners, and length characterizing geometric frameworks. These findings seem in line with previous studies with zebrafish: while short-term memory had flexibility limits in the solution of certain cognitive tasks [66], long-term memory allowed fish to recall a spatial orientation activity up to 10 days after a test [67].

Even though the ability to reorient by geometry is spread among vertebrates [27,68] and invertebrates [29–32], the nature of the spatial relationships beneath has been superficially investigated, most of all in humans [36–40]. In fact, such an issue differs from all the studies concerning the integrated use of conspicuous or local landmarks in association with macrostructural geometry to facilitate the solution of reorientation tasks, where spatial performances of fishes are comparable with those achieved by other land tetrapods [27]. In general, boundary-and landmark-based navigation seems to be supported by two independent cognitive mechanisms: a "spatial system", for the processing of invariable geometric information, and a "landmark system", for the processing of variable nongeometric information [69]. Since homologies between the hippocampus of mammals and the lateral pallium of teleosts have been established [70,71], together with a common organization of the basal ganglia [72], it is plausible that three-dimensional frameworks and features may be differently implemented in the brain of vertebrates (hippocampus versus striatum), including fishes. Furthermore, evidence in chicks [73,74] and fish [75] have highlighted the preferential engagement of the right hemisphere in the encoding of extended layouts during geometry-driven navigation by providing additional contribution to such a functional dissociation.

Besides geometric skills, it has been proven that aquatic organisms possess remarkable learning and memory capacities [76,77], which has led to a comprehensive map of life-spaces being built. Fish species are further supplied with alternative nonvisual sensory modalities, as the "lateral line" [78], whose functionalities may be involved in reorientation spatial phenomena in condition of scant visibility (e.g., visual transparency or blindness conditions) [20,23].

5. Conclusions

To conclude, our study falls within an intriguing and well-explored topic of spatial cognition, that is, the capacity to navigate by means of the geometric properties provided by an environment with a definite shape in terms of extended surfaces. In more detail, we aimed to better disentangle the metric components characterizing such a context by independently dissecting each of these attributes (distance, corners, and length) in relation to left-right positional sense. Not only that, but we were also interested in understanding whether the methodology we applied could have played a crucial part on learning behavior of disoriented zebrafish. Choosing the most suitable procedure, in fact, is an essential step to study animals' behavioral patterns, depending on purposes. Whereas spontaneous activity allows us to corroborate the presence of a given skill as a natural trait, the possibility to train animals to a desired behavior may aid in going beyond the constraints imposed by the phylogenetic history of species, with the aim of understanding the extent to which fish can learn to solve a cognitive task and test their learning skills. Lastly, the use of zebrafish as a model may provide a powerful tool to address navigation issues, with the purpose of drawing a phylogenetic line across remote species as regards the computations underlying geometric reorientation. Moreover, the increasingly widespread engagement of such a species in genomics allows for implementation of combinational approaches [45] targeted to investigate spatial cognition skills at every degree of analysis.

Supplementary Materials: The following is available online at https://www.mdpi.com/article/10.3390/ani11072001/s1, Excel data file: Supplementary Data S1—Learning by doing in zebrafish.

Author Contributions: Conceptualization, G.B. and V.A.S.; methodology, G.B., A.R., and V.A.S.; validation, G.B., A.R., V.A.S., and M.E.M.P.; infographic, G.B.; formal analysis: G.B. and V.A.S.; investigation: G.B., A.R., and V.A.S.; resources: V.A.S.; data curation: G.B., A.R., and V.A.S.; writing—original draft preparation, G.B.; writing—review and editing, G.B., A.R., V.A.S., and M.E.M.P.; visualization, G.B., A.R., V.A.S., and M.E.M.P.; supervision, V.A.S. and M.E.M.P.; project administration, V.A.S.; funding acquisition: V.A.S. All authors have read and agreed to the published version of the manuscript.

Funding: This research was funded by intramural resources financing to V.A.S. from the Center for Mind/Brain Sciences (CIMeC, University of Trento) and by the University of Padua under the STARS Grants program (MetaZeb).

Institutional Review Board Statement: This research was carried out in the Animal Cognition and Neuroscience Laboratory (A.C.N. Lab.) of CIMeC (Center for Mind/Brain Sciences) at the University of Trento (Italy). All husbandry and experimental procedures complied with European Legislation for the Protection of Animals used for Scientific Purposes (Directive 2010/63/EU) and were previously authorized by the University of Trento's Ethic Committee for the Experiments on Living Organisms and by the Italian Ministry of Health (auth. no. 1111-2015-PR and auth. no. 848/2020–PR issued pursuant to art. 31 of Legislative Decree 26/2014).

Acknowledgments: We wish to thank Ciro Petrone, Cristina Pascu, Giampaolo Morbioli, Michela and Erika for the animal care and welfare.

Conflicts of Interest: The authors declare no conflict of interest. The funders had no role in the design of the study; in the collection, analyses, or interpretation of data; in the writing of the manuscript; or in the decision to publish the results.

References

1. Brown, C.; Laland, K.; Krause, J. *Fish Cognition and Behavior*; John Wiley & Sons: Hoboken, NJ, USA, 2011.
2. Odling-Smee, L.; Braithwaite, V.A. The role of learning in fish orientation. *Fish Fish.* **2003**, *4*, 235–246. [CrossRef]
3. Quinn, T.P.; Brannon, E.L. The use of celestial and magnetic cues by orienting sockeye salmon smolts. *J. Comp. Psychol.* **1982**, *147*, 547–552. [CrossRef]
4. Mouritsen, H.; Atema, J.; Kingsford, M.J.; Gerlach, G. Sun compass orientation helps coral reef fish larvae return to their natal reef. *PLoS ONE* **2013**, *8*, e66039. [CrossRef]
5. Hughes, R.N.; Blight, C.M. Two intertidal fish species use visual association learning to track the status of food patches in a radial maze. *Anim. Behav.* **2000**, *59*, 613–621. [CrossRef] [PubMed]
6. Burt de Perera, T.; Macias Garcias, C. Amarillo fish (*Girardinichtys multiradiatus*) use visual landmarks to reorient in space. *Ethology* **2003**, *109*, 341–350. [CrossRef]
7. Reese, E.S. Orientation behavior of butterflyfish (family *Chaetodontidae*) on coral reefs: Spatial learning of route specific landmarks and cognitive maps. *Environ. Biol. Fishes* **1989**, *25*, 79–86. [CrossRef]
8. Rodríguez, F.; Duran, E.; Vargas, J.P.; Torres, B.; Salas, C. Performance of goldfish trained in allocentric and egocentric maze procedures suggests the presence of a cognitive mapping system in fishes. *Learn. Behav.* **1994**, *22*, 409–420. [CrossRef]
9. Sovrano, V.A.; Bisazza, A.; Vallortigara, G. Modularity and spatial reorientation in a simple mind: Encoding of geometric and nongeometric properties of a spatial environment by fish. *Cognition* **2002**, *85*, B51–B59. [CrossRef]
10. Sovrano, V.A.; Bisazza, A.; Vallortigara, G. Modularity as a fish (*Xenotoca eiseni*) views it: Conjoining geometric and nongeometric information for spatial reorientation. *J. Exp. Psychol. Anim. Behav. Process.* **2003**, *29*, 199–210. [CrossRef]
11. Vargas, J.P.; López, J.C.; Salas, C.; Thinus-Blanc, C. Encoding of geometric and featural spatial information by goldfish (*Carassius auratus*). *J. Comp. Psycholo.* **2004**, *118*, 206–216. [CrossRef] [PubMed]
12. Sovrano, V.A.; Dadda, M.; Bisazza, A. Lateralized fish perform better than nonlateralized fish in spatial reorientation tasks. *Behav. Brain Res.* **2005**, *163*, 122–127. [CrossRef]
13. Sovrano, V.A.; Bisazza, A.; Vallortigara, G. Animals' use of landmarks and metric information to reorient: Effects of the size of the experimental space. *Cognition* **2005**, *97*, 121–133. [CrossRef] [PubMed]
14. Brown, A.A.; Spetch, M.L.; Hurd, P.L. Growing in circles: Rearing environment alters spatial navigation in fish. *Psychol. Sci.* **2007**, *18*, 569–573. [CrossRef] [PubMed]
15. Sovrano, V.A.; Bisazza, A.; Vallortigara, G. How fish do geometry in large and in small spaces. *Anim. Cogn.* **2007**, *10*, 47–54. [CrossRef] [PubMed]
16. Lee, S.A.; Vallortigara, G.; Ruga, V.; Sovrano, V.A. Independent effects of geometry and landmark in a spontaneous reorientation task: A study of two species of fish. *Anim. Cogn.* **2012**, *15*, 861–870. [CrossRef]
17. Lee, S.A.; Vallortigara, G.; Flore, M.; Spelke, E.S.; Sovrano, V.A. Navigation by environmental geometry: The use of zebrafish as a model. *J. Exp. Biol.* **2013**, *216*, 3693–3699. [CrossRef]
18. Lee, S.A.; Ferrari, A.; Vallortigara, G.; Sovrano, V.A. Boundary primacy in spatial mapping: Evidence from zebrafish (*Danio rerio*). *Behav. Process.* **2015**, *119*, 116–122. [CrossRef]
19. Sovrano, V.A.; Chiandetti, C. Reorientation ability in redtail splitfin (*Xenotoca eiseni*): Role of environmental shape, rearing in group and exposure time. *Biol. Commun.* **2017**, *62*, 48–56. [CrossRef]
20. Sovrano, V.A.; Potrich, D.; Foà, A.; Bertolucci, C. Extra-visual systems in the spatial reorientation of cavefish. *Sci. Rep.* **2018**, *8*, 17698. [CrossRef]
21. Baratti, G.; Potrich, D.; Sovrano, V.A. The environmental geometry in spatial learning by zebrafish (*Danio rerio*). *Zebrafish* **2020**, *17*, 131–138. [CrossRef]
22. Sovrano, V.A.; Baratti, G.; Lee, S.A. The role of learning and environmental geometry in landmark-based spatial reorientation of fish (*Xenotoca eiseni*). *PLoS ONE* **2020**, *15*, e0229608. [CrossRef]
23. Sovrano, V.A.; Baratti, G.; Potrich, D.; Bertolucci, C. The geometry as an eyed fish feels it in spontaneous and rewarded spatial reorientation tasks. *Sci. Rep.* **2020**, *10*, 8020. [CrossRef] [PubMed]
24. Cheng, K.; Newcombe, N.S. Is there a geometric module for spatial orientation? Squaring theory and evidence. *Psychon. Bull. Rev.* **2005**, *12*, 1–23. [CrossRef] [PubMed]
25. Cheng, K. Whither geometry? Troubles of the geometric module. *Trends Cogn. Sci.* **2008**, *12*, 355–361. [CrossRef] [PubMed]
26. Vallortigara, G. Animals as natural geometers. In *Cognitive Biology*; Tommasi, L., Peterson, M.A., Nadel, L., Eds.; MIT Press: Cambridge, UK, 2009; pp. 83–104.
27. Tommasi, L.; Chiandetti, C.; Pecchia, T.; Sovrano, V.A.; Vallortigara, G. From natural geometry to spatial cognition. *Neurosci. Biobehav. Rev.* **2012**, *36*, 799–824. [CrossRef]
28. Gallistel, C.R.; Matzel, L.D. The neuroscience of learning: Beyond the Hebbian synapse. *Annu. Rev. Psychol.* **2013**, *64*, 169–200. [CrossRef] [PubMed]
29. Wystrach, A.; Beugnon, G. Ants learn geometry and features. *Curr. Biol.* **2009**, *19*, 61–66. [CrossRef] [PubMed]

30. Wystrach, A.; Cheng, K.; Sosa, S.; Beugnon, G. Geometry, features, and panoramic views: Ants in rectangular arenas. *J. Exp. Psychol. Anim. Behav. Process.* **2011**, *37*, 420–435. [CrossRef]
31. Sovrano, V.A.; Rigosi, E.; Vallortigara, G. Spatial reorientation by geometry in bumblebees. *PLoS ONE* **2012**, *7*, e37449. [CrossRef]
32. Sovrano, V.A.; Potrich, D.; Vallortigara, G. Learning of geometry and features in bumblebees (*Bombus terrestris*). *J. Comp. Psychol.* **2013**, *127*, 312–318. [CrossRef]
33. Cheng, K. A purely geometric module in the rat's spatial representation. *Cognition* **1986**, *23*, 149–178. [CrossRef]
34. Cheng, K.; Gallistel, C.R. Testing the geometric power of an animal's spatial representation. In *Animal Cognition*; Roitblat, H., Bever, T.G., Terrace, H., Eds.; Lawrence Erlbaum Associates: Hillsdale, MI, USA, 1984; pp. 409–423.
35. Lee, S.A.; Tucci, V.; Sovrano, V.A.; Vallortigara, G. Working memory and reference memory tests of spatial navigation in mice (*Mus musculus*). *J. Comp. Psychol.* **2015**, *129*, 189–197. [CrossRef] [PubMed]
36. Lee, S.A.; Sovrano, V.A.; Spelke, E.S. Navigation as a source of geometric knowledge: Young children's use of length, angle, distance, and direction in a reorientation task. *Cognition* **2012**, *123*, 144–161. [CrossRef] [PubMed]
37. Gouteux, S.; Spelke, E.S. Children's use of geometry and landmarks to reorient in an open space. *Cognition* **2001**, *81*, 119–148. [CrossRef]
38. Lee, S.A.; Spelke, E.S. Children's use of geometry for reorientation. *Dev. Sci.* **2008**, *11*, 743–749. [CrossRef] [PubMed]
39. Lee, S.A.; Spelke, E.S. A modular geometric mechanism for reorientation in children. *Cogn. Psychol.* **2010**, *61*, 152–176. [CrossRef] [PubMed]
40. Lee, S.A.; Spelke, E.S. Young children reorient by computing layout geometry, not by matching images of the environment. *Psychon. Bull. Rev.* **2011**, *18*, 192–198. [CrossRef] [PubMed]
41. Pecchia, T.; Vallortigara, G. Reorienting strategies in a rectangular array of landmarks by domestic chicks (*Gallus gallus*). *J. Psychol.* **2010**, *124*, 147–158. [CrossRef]
42. Pecchia, T.; Vallortigara, G. View-based strategy for reorientation by geometry. *J. Exp. Biol.* **2010**, *213*, 2987–2996. [CrossRef]
43. Pecchia, T.; Vallortigara, G. Spatial reorientation by geometry with freestanding objects and extended surfaces: A unifying view. *Proc. R. Soc. B Biol. Sci.* **2012**, *279*, 2228–2236. [CrossRef]
44. Lee, S.A.; Spelke, E.S.; Vallortigara, G. Chicks, like children, spontaneously reorient by three-dimensional environmental geometry, not by image matching. *Biol. Lett.* **2012**, *8*, 492–494. [CrossRef]
45. Grunwald, D.J.; Eisen, J.S. Headwaters of the zebrafish—emergence of a new model vertebrate. *Nat. Rev. Genet.* **2002**, *3*, 717–724. [CrossRef]
46. Levin, E.D.; Cerutti, D.T. Behavioral Neuroscience of Zebrafish. In *Methods of Behavior Analysis in Neuroscience*, 2nd ed.; Buccafusco, J.J., Ed.; CRC Press/Taylor & Francis: Boca Raton, FL, USA, 2009.
47. Orger, M.B.; de Polavieja, G.G. Zebrafish behavior: Opportunities and challenges. *Annu. Rev. Neurosci.* **2017**, *40*, 125–147. [CrossRef] [PubMed]
48. Kalueff, A.V.; Stewart, A.M.; Gerlai, R. Zebrafish as an emerging model for studying complex brain disorders. *Trends Pharmacol. Sci.* **2014**, *40*, 63–75. [CrossRef]
49. Salvaggio, A.; Marino, F.; Albano, M.; Pecoraro, R.; Camiolo, G.; Tibullo, D.; Brundo, M.V. Toxic effects of zinc chloride on the bone development in *Danio rerio* (Hamilton, 1822). *Front. Physiol.* **2016**, *7*, 153. [CrossRef] [PubMed]
50. Lo Cascio, P.; Calabrò, C.; Bertuccio, C.; Iaria, C.; Marino, F.; Denaro, M.G. Immunohistochemical characterization of PepT1 and ghrelin in gastrointestinal tract of zebrafish: Effects of Spirulina vegetarian diet on the neuroendocrine system cells after alimentary stress. *Front. Physiol.* **2018**, *9*, 614. [CrossRef]
51. Lauriano, E.R.; Guerrera, M.C.; Laurà, R.; Capillo, G.; Pergolizzi, S.; Aragona, M.; Germanà, A. Effect of light on the calretinin and calbindin expression in skin club cells of adult zebrafish. *Histochem. Cell Biol.* **2020**, *154*, 495–505. [CrossRef] [PubMed]
52. Postlethwait, J.H.; Woods, I.G.; Ngo-Hazelett, P.; Yan, Y.L.; Kelly, P.D.; Chu, F.; Talbot, W.S. Zebrafish comparative genomics and the origins of vertebrate chromosomes. *Genome Res.* **2000**, *10*, 1890–1902. [CrossRef] [PubMed]
53. Aleström, P.; Holter, J.L.; Nourizadeh-Lillabadi, R. Zebrafish in functional genomics and aquatic biomedicine. *Trends Biotechnol.* **2006**, *24*, 15–21. [CrossRef]
54. Kabashi, E.; Brustein, E.; Champagne, N.; Drapeau, P. Zebrafish models for the functional genomics of neurogenetic disorders. *Biochim. Biophys. Acta BBA Mol. Basis Dis.* **2011**, *1812*, 335–345. [CrossRef] [PubMed]
55. Al-Imari, L.; Gerlai, R. Sight of conspecifics as reward in associative learning in zebrafish (*Danio rerio*). *Behav. Brain Res.* **2008**, *189*, 216–219. [CrossRef]
56. Caro, T.M.; Roper, R.; Young, M.; Dank, R. Inter-observer reliability. *Behaviour* **1979**, *69*, 303–315.
57. O'Keefe, J.; Burgess, N. Geometric determinants of the place fields of hippocampal neurons. *Nature* **1996**, *381*, 425–428. [CrossRef]
58. Sovrano, V.A.; Bisazza, A. Recognition of partly occluded objects by fish. *Anim. Cogn.* **2008**, *11*, 161–166. [CrossRef] [PubMed]
59. Darmaillacq, A.S.; Dickel, L.; Rahmani, N.; Shashar, N. Do reef fish, *Variola louti* and *Scarus niger*, perform amodal completion? Evidence from a field study. *J. Comp. Psychol.* **2011**, *125*, 273–277. [CrossRef] [PubMed]
60. Spelke, E.S.; Lee, S.A.; Izard, V. Beyond core knowledge: Natural geometry. *Cogn. Sci.* **2010**, *34*, 863–884. [CrossRef]
61. Izard, V.; Spelke, E.S. Development of sensitivity to geometry in visual forms. *Hum. Evol.* **2009**, *23*, 213–248.
62. Santacà, M.; Agrillo, C. Two halves are less than the whole: Evidence of a length bisection bias in fish (*Poecilia reticulata*). *PLoS ONE* **2020**, *15*, e0233157. [CrossRef] [PubMed]
63. Tran, S.; Gerlai, R. Individual differences in activity levels in zebrafish (*Danio rerio*). *Behav. Brain Res.* **2013**, *257*, 224–229. [CrossRef]

64. Sison, M.; Gerlai, R. Associative learning in zebrafish (*Danio rerio*) in the plus maze. *Behav. Brain Res.* **2010**, *207*, 99–104. [CrossRef] [PubMed]
65. Daggett, J.M.; Brown, V.J.; Brennan, C.H. Food or friends? What motivates zebrafish (*Danio rerio*) performing a visual discrimination. *Behav. Brain Res.* **2019**, *359*, 190–196. [CrossRef] [PubMed]
66. Bloch, S.; Froc, C.; Pontiggia, A.; Yamamoto, K. Existence of working memory in teleosts: Establishment of the delayed matching-to-sample task in adult zebrafish. *Behav. Brain Res.* **2019**, *370*, 111924. [CrossRef] [PubMed]
67. Williams, F.E.; White, D.; Messer, W.S., Jr. A simple spatial alternation task for assessing memory function in zebrafish. *Behav. Process.* **2002**, *58*, 125–132. [CrossRef]
68. Lee, S.A. The boundary-based view of spatial cognition: A synthesis. *Curr. Opin. Behav. Sci.* **2017**, *16*, 58–65. [CrossRef]
69. Lee, S.A.; Spelke, E.S. Two systems of spatial representation underlying navigation. *Exp. Brain Res.* **2010**, *206*, 179–188. [CrossRef] [PubMed]
70. Solstad, T.; Boccara, C.N.; Kropff, E.; Moser, M.B.; Moser, E.I. Representation of geometric borders in the entorhinal cortex. *Science* **2008**, *322*, 1865–1868. [CrossRef]
71. Broglio, C.; Martín-Monzón, I.; Ocaña, F.M.; Gómez, A.; Durán, E.; Salas, C.; Rodríguez, F. Hippocampal pallium and map-like memories through vertebrate evolution. *J. Behav. Brain Sci.* **2015**, *5*, 109–120. [CrossRef]
72. Smeets, W.J.; Marin, O.; Gonzalez, A. Evolution of the basal ganglia: New perspectives through a comparative approach. *J. Anat.* **2000**, *196*, 501–517. [CrossRef]
73. Tommasi, L.; Vallortigara, G. Hemispheric processing of landmark and geometric information in male and female domestic chicks (*Gallus gallus*). *Behav. Brain Res.* **2004**, *155*, 85–96. [CrossRef]
74. Vallortigara, G.; Pagni, P.; Sovrano, V.A. Separate geometric and non-geometric modules for spatial reorientation: Evidence from a lopsided animal brain. *J. Cogn. Neurosci.* **2004**, *16*, 390–400. [CrossRef] [PubMed]
75. Vargas, J.P.; Bingman, V.P.; Portavella, M.; López, J.C. Telencephalon and geometric space in goldfish. *Eur. J. Neurosci.* **2006**, *24*, 2870–2878. [CrossRef] [PubMed]
76. Kieffer, J.D.; Colgan, P.W. The role of learning in fish behaviour. *Rev. Fish Biol. Fish.* **1992**, *2*, 125–143. [CrossRef]
77. Salas, C.; Broglio, C.; Durán, E.; Gómez, A.; Ocaña, F.M.; Jiménez-Moya, F.; Rodríguez, F. Neuropsychology of learning and memory in teleost fish. *Zebrafish* **2006**, *3*, 157–171. [CrossRef] [PubMed]
78. Bleckmann, H.; Zelick, R. Lateral line system of fish. *Integr. Zool.* **2009**, *4*, 13–25. [CrossRef]

animals

Review

Brain Lateralization and Cognitive Capacity

Lesley J. Rogers

School of Science and Technology, University of New England, Armidale, NSW 2351, Australia; lrogers@une.edu.au

Simple Summary: We used to think of brains as symmetrical, functioning in the same way on the left and right sides, but we now know that this is not so. From the small brains of insects to variously sized brains of vertebrates, including humans, the left and right sides process information differently and control different patterns of behaviour. This is known as lateralization. Lateralized brains can carry out different functions simultaneously on the left and right sides (e.g., monitoring for predators while searching for food). Avoiding duplication in this way increases cognitive capacity. This paper considers the cognitive advantages of two kinds of lateralization. The first, known as individual lateralization, means that most individuals in a species are lateralized, roughly half in one direction and the other half in the other direction. The second type of lateralization, known as directional or population lateralization, means that most individuals have the same direction of lateralization. Directional lateralization is important for social behaviour but, as this paper argues, it may not increase cognitive capacity any more than does individual lateralization. Strength of lateralization is discussed and so is the communication between the left and right sides of the brain.

Abstract: One way to increase cognitive capacity is to avoid duplication of functions on the left and right sides of the brain. There is a convincing body of evidence showing that such asymmetry, or lateralization, occurs in a wide range of both vertebrate and invertebrate species. Each hemisphere of the brain can attend to different types of stimuli or to different aspects of the same stimulus and each hemisphere analyses information using different neural processes. A brain can engage in more than one task at the same time, as in monitoring for predators (right hemisphere) while searching for food (left hemisphere). Increased cognitive capacity is achieved if individuals are lateralized in one direction or the other. The advantages and disadvantages of individual lateralization are discussed. This paper argues that directional, or population-level, lateralization, which occurs when most individuals in a species have the same direction of lateralization, provides no additional increase in cognitive capacity compared to individual lateralization although directional lateralization is advantageous in social interactions. Strength of lateralization is considered, including the disadvantage of being very strongly lateralized. The role of brain commissures is also discussed with consideration of cognitive capacity.

Keywords: individual lateralization; directional lateralization; cognitive capacity; parallel processing; social behaviour; visual attention; vertebrates; invertebrates; commissures; strength of lateralization

Citation: Rogers, L.J. Brain Lateralization and Cognitive Capacity. *Animals* **2021**, *11*, 1996. https://doi.org/10.3390/ani11071996

Academic Editor: Andrew Robins

Received: 4 June 2021
Accepted: 30 June 2021
Published: 3 July 2021

1. Introduction

The left and right sides of the brain are specialised to attend to different information, to process sensory inputs in different ways and to control different types of motor behaviour. This is referred to as hemispheric specialization or simply as brain lateralization. Such division of function between the hemispheres has long been known in humans and considered to increase 'neural space' [1] or cognitive capacity [2]. Additionally, in humans, there are clear structural differences between the hemispheres (summarised in [3]). Using functional magnetic resonance imaging of human brains, Gotts et al. [2] provided evidence that the left hemisphere has stronger interaction within itself (intrahemispheric), whereas processing by the right hemisphere involves greater involvement of both hemispheres (interhemispheric) [2].

Lateralization has now been well-documented across species, ranging from the small brains of insects [4] to early vertebrates, birds and mammals [5–7]. This ubiquity of brain lateralization suggests that, regardless of the absolute size of the brain, its capacity and efficiency may be increased by not duplicating all functions on the left and right sides. By avoiding duplication of information processing on the left and right sides of the brain, more "cognitive space" is available to carry out different types of processing. In other words, lateralization of brain function should be able to increase cognitive capacity without the more costly process of increasing brain size.

From studies on a variety of species and using a range of techniques, a general pattern of lateralization has been constructed. As conceptualized by Andrew [8], the left hemisphere controls sustained response to targets, whereas the right hemisphere is specialized for response to potent releasers of innate responses. The left hemisphere directs attention to specific categories of stimuli, often learned categories, and controls feeding responses [9–13]. The right hemisphere has broad attention to a wider variety of stimuli and especially to novel stimuli [14] and to predators (shown in toads [15]; a lizard [16]; a marsupial [17]; and in dogs [18,19]. The right hemisphere is also specialized for expressing intense emotion [18,20,21], for handling geometric information [22] and for dealing with social interactions [23–25]. In line with the latter, and as shown in chicks, attack and copulatory behaviour are functions of the right hemisphere and they can be elicited readily once inhibition of the right hemisphere by the left hemisphere is suppressed or removed [9,26]. Similar right hemisphere activation of attack has been shown also in toads [27,28], frogs [12], lizards [29], penguins [30,31], Australian magpies [32], horses [33] and gelada baboons [34]. It appears, therefore, that these lateralized expressions of behaviour are common to a wide range of vertebrate species.

The first evidence of brain lateralization in non-human species was discovered in avian species; by inhibiting protein synthesis in the left or right forebrain hemisphere of chicks (*Gallus gallus domesticus*) at critical stages of development and then investigating the longer-term effects on behaviour [35], summarised in [36], and by lesioning specific regions of the left or right hemisphere of songbirds and assessing the effects on song production [37]. Then, in rats, it was shown by assessing the effects on behaviour of ablation of the left or right hemisphere [38].

Later, lateralization was revealed simply by testing animals monocularly [39] and this has become a standard method to investigate lateralized behaviour. In species with eyes positioned on the sides of their head, and thus with little overlap of the visual fields, visual information goes mainly to the contralateral side of the brain (discussed further in Section 7). Hence, lateralized differences can be revealed by applying an eye-patch and comparing the performance elicited when the left versus the right eye is seeing [39,40]. The eye-system of the open (seeing) eye, comprising the visual inputs mainly to the contralateral hemisphere, has predominant control of behaviour. Laterality revealed by monocular testing is evident even in small sample sizes, indicating that it must have a significant role in biologically relevant situations.

Within a species, the strength of lateralized responses can vary. Some individuals exhibit strong laterality, while others have weaker laterality or no significant laterality. However, weak or absent laterality expressed in motor behaviour (e.g., hand or limb preference) may not mean that the brain is itself less lateralized for cognitive processing. It is rather a matter of whether one hemisphere alone controls a particular behaviour, as is the case in strong laterality, or that the other hemisphere is involved to some degree.

What does this mean in terms of cognitive capacity? Even when both hemispheres participate in the control of behaviour and interhemispheric control occurs, cognitive capacity is increased as long as each hemisphere performs different computations of available information. Provided each hemisphere is processing information differently and leading to different outcomes, cognitive capacity is increased. By contrast, if both hemispheres are processing information in the same way, and hence duplication is occurring, there is no enhancement of cognitive capacity. In short, more lateralization means greater cognitive capacity.

2. Advantage of Having a Lateralized Brain

Since lateralization is widespread across animal species, it follows that there must be situations in which having a lateralized brain confers an advantage and enhances survival. Logically, this should apply to situations in which the animal has to use both hemispheres in parallel to carry out different functions.

This was shown first in domestic chicks by presenting a silhouette of a predator approaching the chick from its left or right side while the chick was pecking, with focused attention, at a patch of grain and mealworms [41]. Two groups of chicks were tested with both eyes seeing: one group with lateralized visual behaviour, ensured by exposing the eggs to light before hatching [9,42], and the other group lacking visual lateralization, achieved by incubating the eggs in the dark [43]. As discussed above, light-exposed chicks are lateralized for use of the left hemisphere (right eye) in searching for food and for use of the right hemisphere (left eye) to respond to predators. Latency to detect the predator was scored as the time between presentation of the 'predator' and the time when the chick stopped pecking, giving a startle call and, usually, twitching its head. In lateralized chicks, the latency was shorter when the predator's image approached on the chick's left side than when it did so on the right side. The latency of non-lateralized chicks was the same on both sides and not different from the latency of the lateralized chicks on their poorer, right side. The longer time taken by the non-lateralized chicks to detect the predator was not due to reduced levels of fear. In fact, after catching sight of the predator, the non-lateralized chicks were more disturbed by its presence than were the lateralized chicks, as shown by the fact that they produced more distress calls, and continued to do so even after the predator was no longer present [41].

To investigate this difference between lateralized and non-lateralized chicks further, they were tested on a dual task requiring search for grains against a distracting background of pebbles, and at the same time they were presented with a silhouette of a predator moving overhead. Chicks with lateralization of visual function performed both aspects of this task better than chicks lacking lateralization [44]. The lateralized chicks learnt to find grain scattered amongst pebbles, whereas the non-lateralized ones were unable to do so, and the lateralized chicks detected the predator sooner than the non-lateralized chicks. Once they had detected the predator, the non-lateralized chicks were more disturbed by it, as shown by distress calling and being less able to ignore it in order to continue pecking for food [20,44]. Similar results were found also when the lateralized and non-lateralized chicks were tested in groups [45]. Clearly, the lateralized chicks had the capacity to detect the predator while feeding and then to monitor it as they continued to feed. They achieved this increased cognitive capacity by using the different processing abilities of each hemisphere simultaneously; the left hemisphere to discriminate grains from pebbles and the right hemisphere to detect and respond to the predator.

Similar results have been found in both fish and a primate species tested on dual tasks. Topminnow fish, *Girardinus falcatus*, had to feed on shrimps in the presence of an on-looking predatory fish. Topminnow fish with stronger lateralization, assessed by turning bias in a runway, were faster at catching the shrimps than were the fish with no lateralization [46]. Similarly, lateralized female topminnows are able to find food efficiently while avoiding a male attempting to mate with them, whereas non-lateralized females are less able to do so [47].

In marmosets, *Callithrix jacchus,* strength of hand preference for simple reaching was used as an indication of the degree of lateralized *use* of the hemispheres (i.e., not the pattern of brain lateralization *per se* but its expression in motor behaviour). The marmosets were tested on a dual task in which a model predator was introduced to the testing room when the marmoset was performing a discrimination search-task for a favourite food, mealworms [48]. For two types of model predator, a stuffed bird moved overhead or a snake-like model moved on the floor below the marmoset, there was a significant, negative correlation between strength of hand preference and latency to detect the predator. Marmosets with stronger hand preferences detected the presence of the model predator sooner than did marmosets with weaker hand preferences [48]. No difference in latency to detect the predator was found when the marmosets had to perform only one aspect of the task: viz., detection of the predator when they were not feeding at the same time. Hence, the relationship between strength of hand preference and latency to detect the predator emerged only when increased cognitive capacity was needed in the dual task.

A cognitive advantage of being lateralized has even been shown in the invertebrate, larval antlion, *Myrmeleon bore* [49]. Compared to antlions without side-biases, antlions that have significant side-biases in the righting response have enhanced ability to learn to associate a vibrational cue with disappearance of prey. Although motor laterality may not be an accurate measure of brain lateralization, as I have discussed previously for vertebrate species, it is also worth noting a study showing that desert locusts (*Schistocerca gregaria*) with stronger lateralization of forelimb use to reach across a gap perform fewer errors of reaching than do locusts with weaker lateralization [50].

These examples provide evidence that lateralization of the brain increases cognitive capacity, in the sense that it increases the brain's ability to handle more information at any given time. Apparently, this works only for temporarily paired stimuli demanding simultaneous use of the separate specialisations of the hemispheres. If both stimuli needed to be processed within the same hemisphere, interference may occur and cognitive capacity would be reduced. Indeed, when tested in the dual task of pecking at grain while a predator was moved overhead (discussed previously), the behaviour of chicks without visual lateralization indicated that they became increasingly confused, or disturbed, by the dual task since their ability to find food grains scattered among the pebbles deteriorated as the task continued [44]. It appears that the chicks' ability to function well in the dual task was compromised by an inability to separate the required functions into different hemispheres, as Gotts et al. [2] found in a study of humans.

Since animals in the natural environment must be vigilant for predators at the same time as they are feeding, these findings indicate enhanced survival of lateralized animals. A study of larval coral reef fish, *Acanthurus triostegus*, supports this; when exposed to a predatory fish, survival was highest in those larvae that used their left eye to monitor the predator compared to larvae with no eye preference or right eye preference [51].

There is more evidence showing that lateralization influences performance in ways that might be advantageous depending on the context. Cichlid fish, *Geophagus brasiliensis*, for example, are said to be bolder if they are lateralized, as indicated by shorter latency to emerge into an unfamiliar environment [52], and lateralized fish show shorter latency to escape when stimulated by dropping a cylinder into their holding tank [53]. Hence, lateralized fish are not only more exploratory than are non-lateralized fish but they are also faster to respond to danger. Another study on a fish, the convict cichlid, *Archocentrus nigrofasciatus*, showed an association between aggression and strength of laterality: aggression was higher in more strongly lateralized males but the reverse was so in females [54].

Other evidence shows the advantage of having a lateralized brain. Pigeons with stronger laterality, measured as strength of eye preference, learn to find grains among pebbles better than non-lateralized pigeons and there is a linear association between these two factors [55]. Sailfish with stronger laterality for attacking prey are more successful in prey capture [56]. Budgerigars with stronger side bias in preening display enhanced discrimination performance [57]. The latter result has some similarity to that found in cats: testing cats on novel tasks requiring them to open a lid and reach inside to obtain a reward, Isparta et al. [58] found that those using a single preferred paw were better at solving the problems.

Degree of laterality is also associated with behaviour in tasks other than those directly testing cognitive ability. For example, in several tests, including tonic immobility and time to emerge from a box, lateralized chicks are found to be less fearful than non-lateralized chicks [59] and dogs exposed to the sounds of a thunderstorm are less reactive if they have a stronger paw-preference [60]. Across species of parrots, those species with stronger foot preferences for holding food are better able to discriminate seed from pebbles and they also perform better in a string-pulling task [61]. Note that this measure of performance, as a score comparing species, differs from the other studies discussed so far, which have compared variation in laterality within a species. Furthermore, parrot species with stronger foot preferences have larger brains [62], which may be the reason for their better cognitive performance rather than having a lateralized brain per se, although both factors could contribute to better cognitive ability.

Strength of lateralization also affects other cognitive abilities. Lateralized fish, for example, have better numerical abilities than non-lateralized fish [63,64]. As another example, lateralized chicks tested on the dual task mentioned above retained a memory of discrimination between pebbles and grain on the next day, whereas the non-lateralized chicks retained no memory of the task [44]. However, the inability of non-lateralized chicks to remember may have resulted from inability to attend to the pebble-grain task in the presence of the predator (on the previous day of testing) rather than being a direct association between lateralization and memory formation.

In summary, all of these studies show the advantage of being lateralized in such a way that one hemisphere can take charge of performing a particular task while, at the same time, the other hemisphere takes charge of performing another task. This is evidence that lateralization increases cognitive capacity.

3. Tasks Performed Better When Lateralization Is Weak or Absent

Since not all individuals within a species have the same strength of brain lateralization, there may be some contexts in which being less lateralized is an advantage. In contrast to the study on cichlid fish, which reported greater boldness in lateralized fish (see above), Brown and Bilbost [65] found, in rainbow fish, *Melanotaenia nigrans,* that non-lateralized fish are bolder than lateralized fish. As the researchers suggest, boldness is influenced by past experience with predators and may be modulated by fear. Consistent with this, reactivity to stress co-varies with laterality. For example, lateralized sharks react more strongly to stress than do non-lateralized sharks [66] and the same applies to the reaction of lambs to stress [67]. In fact, from research on humans, it seems likely that stress alters interhemispheric integration [68], thereby altering strength of lateralization and cognitive capacity.

These findings indicate that the increased cognitive capacity gained by having a lateralized brain may be associated with heightened stress responses, depending on context. In fact, in the dual task on which lateralized chicks performed better than non-lateralized chicks, it was the non-lateralized ones that were more distressed. Whether this distress translates to higher levels of physiological stress has yet to be determined.

As already mentioned, attack behaviour is a function of the right hemisphere [9]. Nevertheless, as shown in deer, non-lateralized individuals are more likely to engage in successful fights with conspecifics [69]. Additionally, testing damselfish, *Pomacentrus amboinesis,* Chivers et al. [70] found that lateralized individuals were less likely than non-lateralized ones to attack conspecifics when competing for shelter, even though they showed stronger responses to a predator. This result led the researchers to consider that there are costs and benefits of being lateralized. When an animal must attend equally to both sides, they suggest, it would be a disadvantage to be lateralized. In fact, as Dadda et al. [71] showed, in topminnow fish, non-lateralized individuals have an advantage over lateralized individuals in tasks requiring attention to both sides of their body, and hence, requiring the same use of both hemispheres.

4. Balance between Being Lateralized or Not Lateralized

Depending on the type of task and its cognitive demands, performance may be better in individuals with no lateralization or, conversely, better in individuals with lateralization. Overall, however, across and within species, lateralization is more common than non-lateralization. Nevertheless, very strongly lateralized individuals may be at a disadvantage. As an example, pheasants with strong foot-preference have a lower rate of survival than pheasants with a weaker strength of foot-preference [72]. In humans, stronger lateralization provides advantages in some but not all tasks [73]. This illustrates the important point that lateralization is largely specific for each different function. While it may be advantageous for some functions, it may confer no advantage or even a disadvantage for other functions.

As Corballis [74] suggests, there may be a trade-off between symmetry and asymmetry of function, but where the balance point lies depends on the behaviour considered (for research on this issue in humans see [75]), the species, sex, stress levels and possibly other factors, as well as genetic. Within a population the strength of bias is maintained as an evolutionary stable strategy [6,21].

This raises a different question: where does the balance between lateralization and non-lateralization lie within any group of animals? Using game theory analysis of a predator–prey model, Ghirlanda and Vallortigara [76] arrived at the conclusion that most but not all individuals in a group or population are lateralized (see also [77] for a similar result using the analysis of a competition-coordination model). As predicted by game theory, and found in studies of animal populations, the proportion of lateralized individuals in a species ranges from 65 to 90% and such biases in populations are stable, meaning the natural selection restores the proportion of left versus right biased individuals whenever there are slight deviations from the species-typical equilibrium point [78]. Although there are examples of published data in which the group bias is greater than 90% (e.g., footedness in some species of cockatoo [61,62]), the sample size tested needs to be considered.

5. Population Versus Individual Lateralization

The increased cognitive capacity of brains that carry out different computational or neural processes on each side could be achieved regardless of the direction of the laterality. Despite this, most examples of lateralization discussed so far in this paper are directional, meaning that the direction of the laterality is the same in the majority of individuals in the group or species. In other words, lateralization is not only present at the individual level but also at the level of the population.

There may be ontogenetic reasons for this situation. For example, in the final stages of incubation before hatching, the chick embryo is oriented within the egg so that its right eye is next to the shell and the left eye is occluded by the chick's body. This posture determines the direction of structural differences in developing visual pathways as a consequence of light stimulation of the right eye only [79]. Hence, light exposure at this critical stage of development leads to a population bias for asymmetry of visual behaviour [9,42].

Whatever the reason for individuals having the same direction of asymmetry, because it is widespread across species, the advantage that it confers must over-ride any potential disadvantages. Population lateralization to detect and respond more readily to predators on the left seems to be disadvantageous since predators are just as likely to approach on the right or the left, unless the predators themselves have population-level lateralization that predisposes them to approach prey from behind and capture them on the predator's right side. There are examples of such right-side bias in predatory response: the cane toad, for example, strikes at prey once the prey has moved into the toad's right visual field, whereas prey items are ignored when they are in the toad's left visual field [11]. A similar result has been found in the music frog [80]. Such preferential use of the right eye in feeding, or predation, originally shown in chicks [35], has also been reported to occur in humpback whales [81] and blue whales [82]. A right-side preference for prey capture by wild stilts has also been reported [83]. Other predators, however, may attack prey to their left or right (individual bias but no population bias), as found to be the case in sailfish [56].

Interestingly, although individual sailfish showed more success in prey capture on their preferred side and those with stronger laterality were more successful, the population showed no side-difference in success of capturing prey, which also implies that the prey were not better at escaping when attacked from their left side, although this has been found in some tests of amphibians [15], a species of marsupial [17] and another species of fish [84].

If population-level lateralization, also called directional lateralization, does have certain disadvantages, these must be less important than the advantages which it bestows. From the evidence discussed so far, it seems that the advantage of population-level lateralization must have something to do with social behaviour because it is in social interactions that it is manifested, as discussed next.

6. Social Cognition

The first evidence that directional lateralization is associated with social behaviour came from the study of social hierarchies in groups of chicks with lateralized brains for visual function compared to groups of chicks without this lateralization [85]. Quite rapidly, groups of chicks with population-level lateralization established stable social hierarchies, measured by scoring access to a limited food source, whereas those without laterality failed to form stable hierarchies. In fish also, being lateralized at the population level is associated with social group formation and maintenance. Bisazza et al. [86] tested 16 species of fish, some known to form shoals and others not so. They measured turning behaviour of each fish individually and determined whether each species had individual lateralization or directional lateralization. All of the species that displayed shoaling behaviour were lateralized at the population level, whereas this occurred in less than half of the species that did not shoal [86]. Clearly, a shoal is maintained if individual fish turn together in the same direction. The directional lateralization is essential for this particular aspect of social behaviour.

With particular relevance to lateralized chicks forming stable social hierarchies, young chicks have a quality called transitive inference, a cognitive ability that permits a chick to infer the social rank of another chick by observing it rather than interacting directly with it [87]. Using transitive inference, the animal is able to predict its position in the social group and thereby avoid fights and, as found by Daisley et al. [88], they can make such inferences when they use their left eye but not when they use their right eye. These examples demonstrate that the right hemisphere is specialised for functions essential to social behaviour.

The ability recognise familiar from unfamiliar conspecifics is a function of the left eye and right hemisphere, as shown in chicks [23,89] and fish [90]. Chicks have been shown to respond consistently to social signals when using their left eye [87,88] and, provided they are using their left eye, they can learn to avoid pecking a distasteful bead by observation of the behaviour of a conspecific [91]. These are aspects of social cognition and they involve specialisations of the right hemisphere.

Specialisation of the right hemisphere to assess and respond in social situations is seen also in mammals. Sheep can recognise faces of other sheep using the right but not the left hemisphere [92] and they also respond to the emotional expression of those faces using the right hemisphere [93]. Considered from the other perspective, the right hemisphere has a dominant role in producing expressions of fear, as shown in rhesus monkeys [94] and marmosets [95]. Moreover, in a wide range of mammalian species, including bats, walruses, whales, dolphins, horses, kangaroos, sheep, deer and bison, maternal animals position their offspring on their left side [96–98], meaning that the maternal animal uses her right hemisphere to monitor the offspring. A similar left-side/right-hemisphere preference has been shown in gorillas and chimpanzees for monitoring conspecifics [99]. These examples show that directional bias is a widespread characteristic of social behaviour.

One way to investigate the association between directional lateralization and social behaviour is to compare laterality in social and non-social species. Comparison of different lateralities in several species of bee has provided some information on this topic. The hypothesis tested was that social species should show directional laterality, whereas asocial species should not, although they may show individual laterality. This line of research began with the discovery of directional lateralization in the honeybee, *Apis mellifera*, a highly social species [100]. It has long been known that honeybees can be conditioned to associate a specific odour (e.g., lemon or vanilla) with a food reward (sugar solution). They detect the odour with receptors on their antennae and respond by extending the proboscis. Letzkus et al. [100] discovered that honeybees can learn this task when they use their right antenna but not when they are forced to use their left antenna. Not surprisingly, they can recall memory of the task when the odour is presented to the right, but not the left, antenna, at least, in the short-term, up to an hour after learning [101]. When tested for recall of the memory 6 h or more after learning, they can do so using their left but not their right antenna [101]. Hence, short-term and long-term memory is laid down in different sites, and likely on different sides, of the brain and accessed by the right or left antenna, respectively. These are directional lateralities of olfactory learning and memory recall.

The population bias for olfactory learning is also present in three species of Australian stingless bees [102], all of which are social species, but not in the largely asocial mason bee, *Osmia sp.* [103]. However, what about asymmetry in mason bees when they do happen to interact socially? Discovery of directional lateralization of agonistic responses in honeybees [104] suggested a way to test this since mason bees do interact with each other when they fight. In aggressive encounters, which are particularly intense between females, mason bees do show directional lateralization: both males and females interact more aggressively when they use the left antenna but not when they use the right antenna [105]. Hence, aggressive interaction, which is an aspect of social behaviour, is directionally lateralized. This does not imply anything about a role of lateralization in an individual's likelihood of success in aggressive interactions, examples of which are discussed below.

This finding supports the hypothesis that population or directional lateralization evolves, or develops, in social interactions and it is, of course, evident in a wider range of behaviour in social species than it is in species that are largely solitary. Even social species may show individual but not population lateralization in behaviour that does not involve social interaction. As an example, Ong et al. [106] found that this is so for direction choice when honeybees are flying through holes in a barrier and, since the bees were tested without other bees being nearby, they were not interacting socially while they made a choice. Individual bees were found to have lateral preferences but about half preferred the right side and the other half preferred the left side. Overall, therefore, population-level asymmetry is present only in social behaviour.

During evolution, social interaction could have selected for alignment of laterality in most individuals. Could this lead to any increase in cognitive capacity? It has long been hypothesised that social interaction contributes to the evolution of increased brain size and cognitive capacity in primates [107] and this hypothesis is supported by evidence that neocortical size correlates with social group size in primates [108] and also in insectivores and carnivores [109]. It is not obvious, however, that aligning the direction of laterality (i.e., directional lateralization) in social animals provides any further increase in cognitive capacity than already gained by having individual lateralization.

In fact, it seems that individual-level lateralization almost certainly evolved first and there was selection for it because it enhanced cognitive ability. Then, as sociality evolved, directional lateralization did so along with it, not because it further enhanced cognitive ability but because it conferred an advantage in social interactions.

Returning to the original research addressing this question, we can see one reason for aligning laterality within a population: in groups of chicks, it pays to be lateralized at the population level because it stabilises the social hierarchy, as discussed above [85]. It also maintains the coherency of a shoal of fish, as also mentioned above [86], and as reported recently, more strongly lateralized fish (measured as eye preference) are more likely to be in the safest, less exposed position in the shoal [110]. In many mammalian species left-side preference for positioning of the young next to the mother, presumably, ensures optimum interaction between mother and young [96,97]. These are social advantages but not necessarily cognitive advantages. In other words, there can be social advantages for aligning direction of laterality [78], without evidence of any cognitive advantage.

In some species directional laterality increases success in aggressive interactions (e.g., in flies [111]) and in other species it decreases the likelihood of success (in deer [69]). As another example of the latter, Schnell et al. [112] found that the majority of giant Australian cuttlefish have a preference to use their left eye during escalated fighting but the minority with a right-eye preference have more success in the outcomes of these fights. However, in mating, most male cuttlefish use their left eye as they approach the female on her right side and it is these males that achieve higher mating success [112], which illustrates that there must be a selective balance between advantage in one behaviour and disadvantage in another. In sage-grouse also, males lateralized with a left-eye preference in aggressive encounters are more successful in mating [113].

Again, these are examples of social advantage of population lateralization but not of superior cognitive capacity. Nevertheless, success in aggressive interactions depends on tactical decisions, and the same may apply to success in mating, and this could reflect higher cognitive capacity. Whether or not this is so must depend on the degree of sensory processing needed to make these decisions and we do not yet know this.

7. Interaction between the Hemispheres

So far, I have discussed lateralization revealed by testing animals monocularly and followed the general interpretation, common to most papers in this area of research, that behaviour performed when using the right eye reflects specialisations of the left hemisphere and vice versa. Although this is largely correct, the situation is a little more complex than that.

In species with small binocular fields, visual inputs from one eye are, indeed, processed primarily in the contralateral hemisphere. However, this is does not mean the contralateral hemisphere is used exclusively, although that is often assumed to be the case because it is the parsimonious way to interpret results from monocular tests. This interpretation ignores the fact that some visual inputs go to the ipsilateral hemisphere too, even though these inputs are much less than the visual inputs to the contralateral hemisphere. Additionally, brains have interhemispheric commissures and this is the case in almost all vertebrate species, even though these commissures are considerably smaller than the major interhemispheric commissure, the corpus callosum, in humans [114,115].

The avian brain, for example, has an interhemispheric commissure, the anterior commissure, and two commissures in the thalamus, the tectal and posterior commissures. In addition, each eye sends at least some inputs that recross the midline (decussate) to the ipsilateral hemisphere. Although we tend to interpret the behavioural differences expressed when the left eye or the right eye is used as reflecting the specialisation of the right or left hemisphere, respectively, in each case the other hemisphere is not without some input and, potentially, has some role in processing sensory information, albeit less than the hemisphere contralateral to the seeing eye. It follows, therefore, that lateralization not only enhances the brain's cognitive capacity by allocating different processing of information to each hemisphere but also by utilising at the same time, and to some degree, the differing cognitive abilities of each hemisphere. This means that, even though one hemisphere may take a dominant role in processing and controlling response, the other hemisphere is not "silent" and may provide some important contrasting analysis of inputs.

Recent research in Xiao and Güntürkün [116] has studied the role of the anterior commissure in the pigeon tested on a colour-discrimination task. This commissure connects the arcopallial regions of the left and right hemispheres. The researchers blocked the activity of the arcopallium of one hemisphere temporarily and recorded the activity patterns of the neurons in the acrcopallium of the other hemisphere while the bird performed the colour discrimination task. A clear asymmetric effect was found: blocking the activity of the left arcopallium had a greater effect on recordings obtained from the right acropallial neurones than vice versa. Hence, interhemispheric transfer via the anterior commissure is greater from the left to the right than from the right to the left. As Xiao and Güntürkün [116] argue, the neurones in this commissure are largely excitatory and, hence, the left side should step up activity of the right side, thereby acting as a balance against the dominant role of the left hemisphere in colour discrimination. To understand this more completely, it is necessary to know that, in pigeons, the left hemisphere receives visual inputs from both eyes (the tectofugal visual system), whereas the right hemisphere receives inputs mostly from the left eye [117,118]. In domestic chicks, it is the other visual system (the thalamofugal system) that is lateralized: in this visual system, the Wulst region of the right hemisphere receives strong inputs from both eyes, whereas the left Wulst receives input mainly from the right eye [79]. A recent study [119] has shown lateralized integration of visual information in the Wulst regions of the chick brain.

Returning to the study on pigeons [116], although the left hemisphere of the pigeon has a dominant role in colour discrimination, the anterior commissure may provide some tempering of this asymmetry before the animal makes a response. Further research is needed to confirm this interpretation.

The tectal commissure, connecting the optic tectum on one side to its counterpart on the other side, and the posterior commissure may have similar tempering roles on the input of visual information, at least in the chick. When the tectal and posterior commissures of the chick are sectioned, asymmetry of behaviour emerges [120]. Presented with a small red bead, chicks with the commissures sectioned pecked much more when they were using their right eye than when they were using their left eye. In fact, when using their right eye, the chicks pecked more and more each time the red bead was presented to them. No such asymmetry was seen in sham operated chicks or unoperated controls. Hence, without the commissures the brain functions asymmetrically and with the commissures this asymmetry is not seen, presumably because the left hemisphere suppresses the right via the tectal or posterior commissure. That release from control and expression of asymmetry occurs also for attack and copulation behaviour in chicks. Both attack and copulation responses are controlled by the right hemisphere and released from inhibition by the left hemisphere following treatment of the left hemisphere with cycloheximide [9] or when the right, but not the left, eye is occluded [121].

These studies suggest that, although the hemispheres have different ways of processing information and, indeed, each handles a different package of information, the brain can make some adjustments for these differences via the commissures. These adjustments may go some of the way in explaining why individuals vary in the strength of lateralization. It could be these commissures which set the balance discussed above in Section 4.

We know that this balance changes during early development [122] and it is altered by stress [66–68,123,124]. More research will, hopefully, elucidate the mechanism(s) by which one hemisphere can assume dominant control of a particular behaviour (strong lateralization) or, as Reddon and Hurd [125] explain it, there is a "consensus of action of the two hemispheres" (weak lateralization). Reddon and Hurd [125] suggest an alternative route by which one hemisphere may override the other and so assume dominance, or fail to do so, as the case may be, and that route involves the habenula nuclei of the epithalamus.

The point relevant to discussion of cognitive capacity is that different sensory inputs to each side of the brain, followed by different types of neural processing are mechanisms of achieving increased cognitive capacity, and interhemispheric and other commissures play a role in balancing or coordinating these left-right differences.

8. Conclusions

Cognitive capacity is increased when each hemisphere can be used independently, at the same time. Examples of this ability in different species have been discussed. Other research on rhesus macaques has led to the conclusion that the two hemispheres have independent capacities, which are limited by competition for sensory encoding rather than by a failure of memory formation or recall [126].

Strength of lateralization varies between individuals, as does cognitive capacity. This variation must depend on the task considered and a range of factors play a role, including developmental processes and the selective advantage/disadvantage of having a lateralized brain.

There is a need to investigate a broad range of lateralities within individuals to see what functions are associated in terms of lateralization and what functions are lateralized independently. I realise that this is being studied in humans [127–129] and there are some studies of it in non-human species, discussed in this paper, but more examples are needed. Furthermore, cognitive capacity related to laterality in sensory modalities other than vision needs to be investigated, as well as potential interaction between laterality in the different sensory modalities.

Whereas being lateralized confers cognitive advantage to individuals, aligning the direction of lateralization in the majority of individuals of a population, or species, seems unlikely to enhance cognitive capacity to any greater degree than does individual lateralization. Hence, selection for directional (population) lateralization is dependent on social interactions, and not enhancement of cognitive processing. While social behaviour might increase cognitive capacity of species via evolution of increased brain size, occurring with larger group size, as hypothesised by Dunbar [107–109], or by the possibility of group-consensus decisions [130], it does not do so by brain lateralization *per se*.

Funding: This research received no external funding.

Institutional Review Board Statement: Not applicable.

Data Availability Statement: No need.

Conflicts of Interest: The author declares no conflict of interest.

References

1. Corballis, M.C. Bilaterally Symmetrical: To be or not to be? *Symmetry* **2020**, *12*, 326. [CrossRef]
2. Gotts, S.J.; Jo, H.J.; Wallace, G.; Saad, Z.S.; Cox, R.; Martin, A. Two distinct forms of functional lateralization in the human brain. *Proc. Natl. Acad. Sci. USA* **2013**, *110*, 3435–3444. [CrossRef]
3. Ocklenburg, S.; Güntürkün, O. *The Lateralized Brain: The Neuroscience and Evolution of Hemispheric Asymmetries*; Academic Press: London, UK, 2018.
4. Frasnelli, E.; Vallortigara, G.; Rogers, L.J. Left–right asymmetries of behaviour and nervous system in invertebrates. *Neurosci. Biobehav. Rev.* **2012**, *36*, 1273–1291. [CrossRef] [PubMed]
5. Rogers, L.J.; Vallortigara, G.; Andrew, R.J. *Divided Brains: The Biology and Behaviour of Brain Asymmetries*; Cambridge University Press: Cambridge, UK, 2013.
6. Vallortigara, G.; Rogers, L.J. A function for the bicameral mind. *Cortex* **2020**, *124*, 274–285. [CrossRef] [PubMed]
7. Güntürkün, O.; Ströckens, F.; Ocklenburg, S. Brain Lateralization: A Comparative Perspective. *Physiol. Rev.* **2020**, *100*, 1019–1063. [CrossRef]
8. Andrew, R. Origins of asymmetry in the CNS. *Semin. Cell Dev. Biol.* **2009**, *20*, 485–490. [CrossRef]
9. Rogers, L.J. Light experience and asymmetry of brain function in chickens. *Nature* **1982**, *297*, 223–225. [CrossRef] [PubMed]
10. Robins, R.; Rogers, L.J. Complementary and lateralized forms of processing in *Bufo marinus* for novel and familiar prey. *Neurobiol. Learn. Mem.* **2006**, *86*, 214–227. [CrossRef]
11. Robins, A.; Rogers, L.J. Lateralized prey-catching responses in the cane toad, *Bufo marinus*: Analysis of complex visual stimuli. *Anim. Behav.* **2004**, *68*, 767–775. [CrossRef]
12. Robins, A.; Rogers, L.J. Lateralized visual and motor responses in the green tree frog, *Litoria caerulea*. *Anim. Behav.* **2006**, *72*, 843–852. [CrossRef]
13. Liu, Y.; Shen, J.; Fang, K.; Song, J.; Fan, Y.; Yang, J.; Shen, D.; Lu, F.; Fang, G. Behavior and approximate entropy of right-eye lateralization during predation in the music frog. *Asian Herpetol. Res.* **2020**, *11*, 115–123.
14. MacNeilage, P.F.; Rogers, L.J.; Vallortigara, G. Origins of the left and right brain. *Sci. Am.* **2009**, *301*, 60–67. [CrossRef]
15. Lippolis, G.; Bisazza, A.; Rogers, L.J.; Vallortigara, G. Lateralization of predator avoidance responses in three species of toads. *Laterality* **2002**, *7*, 163–183. [CrossRef]
16. Martín, J.; López, P.; Bonati, B.; Csermely, D. Lateralization when monitoring predators in the wild: A left eye control in the common wall lizard (*Podarcis muralis*). *Ethology* **2010**, *116*, 1226–1233. [CrossRef]
17. Lippolis, G.; Westman, W.; McAllan, B.M.; Rogers, L.J. Lateralization of escape responses in the striped-faced dunnart, *Sminthopsis macroura* (Dasyuridae: Marsupalia). *Laterality* **2005**, *10*, 457–470. [CrossRef] [PubMed]

18. Siniscalchi, M.; Sasso, R.; Pepe, A.M.; Vallortigara, G.; Quaranta, A. Dogs turn left to emotional stimuli. *Behav. Brain Res.* **2010**, *208*, 516–521. [CrossRef]
19. Siniscalchi, M.; Bertino, D.; D'Ingeo, S.; Quaranta, A. Relationship between motor laterality and aggressive behavior in sheepdogs. *Symmetry* **2019**, *11*, 233. [CrossRef]
20. Dharmaretnam, M.; Rogers, L. Hemispheric specialization and dual processing in strongly versus weakly lateralized chicks. *Behav. Brain Res.* **2005**, *162*, 62–70. [CrossRef]
21. Vallortigara, G.; Rogers, L.J. Survival with an asymmetrical brain: Advantages and disadvantages of cerebral lateralization. *Behav. Brain Sci.* **2005**, *28*, 575–589. [CrossRef]
22. Tommasi, L.; Vallortigara, G. Hemispheric processing of landmark information and geometric information in the domestic chick (*Gallus gallus*). *Behav. Brain Res.* **2000**, *112*, 119–126. [CrossRef]
23. Vallortigara, G.; Andrew, R. Differential involvement of right and left hemisphere in individual recognition in the domestic chick. *Behav. Process.* **1994**, *33*, 41–57. [CrossRef]
24. Giljov, A.; Karenina, K. Differential roles of the right and left brain hemispheres in the social interactions of a free-ranging ungulate. *Behav. Process.* **2019**, *168*. [CrossRef]
25. Karenina, K.; Giljov, A.; Malashichev, Y. Lateralization of mother-infant interactions in wild horses. *Behav. Process.* **2018**, *148*, 49–55. [CrossRef]
26. Bullock, S.P.; Rogers, L.J. Glutamate-induced asymmetry in the sexual and aggressive behavior of young chickens. *Pharmcol. Biochem. Behav.* **1985**, *24*, 549–554. [CrossRef]
27. Robins, A.; Lippolis, G.; Bisazza, A.; Vallortigara, G.; Rogers, L.J. Lateralized agonistic responses and hindlimb use in toads. *Anim. Behav.* **1998**, *56*, 875–881. [CrossRef] [PubMed]
28. Vallortigara, G.; Rogers, L.J.; Bisazza, A.; Lippolis, G.; Robins, A. Complementary right and left hemifield use for predatory and agonistic behaviour in toads. *Neuroreport* **1998**, *9*, 3341–3344. [CrossRef] [PubMed]
29. Deckel, A.W. Laterality of aggressive responses in *Anolis*. *J. Exp. Zoöl.* **1995**, *272*, 194–200. [CrossRef]
30. Stor, T.; Rebstock, G.A.; Borboroglu, P.G.; Boersma, P.D. Lateralization (handedness) in Magellanic penguins. *PeerJ* **2019**, *7*, e6936. [CrossRef] [PubMed]
31. Lemaire, B.S.; Viblanc, V.A.; Jozet-Alves, C. Sex-specific lateralization during aggressive interactions in breeding king penguins. *Ethology* **2019**, *125*, 439–449. [CrossRef]
32. Koboroff, A.; Kaplan, G.; Rogers, L.J. Hemispheric specialization in Australian magpies (*Gymnorhina tibicen*) shown as eye preferences during response to a predator. *Brain Res. Bull.* **2008**, *76*, 304–306. [CrossRef]
33. Austin, N.; Rogers, L. Limb preferences and lateralization of aggression, reactivity and vigilance in feral horses, *Equus caballus*. *Anim. Behav.* **2012**, *83*, 239–247. [CrossRef]
34. Casperd, J.M.; Dunbar, R. Asymmetries in the visual processing of emotional cues during agonistic interactions by gelada baboons. *Behav. Process.* **1996**, *37*, 57–65. [CrossRef]
35. Rogers, L.; Anson, J. Lateralisation of function in the chicken fore-brain. *Pharmacol. Biochem. Behav.* **1979**, *10*, 679–686. [CrossRef]
36. McCabe, B. Pharmacological agents and electrophysiological techniques. In *Lateralized Brain Functions: Methods in Human and Non-Human Species. Neuromethods*; Rogers, L.J., Vallortigara, G., Eds.; Humana Press, Springer-Nature: New York, NY, USA, 2017; Volume 122, pp. 251–276.
37. Nottebohm, F. Asymmetries in neural control of vocalization in the canary. In *Lateralization of the Nervous System*; Harnard, S., Doty, R.W., Goldstein, L., Jaynes, J., Krauthamer, G., Eds.; Academic Press: New York, NY, USA, 1977; pp. 23–44.
38. Denenberg, V.H. Hemispheric laterality in animals and the effects of early experience. *Behav. Brain Sci.* **1981**, *4*, 1–49. [CrossRef]
39. Mench, J.; Andrew, R.J. Lateralisation of a food search task in the domestic chick. *Behav. Neural Biol.* **1986**, *46*, 107–114. [CrossRef]
40. Vallortigara, G.; Andrew, R. Lateralization of response by chicks to change in a model partner. *Anim. Behav.* **1991**, *41*, 187–194. [CrossRef]
41. Rogers, L.J. Evolution of hemispheric specialisation: Advantages and disadvantages. *Brain Lang.* **2000**, *73*, 236–253. [CrossRef]
42. Rogers, L.J. Light input and the reversal of functional lateralization in the chicken brain. *Behav. Brain Res.* **1990**, *38*, 211–221. [CrossRef]
43. Rogers, L.J. Early experiential effects on laterality: Research on chicks has relevance to other species. *Laterality* **1997**, *2*, 199–219. [CrossRef] [PubMed]
44. Rogers, L.J.; Zucca, P.; Vallortigara, G. Advantages of having a lateralized brain. *Proc. Biol. Sci.* **2004**, *271* (Suppl. 6), 420–422. [CrossRef]
45. Wichman, A.; Freire, R.; Rogers, L.J. Light exposure during incubation and social and vigilance behaviour of domestic chicks. *Laterality* **2009**, *14*, 381–394. [CrossRef] [PubMed]
46. Dadda, M.; Bisazza, A. Does brain asymmetry allow efficient performance of simultaneous tasks? *Anim. Behav.* **2006**, *72*, 523–529. [CrossRef]
47. Dadda, M.; Bisazza, A. Lateralized female topminnows can forage and attend to a harassing male simultaneously. *Behav. Ecol.* **2006**, *17*, 358–363. [CrossRef]
48. Piddington, T.; Rogers, L.J. Strength of hand preference and dual task performance by common marmosets. *Anim. Cogn.* **2012**, *16*, 127–135. [CrossRef]

49. Miler, K.; Kuszewska, K.; Woyciechowski, M. Larval antlions with more pronounced behavioural asymmetry show enhanced cognitive skills. *Biol. Lett.* **2017**, *13*. [CrossRef]
50. Bell, A.T.A.; Niven, J.E. Strength of forelimb lateralization predicts motor errors in an insect. *Biol. Lett.* **2016**, *12*. [CrossRef]
51. Besson, M.; Gache, C.; Bertucci, F.; Brooker, R.M.; Roux, N.; Jacob, H.; Berthe, C.; Sovrano, V.A.; Dixson, D.L.; Lecchini, D. Exposure to agricultural pesticide impairs visual lateralization in a larval coral reef fish. *Sci. Rep.* **2017**, *7*. [CrossRef]
52. Reddon, A.R.; Hurd, P.L. Individual differences in cerebral lateralization are associated with shy–bold variation in the convict cichlid. *Anim. Behav.* **2009**, *77*, 189–193. [CrossRef]
53. Dadda, M.; Koolhaas, W.H.; Domenici, P. Behavioural asymmetry affects escape performance in a teleost fish. *Biol. Lett.* **2010**, *6*, 414–417. [CrossRef] [PubMed]
54. Reddon, A.; Hurd, P.L. Aggression, sex and individual differences in cerebral lateralization in a cichlid fish. *Biol. Lett.* **2008**, *4*, 338–340. [CrossRef]
55. Güntürkün, O.; Diekamp, B.; Manns, M.; Nottelmann, F.; Prior, H.; Schwarz, A.; Skiba, M. Asymmetry pays: Visual lateralization improves discrimination success in pigeons. *Curr. Biol.* **2000**, *10*, 1079–1081. [CrossRef]
56. Kurvers, R.H.; Krause, S.; Viblanc, P.E.; Herbert-Read, J.; Zaslansky, P.; Domenici, P.; Marras, S.; Steffensen, J.F.; Svendsen, M.B.; Wilson, A.; et al. The Evolution of Lateralization in Group Hunting Sailfish. *Curr. Biol.* **2017**, *27*, 521–526. [CrossRef] [PubMed]
57. D'Antonio-Bertagnolli, A.J.; Anderson, M.J. Lateral asymmetry in the freely occurring behaviour of budgerigars (*Melopsittacus undulatus*) and its relation to cognitive performance. *Laterality* **2017**, *23*, 344–363. [CrossRef] [PubMed]
58. Isparta, S.; Demirbas, Y.S.; Bars, Z.; Kul, B.C.; Güntürkün, O.; Ocklenburg, S.; Pereira, G.D.G. The relationship between problem-solving ability and laterality in cats. *Behav. Brain Res.* **2020**, *391*. [CrossRef]
59. Archer, G.S.; Mench, J.A. Exposing avian embryos to light affects post-hatch anti-predator fear responses. *Appl. Anim. Behav. Sci.* **2017**, *186*, 80–84. [CrossRef]
60. Branson, N.J.; Rogers, L.J. Relationship between paw preference strength and noise phobia in *Canis familiaris*. *J. Comp. Psychol.* **2006**, *120*, 176–183. [CrossRef]
61. Magat, M.; Brown, C. Laterality enhances cognition in Australian parrots. *Proc. R. Soc. B Biol. Sci.* **2009**, *276*, 4155–4162. [CrossRef]
62. Kaplan, G.; Rogers, L. Brain size associated with foot preferences in Australian parrots. *Symmetry* **2021**, *13*, 867. [CrossRef]
63. Dadda, M.; Agrillo, C.; Bisazza, A.; Brown, C. Laterality enhances numerical skills in the guppy, *Poecilia reticulata*. *Front. Behav. Neurosci.* **2015**, *9*. [CrossRef]
64. Gatto, E.; Agrillo, C.; Brown, C.; Dadda, M. Individual differences in numerical skills are influenced by brain lateralization in guppies (*Poecilia reticulata*). *Intelligence* **2019**, *74*, 12–17. [CrossRef]
65. Brown, C.; Bilbost, A.-L. Laterality is linked to personality in the black-lined rainbowfish, *Melanotaenia nigrans*. *Behav. Ecol. Sociobiol.* **2014**, *68*, 999–1005. [CrossRef]
66. Byrnes, E.E.; Pouca, C.V.; Brown, C. Laterality strength is linked to stress reactivity in Port Jackson sharks (*Heterodontus portusjacksoni*). *Behav. Brain Res.* **2016**, *305*, 239–246. [CrossRef]
67. Barnard, S.; Matthews, L.; Messori, S.; Podaliri-Vulpiani, M.; Ferri, N. Laterality as an indicator of emotional stress in ewes and lambs during a separation test. *Anim. Cogn.* **2015**, *19*, 207–214. [CrossRef]
68. Berretz, G.; Packheiser, J.; Wolf, O.T.; Ocklenburg, S. Dichotic listening performance and interhemispheric integration after stress exposure. *Sci. Rep.* **2020**, *10*. [CrossRef]
69. Jennings, D.J. Information gathering during contests: The relationship between lateralisation and contestant behaviour during fallow deer fights. *Behav. Process.* **2014**, *103*, 278–282. [CrossRef]
70. Chivers, D.P.; McCormick, M.I.; Warren, D.T.; Allan, B.J.M.; Ramasamy, R.A.; Arvizu, B.K.; Glue, M.; Ferrari, M.C.O. Competitive superiority versus predation savvy: The two sides of behavioural lateralization. *Anim. Behav.* **2017**, *130*, 9–15. [CrossRef]
71. Dadda, M.; Zandona, E.; Agrillo, C.; Bisazza, A. The costs of hemispheric specialization in a fish. *Proc. R. Soc. B Biol. Sci.* **2009**, *276*, 4399–4407. [CrossRef] [PubMed]
72. Whiteside, M.A.; Bess, M.M.; Frasnelli, E.; Beardsworth, C.; Langley, E.J.G.; Van Horik, J.O.; Madden, J.R. Low survival of strongly footed pheasants may explain constraints on lateralization. *Sci. Rep.* **2018**, *8*. [CrossRef] [PubMed]
73. Boles, D.B.; Barth, J.M.; Merrill, E.C. Asymmetry and performance: Toward a neurodevelopmental theory. *Brain Cogn.* **2008**, *66*, 124–139. [CrossRef]
74. Corballis, M.C. The Evolution of lateralized brain circuits. *Front. Psychol.* **2017**, *8*. [CrossRef]
75. Hirnstein, M.; Hugdahl, K.; Hausmann, M. How brain asymmetry relates to performance—A large-scale dichotic listening study. *Front. Psychol.* **2014**, *4*. [CrossRef]
76. Ghirlanda, S.; Vallortigara, G. The evolution of brain lateralization: A game-theoretical analysis of population structure. *Proc. Biol. Sci.* **2004**, *271*, 853–857. [CrossRef] [PubMed]
77. Ghirlanda, S.; Frasnelli, E.; Vallortigara, G. Intraspecific competition and coordination in the evolution of lateralization. *Philos. Trans. R. Soc. Lond. B Biol. Sci.* **2008**, *364*, 861–866. [CrossRef]
78. Vallortigara, G. The evolutionary psychology of left and right: Costs and benefits of lateralization. *Dev. Psychobiol.* **2006**, *48*, 418–427. [CrossRef] [PubMed]
79. Rogers, L.J.; Sink, H.S. Transient asymmetry in the projections of the rostral thalamus to the visual hyperstriatum of the chicken, and reversal of its direction by light exposure. *Exp. Brain Res.* **1988**, *70*, 378–384. [CrossRef]

80. Shen, J.; Fang, K.; Fan, Y.; Song, J.; Yang, J.; Shen, D.; Liu, Y.; Fang, G. Dynamics of electroencephalogram oscillations underlie right-eye preferences in predatory behavior of the music frogs. *J. Exp. Biol.* **2019**, *222*. [CrossRef]
81. Canning, C.; Crain, D.; Eaton, T.S.; Nuessly, K.; Friedlaender, A.; Hurst, T.; Parks, S.; Ware, C.; Wiley, D.; Weinrich, M. Population-level lateralized feeding behaviour in North Atlantic humpback whales, *Megaptera novaeangliae*. *Anim. Behav.* **2011**, *82*, 901–909. [CrossRef]
82. Friedlaender, A.S.; Herbert-Read, J.E.; Hazen, E.L.; Cade, D.E.; Calambokidis, J.; Southall, B.L.; Stimpert, A.K.; Goldbogen, J.A. Context-dependent lateralized feeding strategies in blue whales. *Curr. Biol.* **2017**, *27*, 1206–1208. [CrossRef] [PubMed]
83. Ventolini, N.; Ferrero, E.A.; Sponza, S.; Della Chiesa, A.; Zucca, P.; Vallortigara, G. Laterality in the wild: Preferential hemifield use during predatory and sexual behaviour in the black-winged stilt. *Anim. Behav.* **2005**, *69*, 1077–1084. [CrossRef]
84. Cantalupo, C.; Bisazza, A.; Vallortigara, G. Lateralization of predator-evasion response in a teleost fish (*Girardinus falcatus*). *Neuropsychologia* **1995**, *33*, 1637–1646. [CrossRef]
85. Rogers, L.; Workman, L. Light exposure during incubation affects competitive behaviour in domestic chicks. *Appl. Anim. Behav. Sci.* **1989**, *23*, 187–198. [CrossRef]
86. Bisazza, A.; Cantalupo, C.; Capocchiano, M.; Vallortigara, G. Population lateralisation and social behaviour: A study with 16 species of fish. *Laterality* **2000**, *5*, 269–284. [CrossRef]
87. Daisley, J.N.; Mascalzoni, E.; Salva, O.R.; Rugani, R.; Regolin, L. Lateralization of social cognition in the domestic chicken (*Gallus gallus*). *Philos. Trans. R. Soc. B: Biol. Sci.* **2008**, *364*, 965–981. [CrossRef] [PubMed]
88. Daisley, J.N.; Vallortigara, G.; Regolin, L. Logic in an asymmetrical (social) brain: Transitive inference in the young domestic chick. *Soc. Neurosci.* **2010**, *5*, 309–319. [CrossRef]
89. Deng, C.; Rogers, L.J. Prehatching visual experience and lateralization of the visual Wulst. *Behav. Brain Res.* **2002**, *134*, 375–385. [CrossRef]
90. Sovrano, V.; Andrew, R. Eye use during viewing a reflection: Behavioural lateralisation in zebrafish larvae. *Behav. Brain Res.* **2006**, *167*, 226–231. [CrossRef]
91. Salva, O.R.; Daisley, J.N.; Regolin, L.; Vallortigara, G. Lateralization of social learning in the domestic chick, *Gallus gallus domesticus*: Learning to avoid. *Anim. Behav.* **2009**, *78*, 847–856. [CrossRef]
92. Kendrick, K.M. Brain asymmetries for face recognition and emotion control in sheep. *Cortex* **2006**, *42*, 96–98. [CrossRef]
93. Tate, A.J.; Fischer, H.; Leigh, A.E.; Kendrick, K.M. Behavioural and neurophysiological evidence for face identity and face emotion processing in animals. *Philos. Trans. R. Soc. Lond. B Biol. Sci.* **2006**, *361*, 2155–2172. [CrossRef]
94. Hauser, M.D. Right hemisphere dominance for the production of facial expression in monkeys. *Science* **1993**, *261*, 475–477. [CrossRef] [PubMed]
95. Hook-Costigan, M.A.; Rogers, L.J. Lateralized use of the mouth in production of vocalizations by marmosets. *Neuropsychologia* **1998**, *36*, 1265–1273. [CrossRef]
96. Giljov, A.; Karenina, K.; Malashichev, Y. Facing each other: Mammal mothers and infants prefer the position favouring right hemisphere processing. *Biol. Lett.* **2018**, *14*. [CrossRef] [PubMed]
97. Karenina, K.; Giljov, A.; Ingram, J.; Rowntree, V.J.; Malashichev, Y. Lateralization of mother–infant interactions in a diverse range of mammal species. *Nat. Ecol. Evol.* **2017**, *1*. [CrossRef]
98. Karenina, K.; Giljov, A.; Ivkovich, T.; Burdin, A.; Malashichev, Y. Lateralization of spatial relationships between wild mother and infant orcas, *Orcinas orca*. *Anim. Behav.* **2013**, *86*, 1225–1231. [CrossRef]
99. Quaresmini, C.; Forrester, G.S.; Spiezio, C.; Vallortigara, G. Social environment elicits lateralized behaviors in gorillas (*Gorilla gorilla gorilla*) and chimpanzees (Pan troglodytes). *J. Comp. Psychol.* **2014**, *128*, 276–284. [CrossRef] [PubMed]
100. Letzkus, P.; Ribi, W.A.; Wood, J.T.; Zhu, H.; Zhang, S.; Srinivasan, M.V. Lateralization of Olfaction in the Honeybee *Apis mellifera*. *Curr. Biol.* **2006**, *16*, 1471–1476. [CrossRef] [PubMed]
101. Rogers, L.J.; Vallortigara, G. From Antenna to antenna: Lateral shift of olfactory memory recall by honeybees. *PLoS ONE* **2008**, *3*, e2340. [CrossRef] [PubMed]
102. Frasnelli, E.; Vallortigara, G.; Rogers, L.J. Origins of brain asymmetry: Lateralization of odour memory recall in primitive Australian stingless bees. *Behav. Brain Res.* **2011**, *224*, 121–127. [CrossRef] [PubMed]
103. Anfora, G.; Frasnelli, E.; Maccagnani, B.; Rogers, L.J.; Vallortigara, G. Behavioural and electrophysiological lateralization in a social (*Apis mellifera*) but not in a non-social (*Osmia cornuta*) species of bee. *Behav. Brain Res.* **2010**, *206*, 236–239. [CrossRef]
104. Rogers, L.J.; Rigosi, E.; Frasnelli, E.; Vallortigara, G. A right antenna for social behaviour in honeybees. *Sci. Rep.* **2013**, *3*. [CrossRef]
105. Rogers, L.J.; Frasnelli, E.; Versace, E. Lateralized antennal control of aggression and sex differences in red mason bees, *Osmia bicornis*. *Sci. Rep.* **2016**, *6*. [CrossRef]
106. Ong, M.; Bulmer, M.; Groening, J.; Srinivasan, M.V. Obstacle traversal and route choice in flying honeybees: Evidence for individual handedness. *PLoS ONE* **2017**, *12*, e0184343. [CrossRef] [PubMed]
107. Dunbar, R.I.M. The social brain hypothesis. *Evol. Anthropol.* **1998**, *6*, 178–190. [CrossRef]
108. Dunbar, R. Neocortex size and group size in primates: A test of the hypothesis. *J. Hum. Evol.* **1995**, *28*, 287–296. [CrossRef]
109. Dunbar, T.I.M.; Bever, J. Neocortex size determines group size in insectivores and carnivores. *Ethology* **1998**, *104*, 695–708. [CrossRef]
110. Middlemiss, K.L.; Cook, D.G.; Jaksons, P.; Jerrett, A.R.; Davison, W. Lateralisation of visual function in yellow-eyed mullet (*Aldrichetta forsteri*) and its role in schooling behaviour. *Mar. Freshw. Behav. Physiol.* **2018**, *51*, 15–29. [CrossRef]

111. Romano, D.; Canale, A.; Benelli, G. Do right-biased boxers do it better? Population-level asymmetry of aggressive displays enhances fighting success in blowflies. *Behav. Process.* **2015**, *113*, 159–162. [CrossRef]
112. Schnell, A.K.; Jozet-Alves, C.; Hall, K.C.; Radday, L.; Hanlon, R.T. Fighting and mating success in giant Australian cuttlefish is influenced by behavioural lateralization. *Proc. R. Soc. B: Biol. Sci.* **2019**, *286*. [CrossRef] [PubMed]
113. Krakauer, A.H.; Blundell, M.A.; Scanlan, T.N.; Wechsler, M.S.; McCloskey, E.A.; Yu, J.H.; Patricelli, G. Successfully mating male sage-grouse show greater laterality in courtship and aggressive interactions. *Anim. Behav.* **2016**, *111*, 261–267. [CrossRef]
114. Letzner, S.; Simon, A.; Güntürkün, O. Connectivity and neurochemistry of the commissura anterior of the pigeon (Columba livia). *J. Comp. Neurol.* **2015**, *524*, 343–361. [CrossRef]
115. Suárez, R.; Ao, L.R.; Fenlon, L.; Morcom, L.; Kozulin, P.; Kurniawan, N.D.; Richards, L.J. A pan-mammalian map of interhemispheric brain connections predates the evolution of the corpus callosum. *Proc. Natl. Acad. Sci. USA* **2018**, *115*, 9622–9627. [CrossRef]
116. Xiao, Q.; Güntürkün, O. The commissura anterior compensates asymmetries of visual representation in pigeons. *Laterality* **2021**, *26*, 213–237. [CrossRef] [PubMed]
117. Gùntùkùn, O.; Hellmann, B.; Melsbach, G.; Prior, H. Asymmetries of representation in the visual system of pigeons. *NeuroReport* **1998**, *9*, 4127–4130. [CrossRef] [PubMed]
118. Letzner, S.; Manns, M.; Güntürkün, O. Light-dependent development of the tectorotundal projection in pigeons. *Eur. J. Neurosci.* **2020**, *52*, 3561–3571. [CrossRef] [PubMed]
119. Costalunga, G.; Kobylkov, D.; Rosa-Salva, O.; Vallortigara, G.; Mayer, U. Light-incubation effects on lateralisation of single unit responses in the visual Wulst of domestic chicks. *Brain Struct. Funct.* **2021**, 1–17. [CrossRef]
120. Parsons, C.; Rogers, L. Role of the tectal and posterior commissures in lateralization of the avian brain. *Behav. Brain Res.* **1993**, *54*, 153–164. [CrossRef]
121. Rogers, L.J.; Zappia, J.V.; Bullock, S.P. Testosterone and eye-brain asymmetry for copulation in chickens. *Experientia* **1985**, *41*, 1447–1449. [CrossRef]
122. Andrew, R.J. Behavioural development and lateralization. In *Comparative Vertebrate Lateralization*; Rogers, L.J., Andrew, R.J., Eds.; Cambridge University Press: Cambridge, UK, 2002; pp. 157–205.
123. Archer, G.S.; Mench, J.A. Natural incubation patterns and the effects of exposing eggs to light at various times during incubation on post-hatch fear and stress responses in broiler (meat) chickens. *Appl. Anim. Behav. Sci.* **2014**, *152*, 44–51. [CrossRef]
124. Brown, C.; Western, J.; Braithwaite, V.A. The influence of early experience on, and inheritance of, cerebral lateralization. *Anim. Behav.* **2007**, *74*, 231–238. [CrossRef]
125. Reddon, A.R.; Hurd, P.L. Acting unilaterally: Why do animals with strongly lateralized brains behave differently than those with weakly lateralized brains? *Biosci. Hypotheses* **2009**, *2*, 383–387. [CrossRef]
126. Buschman, T.J.; Siegel, M.; Roy, J.; Miller, E.K. Neural substrates of cognitive capacity limitations. *Proc. Natl. Acad. Sci. USA* **2011**, *108*, 11252–11255. [CrossRef] [PubMed]
127. Innes, B.; Burt, D.M.; Birch, Y.; Hausmann, M. A leftward bias however you look at it: Revisiting the emotional chimeric face task as a tool for measuring emotion lateralization. *Laterality* **2016**, *21*, 643–661. [CrossRef] [PubMed]
128. Karlsson, E.M.; Johnstone, L.; Carey, D.P. The depth and breadth of multiple perceptual asymmetries in right handers and non-right handers. *Laterality* **2019**, *24*, 707–739. [CrossRef]
129. Cai, Q.; Van der Haegen, L.; Brysbaert, M. Complementary hemispheric specialization for language production and visuospatial attention. *Proc. Natl. Acad. Sci. USA* **2013**, *110*, 322–330. [CrossRef] [PubMed]
130. Sasaki, T.; Pratt, S.C. Groups have a larger cognitive capacity than individuals. *Curr. Biol.* **2012**, *22*, 827–829. [CrossRef] [PubMed]

animals

MDPI

Review

A Review of the Model/Rival (M/R) Technique for Training Interspecies Communication and Its Use in Behavioral Research

Irene M. Pepperberg [1,2]

[1] Department of Psychology, Harvard University, Cambridge, MA 02138, USA; impepper@media.mit.edu
[2] The Alex Foundation, 30 Curry Circle, Swampscott, MA 01907, USA

Simple Summary: Comparing the cognitive capacities of nonhumans to those of humans can be quite difficult, particularly given that humans can be questioned directly (e.g., "How many?", "What color?") but that most nonhumans must be tested by various indirect means that might not demonstrate the full range of their capacities. A few nonhumans, however, have acquired some level of symbolic representation (e.g., labels for items such as physical objects and their attributes, for concepts and relations among these items and concepts, and for actions that can be carried out on or with these items), which allows for a limited form of interspecies communication that can be used for direct questioning. Why have so few nonhumans acquired this skill, and what are the advantages of having it? I describe a specific training procedure, the Model/Rival (M/R) protocol, that enabled several Grey parrots to learn some level of referential communication; I discuss the specific elements of such training that are both necessary and sufficient for successful acquisition and how lack of any of these elements can cause failure. I also describe some experiments that were facilitated by interspecies communication, and how acquisition of this ability might affect the extent to which nonhumans can process information.

Abstract: In this paper, I will review the Model/Rival (M/R) technique that has been used to establish interspecies communication with Grey parrots (*Psittacus erithacus*). I will describe the original format developed by Todt, the relationship to other forms of observational learning outlined by other researchers, and the adaptations that I devised. I will describe how my undergraduate trainers and I isolated the various components that constitute the technique and explain how each is necessary, but how only the combination of *all* components is sufficient for successful implementation—and how improper implementation can lead to failure. I will briefly summarize the results of proper implementation—including the importance of interspecies communication itself as a technique for studying animal cognition.

Keywords: animal–human communication; interspecies communication; vocal learning; animal cognition; model/rival training; symbolic reference; grey parrots

Citation: Pepperberg, I.M. A Review of the Model/Rival (M/R) Technique for Training Interspecies Communication and Its Use in Behavioral Research. *Animals* **2021**, *11*, 2479. https://doi.org/10.3390/ani11092479

Academic Editors: Lucia Regolin and Maria Loconsole

Received: 26 July 2021
Accepted: 18 August 2021
Published: 24 August 2021

1. Introduction

Determining the cognitive capacities of nonhumans is not a simple task; just the act of devoting a special issue to the topic makes the problem abundantly clear. A form of the Heisenberg uncertainty principle always exists when performing such studies—i.e., the process by which measuring occurs can affect the result. Moreover, given that most such experiments are comparative in nature and that human cognition is the standard against which nonhuman abilities are most often measured, the fact that humans and nonhumans are generally tested via very different techniques complicates the issue even further. For example: other than in studies with pre-verbal infants, humans are most often tested using their language abilities—by simply being asked pertinent questions—whereas nonhumans are mostly incapable of being tested in that manner. However, what about those nonhumans who actually can be questioned in the same manner as humans? Specifically, what about nonhumans that have acquired some level of symbolic representation (a term I use interchangeably with symbolic reference): that is, what about nonhumans who understand the semantic and pragmatic use of noniconic symbols—be they auditory/vocal, manual, or lexical—to stand for (but not be limited to) items such as physical objects and their attributes, various concepts, relations among these items and concepts, and actions that can be done to or with these items? Here I am utilizing Deacon's assertions about reference [1]; that reference is not present when a label or sign is simply *associated* with something (e.g., as is a green button that, when pressed in a specific situation, delivers various food items to relieve hunger, whereas a red one does not), but is present if the label actually *represents* something independent of context (e.g., "green" describes the color of grass, a certain wavelength of light, or the skin color of a famous fictional entity from a far-off planet). Thus, any speaker of English would instantly understand what is being communicated by the label "green" in the latter case—as would a nonhuman who has also acquire some level of symbolic reference.

Such nonhumans can be tested via the use of interspecies communication [2]. Interspecies communication (a) directly states the precise content of questions to be asked—animals need not determine the nature of a question through hundreds (if not thousands) of instances of trial-and-error learning, thus making the task efficient; (b) incorporates research showing that social animals may respond more readily and accurately within an ecologically valid social context; (c) allows straightforward data comparisons among species, including humans; (d) is an open, arbitrary, creative code with enormous signal variety, enabling an animal to respond in novel, possibly innovative ways that demonstrate greater competence than the required responses of operant paradigms, and allows researchers to examine the exact nature and extent of information an animal perceives; (e) allows rigorous testing that avoids expectation cuing: subjects can be made to choose responses from their entire repertoire rather than from a subset relevant only to a particular topic. In some instances, interspecies communication via symbolic reference may simply be a means to more facilely demonstrate nonhumans' inherent capacities; in other instances, however, it may also significantly affect their abilities to process information [3–5].

Although I will spend some time discussing how acquisition of symbolic representation may actually affect cognitive capacities (see Section 7), the main portion of this paper will be devoted to reviewing exactly how at least one species, the Grey parrot (*Psittacus erithacus*), was taught such a form of interspecies communication. I will review the various components that constitute the training technique, explain how each is necessary, and demonstrate how only the combination of *all* components is sufficient for successful implementation.

2. Choice of Subject Species

In the late 1960s and early 1970s, several projects were initiated in order to train nonhumans to communicate with humans. Their choices of species were based on the idea that subjects should, like the great apes, have a close phylogenetic relationship to humans (reviewed in [6]) or, like the cetaceans, at least have large brains relative to their body size (reviewed in [7]); the idea was that possibly latent abilities in their own communication systems would be stimulated or that their cognitive capacities would provide the bases for training to succeed. At the time, the idea of using a bird—a creature evolutionarily far removed from humans and with a brain the size of a shelled-walnut—was considered absurd. Nevertheless, considerable data existed on the vocal communication abilities of birds in general—with respect to usage and especially for parallels between avian vocal learning and young children's early speech acquisition—and on Grey parrots in particular (reviewed in [5]). Other papers commented upon the exceptional clarity of Grey parrots' reproduction of human speech [8,9], and still other studies documented the intelligence of this particular species (reviewed in [2,5]). Thus, I decided to obtain a Grey parrot and begin its training in a laboratory setting [2].

3. The First Model/Rival Technique and Social Modeling Theory—Some History

Despite the material discussed above, until my research birds such as Grey parrots were still consistently dismissed as mindless mimics [10], and early studies attempting to establish interspecies communication with mimetic birds, having had negligible success, did little to alter that view (reviewed in detail in [11]). Such projects, using operant conditioning and nonreferential food rewards [12,13], generally confounded the label of the object or action being trained with that of the unrelated food reward, preventing the subject from making appropriate associations between the human labels being taught and their actual meaning [2,5,11]. With such training, all labels being taught simply became requests for food by a hungry subject—i.e., one that was usually kept below its free-feeding weight. Birds in these studies not only failed to learn referential speech but also failed to learn much speech at all. A subsequent study [8], primarily interested in examining antiphonal duetting in Grey parrots, instead used a modeling technique (a form of observational learning), in which one human was exclusively a cooperative partner (the trainer) of the parrot, while another human acted both as a model (M) for the bird's responses and as a rival (R) for the attention of the trainer (hence the term M/R training). The trainer interacted with both the parrot and human acting as a model, but the model never interacted with the parrot. The vocal patterns to be learned were not meaningful, although they gave the appearance of conversation (e.g., "What is your name?" "My name is Lora"). As a consequence of this procedure, each parrot engaged in verbal interaction solely with its particular trainer; learning of verbal patterns from other trainers was inhibited, as was the ability of the bird to transfer its responses to someone other than the principal trainer. However, birds in this program did acquire novel vocalizations rapidly (generally less than one month after introduction, usually within 3–6 days). The important difference between this study and previous ones was the introduction of social interaction—a not entirely surprising addition, given that communication is a social act and that in nature birds primarily learn their communication skills from other birds [8].

This study caused me to delve deeper into the various issues involved in determining how to establish referential communication and led to my examining research on social modeling theory and its applications [11,14,15]. Social modeling theory was the outcome of social psychologists' attempts to determine the underlying mechanisms of what they considered real-world learning (including observational learning), in contrast to the laboratory-based behavioristic, operant associationist paradigm common at the time [16]. Social modeling theorists proposed that "provision of models not only serves to accelerate the learning process, but also, in cases where errors are dangerous or costly, becomes an essential means of transmitting behavior patterns" [16], p. 54. In an attempt to devise procedures to enable humans to overcome strong inhibitions or phobias, they derived additional aspects of the theory; the idea was that, by determining how the toughest learning problems were resolved (e.g., showing how a subject could overcome a snake phobia by a combination of observing others handling snakes, watching someone else model the process of overcoming *their* initial hesitation about handling snakes, and then imitating the actions of this successful person), they could make some inferences about general learning mechanisms [17]. Overall, these researchers developed a theory that emphasized how attention, comprehension, and motivation affected learning. The theory thus consists of a set of principles that describe the optimal form of social input for any type of learning [14]. My understanding of the theory led me to believe that by modifying the original M/R protocol so as to incorporate aspects of a subset of four of these principles, I would succeed in enabling nonhumans—in my case, Grey parrots—to acquire some form of interspecies communication [2,11]. I describe each of these principles in turn, and then explain how I used them to revise the M/R procedure. (Note that in the following general discussion, the term "student" refers to any individual that is learning a task, whether human or nonhuman).

3.1. Principle 1

One principle states that the student's level of competence must be taken into account during all aspects of training. Researchers had already determined, for example, that human children most easily and most often acquire whatever is just slightly beyond their current abilities—whether it be new labels, actions, etc. (e.g., [15,18–21]). Specifically, with respect to modeling, interactions that model a new behavior that differs only slightly from an existing behavior or that encode only slightly unfamiliar information are most easily learned [11].

A corollary of this principle is that for learning to proceed efficiently, both tutor and model must constantly adjust their demonstration so that it will adapt to—and thus continue to challenge—a student's increasing knowledge (note [20]). Finding the appropriate level of input requires careful balancing. Input that is too simple may be ignored because the learner loses interest; input that is too advanced for the learner's level may similarly be ignored (e.g., as being irrelevant). As trainers interact with the student, they must therefore consistently determine the student's current level, and make upward or downward adjustments; the process is recursive and continuous, with all actors working in concert [11].

3.2. Principle 2

A second principle of this theory is that the modeling must help the student understand how new material relates to current problems and what advantage is conferred by learning the new material. Practically, training is therefore most effective when: (a) the student sees and then practices the targeted behavior under conditions similar to those in its regular environment, and (b) the appropriate use and consequence of the behavior are explicitly demonstrated (e.g., [14,17,22,23]). Thus, the trainers must provide some motivation for the subjects to acquire the requisite knowledge (note [20]). For example, subjects who are shown exactly how a novel utterance can be used to request a desired object are likely to learn more readily than if they were in situations without such demonstrations. This principle ties into functionality and contextual applicability, which will be discussed in detail below.

3.3. Principle 3

A third principle states that the effectiveness of the training is related to the intensity of the interaction between a student and the models. Here, intensity is viewed as the extent to which tutors arouse a response in a student [17] and, again, requires a careful balance between extremes [24]. Too little intensity, and the student will likely ignore the training. Moreover, overly nurturant models may inhibit learning by preventing students from experimenting or attempting to go beyond their current levels (note the tie-in to the principle concerning challenging a student intellectually). However, increasing the intensity of interaction may not always increase learning. Overly aggressive models may arouse fear or counter-aggression strong enough to block processing of any input.

3.4. Principle 4

The fourth principle states that the more that inhibition or resistance exists towards learning, the more important are the first three principles. The fourth principle is particularly relevant when the task involves acquiring what I call *exceptional* communication [25]—communication characterized by vocal learning that is unlikely to occur in the normal course of development, such as nonhuman acquisition of symbolic reference, particularly with respect to the likelihood of nonhuman acquisition of the more complex elements of a human system. The term "exceptional" may also imply some resistance toward acquiring the targeted behavior. Thus, for exceptional learning to occur, social modeling theory predicts that tutor/models be even more attuned to the student's level, that interactions be even more balanced with respect to intensity, and that demonstrations be even more explicit as to the real world uses and consequences of a targeted behavior than during normal learning [2,11]. This principle was especially crucial for understanding how I adapted social modeling theory to teach a Grey parrot to communicate with humans and to use this communication to examine his cognitive abilities.

3.5. Adapting the Principles to Training

Social modeling theory overall provides a framework for devising optimal training procedures—i.e., deciding upon optimal forms of input—but this framework must be put into practice. To wit, what factors characterize such optimal input? From the principles described above, optimal input should (a) correlate well with specific aspects of an individual's environment (i.e., be "referential" [26]), (b) have functional meaning relevant to the individual's environment (also known as "contextual applicability" or utility), and (c) be socially interactive. I describe these factors in some detail elsewhere (e.g., [11,27,28]); however, they bear re-examination here.

Reference—reference, for the most part, involves content; that is, what signals "are about" [26]. Reference, although generally defined as the direct relationship between a signal and an object or action, is not always easily determined. Signals, for example, may have multiple (if somewhat related) cultural meanings. Thus, "parrot" generally refers to one of a set of brightly colored birds (what [26] describes as an "external referent"), but may also refer to an action, as to "parrot", or mindlessly mimic, a behavior. Similarly, in nature an alarm call may refer to either the predator or the action that the emitter is about to take—or both. The more explicit the referent of a signal, however, the more easily the signal appears to be learned.

Functionality—functionality (also known as contextual applicability) involves the pragmatics of signal use: when a signal is to be used and the effects of using information in the signal [11,25]. Functionality is demonstrated by showing when using a signal is advantageous and the specific advantage gained by its use. Environmental context (e.g., the situation, intonation) also factors into the manner in which a signal functions. For example, a comment such as "My, don't we look nice today" said with one type of intonation will have one meaning and effect for a little girl in a party dress, but a different meaning and effect when said with a different intonation to a hungover friend [11]. Functionality also helps define reference; that is, context defines "parrot" as either a noun or a verb. As with reference, the more explicit a signal's functionality, the more readily the signal appears to be learned.

Social interaction—the three major functions of social interaction can be clarified by examples [11]. First, social interaction can highlight which of several environmental components should be noted; e.g., a subject can be directed to an object's shape to learn shape labels ("Look at the blocks. The *shape* of this one is *square*; the *shape* of that one is *round*"). Next, social interaction can emphasize common attributes—and thus possible underlying rules—of diverse actions (i.e., "*Give* me the *ball*" versus "*Give* me the *block*" versus "*Take* the *ball*"). Finally, social interaction is what allows input to be continuously adjusted to match the receiver's level ("Yes, you picked a ball in this pile of toys! Now can you find the *red* ball?"). Interaction may also provide a contextual explanation for an action and demonstrate its consequences ("There are lots of toys in the box! Tell me which one you want, and you can have it."). Interactive input of various forms thus facilitates learning.

In sum, reference and functionality concern how a label is used in the world, social interaction highlights various components of its use, and all three are necessary for meaningful learning [11]. Importantly, reference and functionality are distinct from what associative learning theory calls "reinforcement"—reinforcement is some general, positive outcome that is associated with an action (e.g., food received for hitting a button, whether the button-press in the task involves distinguishing colors, shapes, or grids), whereas reference and functionality are specific to a particular outcome out of many possibilities (e.g., "want red apple" specifies the action, a particular type of fruit and a particular varietal). I thus reasoned that, in order to teach a parrot to communicate with humans, my training procedure needed to take these factors into account. My most important insight, however, was the hypothesis that a parrot's acquisition of a human-based code was a form of exceptional learning. Given the earlier failures, I believed that despite these birds' abilities to reproduce all sorts of sounds, some strong inhibition existed towards learning to use allospecific sounds in a functional manner. I further believed that, to overcome this inhibition, training would have to be carefully adjusted to the parrot's abilities and include intense interactions and extremely clear demonstrations of reference and functionality [11]. Could I modify Todt's [8] technique appropriately? By using a form of observational learning, he had demonstrated the effectiveness of social interaction; would I succeed if I figured out how to incorporate referentiality and functionality, and extend and improve the level of social interaction?

4. The Second M/R Protocol

In honor of Todt's breakthrough, I continued to call the revised training protocol the M/R procedure. However, several modifications were made in this new design. The major changes are the demonstration of referential and contextual use of labels, and of corrective feedback among two human trainers and the parrot in order to demonstrate the targeted vocal behavior [2]. The parrot watches and listens as one trainer presents objects to, and queries the other trainer about, various items (e.g., "What's here?", "What color?")—items in which the bird has already demonstrated some interest. Because the reward for uttering "x" is x and only x (an *intrinsic* reward), and not an irrelevant item like a piece of food (an *extrinsic* reward), the bird thereby observes the closest possible connection between the object and the label to be learned. The trainer gives praise and transfers the named object to the human partner to reward correct answers. Incorrect responses are punished by scolding and temporarily removing items from sight. Thus, the second human is a model for the parrot's responses, and its rival for both the trainer's attention and acquisition of the item. The second human also illustrates consequences of errors. The model must try again or talk more clearly if the response was deliberately incorrect or garbled; that is, the model is subject to corrective feedback, which the bird observes. The parrot is included in these interactions, being queried and rewarded for successive approximations to correct responses; training is adjusted to its performance level. If a bird is inattentive or accuracy regresses, trainers threaten to leave. Note that the procedure also clearly demonstrates how the two trainers jointly attend to the object in question, another aspect of social interaction input thought to be important for acquisition of referential labels [29].

Unlike M/R procedures used by other researchers (see [30]), ours also exchanges roles of trainer and model to emphasize the importance of three-way interactions. The parrot thus sees how questioner and respondent inhabit both roles, and how the procedure causes environmental change (i.e., the transfer of the designated item). Role reversal also counteracts an earlier methodological problem [8]. Birds whose trainers always maintained their respective roles responded only to that human questioner. After several demonstrations and role reversals, the parrot itself is questioned—by both humans in their turn as trainer—and attempts at the label are rewarded with the object. The humans will then use these attempts themselves, with the person acting as trainer responding with phrases such as "That's close, say better!" or "Talk clearly", and giving the model another chance to respond appropriately. With our system, birds respond to, interact with, and learn from any human and, importantly, acquire the ability to ask questions themselves [11].

Although giving the parrot the object it had just labeled emphasized the referentiality and functionality of the label, this procedure also, of course, confounded identification of an object with the request for the object. We thus subsequently had to teach the birds (again through the M/R technique) to use "I want x", which enabled them to request unrelated, preferred rewards while learning labels for various items that they found of little or no inherent interest [31]. Specifically, the bird could identify object *y* with label "y", receive *y*, toss it, and then request *x* as the reward.

We maintained strict controls during training and testing (discussed in [5,13]), but outside of these formal sessions, we tried to provide as much vocal and social stimulation as possible, providing human interaction to substitute for that which this single bird would have received from his flock-mates in nature. He was allowed free access (contingent upon his vocal requests; e.g., "Wanna go gym") to all parts of the laboratory for the ~8 h/day that trainers were present; in fact, trials could occur at various locations. He was confined to a standard cage (~62 × 62 × 73 cm) only during sleeping hours. Water and a standard seed mix for psittacids (sunflower seeds, dried corn, kibble, oats, safflower, etc.) were available continuously; fresh fruits, vegetables, specialty nuts (cashews, almonds, pecans, walnuts) and toys (keys, pieces of wood, paper, rawhide, etc.) were provided at his vocal requests (e.g., "I want cork"). Once he learned how to question his trainers, he could also ask them for the labels of novel objects/colors/shapes in the laboratory (e.g., "What's here?", "What toy/matter/color/shape?").

With the help of this modified technique and these housing conditions, my most successful subject, Alex, acquired labels for over 100 different objects, seven colors, five shapes, exact numbers (up to eight), and categories; he acquired concepts of same/different, bigger/smaller, addition, numerosity, something akin to 'zero', and functional use of phrases such as "I want x" and "Wanna go y" where x and y were appropriate object and location labels (e.g., [4,11,32–34]). In a subsequent section (Section 7), I will discuss some of his capacities and those of other Grey parrots. For now, I concentrate on training issues, and one issue in particular: how the initial studies did not enable us to learn exactly why the M/R technique was so successful; we could not yet specify which training elements were necessary and sufficient. Thus, additional, different types of experiments, each of which carefully eliminated specific aspects of training, were needed to determine these criteria.

5. Experiments That Eliminate Aspects of M/R Training

These different experiments had to wait, for the most part, for the acquisition of additional subjects. The rationale was that, had I significantly changed Alex's training protocols, he might have ceased learning simply because I had made any changes, not because of the quality of the changes. New subjects, lacking a history of training, would not be influenced by such prior experiences. Hence, the introduction into the lab, at various times, of the (then juvenile) Grey parrots Kyo, Alo, Griffin, and Arthur. For these new subjects, I could provide input that varied with respect to various amounts of reference, functionality, and social interaction. I therefore contrasted sessions involving standard M/R tutoring on some labels with sessions on other labels that used audiotapes, videos, as well as variants of M/R and video input that involved different levels of human interaction. Moreover, to see what could happen if we did make changes in Alex's learning environment, I also exposed him to a variant of M/R training.

5.1. *Audiotape Instruction*

To test how learning might be affected if input totally lacked reference, context, and social interaction, I used two juvenile parrots, Alo and Kyaaro, contrasting their training via audiotapes with the standard M/R protocol. In the audiotape sessions, juveniles sat on a perch and listened to tapes while isolated from each other and the rest of the laboratory, and no objects were associated with the sounds presented over the speaker [35]. The two labels used for one bird in the audiotape condition were the two labels used for the other bird with the M/R procedure, to ensure that any difference in acquisition would not be a consequence of the form of the label. Notably, all four labels had been acquired by Alex, so I knew that they could be learned by a Grey parrot [11,35]. Both Kyaaro and Alo had already demonstrated that they found the objects to which these labels referred to be of considerable interest. To ensure that sessions paralleled M/R procedures, audiotapes consisted of the audio portion of a video of one of Alex's M/R sessions for each label. Although Alex already knew the labels being used in these videos [11], his M/R sessions were not structured as reviews, but replicated actual training. For him, however, the sessions *were* reviews and thus were not engaging, and he occasionally erred or interrupted with requests for other items and changes of location, which enabled us to demonstrate the same corrective feedback as usual [2,11]. As in regular live M/R sessions, trainers switched roles and occasionally deliberately erred. The audio was analyzed (Kay 5500 DSP Sona-Graph, Kay Electric Company, Lincoln Park, NJ, USA) to ensure that it was not degraded compared to that of Alex "live". Alo and Kyaaro each learned the two labels trained via the M/R procedure but neither label trained via audiotape. This study clearly showed that absence of all three aspects of input hindered acquisition, but not whether the presence of some subset or some intermediate amount of these types of input would be sufficient to engender learning.

5.2. M/R Variant 1

What would happen if we provided plenty of social interaction (more than in Todt's original study [8]), but eliminated reference and functionality? The experiment to test this with Alex was part of research on ordinality, counting, and serial learning [11]. We wanted to teach him a number line—that is, to recite a set of number labels in order of their value, in ways similar to the behavior of children in their early stages of numerical education [36]—but did not want this training to affect our other numerical studies. Thus, while Alex watched his trainers, one human asked a question, "Say number?" and the other uttered a string of foreign language number labels without reference either to specific objects in the laboratory or to Alex's previously acquired English number labels [11,35]. The set, *il ee bam ba oo yuk chil gal*, was derived from Korean count labels both to permit comparisons with children [36] and to be maximally different from English numbers he already knew. (That is, his number labels used to distinguish actual quantities (i.e., cardinal numbers), "one", "two", etc. were to be kept distinct from his number labels used to distinguish position in a series (i.e., ordinal numbers), in a system not identical with, but related to, humans' use of "first", "second", etc.). *Bam* (pronounced /baem/) and *ba* were substituted for the Korean *sam* and *sa* because of Alex's occasional difficulty in producing "ss". Alex thus did not initially observe any modeled connection between labels and their referents: that is, his input would lack as much reference and functionality as possible compared with his standard training [2,35]. Correct replies would receive vocal praise and the opportunity to request anything he desired (i.e., he would receive *x* for stating "I want x" [31], but *x* would have no referential relation to the count labels on which he was being trained); errors in the string (omissions, interchanging the order of elements) would elicit scolding, time-outs, and corrective feedback [35].

The training failed before we could even begin to collect data [35]. Although the plan was to have us train Alex in M/R variant 1 without any reference whatsoever, he would not attend (e.g., would turn his back and preen) until we included a minimal point of reference. We decided to use a sheet of paper with the symbols 1–8 printed along the diagonal. (NB: at that time, he did not know that his English number labels corresponded to these symbols; such training came later). In a typical session, a trainer showed the paper to the model and stated, "Say number?" (In contrast, queries in all previous number-related studies had been "How many?"). If the model correctly uttered the altered Korean labels in order, s/he was allowed to request a toy or food; errors resulting in mild scolding (e.g., "No, you're wrong!") and corrective feedback. As in basic M/R sessions, we routinely reversed roles of model and trainer; Alex was also given a turn. Alex's only reward was the opportunity to request a favored item ("I want *x*") for a correct response. Training, therefore, lacked functional meaning and all but minimal reference. The procedure did, however, maintain joint attention among bird, humans, and the pictured numbers.

Alex eventually did acquire the modeled string of number labels, but his results differed in two striking ways from those for utterances learned in standard M/R sessions [2,35]. First, whereas with standard input he would generally attempt new labels in just a few days of training and rarely needed more than a few weeks to achieve a recognizable reproduction, here he needed 9 months of training to learn the elements of the string. Second, and most interesting, was that he could not immediately use, nor subsequently learn to use, these labels referentially in any manner. He could merely repeat the labels by rote.

Whether or not the paper with the printed numerals was present, he responded with the string whenever he heard "Say number?" Even after we modeled a correspondence of the string with various sets of items, he could not learn any 1:1 relationship between a set of eight objects and the string of labels, nor could he use elements in the string to refer to smaller quantities, e.g., to say "il ee bam" when presented with three items and queried "Say number?" Alex had thus learned to reproduce—that is, mimic—but not comprehend these human vocalizations [35]. I doubted that the task was too difficult, given his previous success on both production and comprehension of human labels after M/R training (e.g., [37,38]) and his ability to refer to sets of objects with his prior number labels [11]. I believed that his failure was, as predicted by social modeling theory, a consequence of the type of training—one lacking reference and functionality—and not a lack of general cognitive capacity [11].

5.3. M/R Variant 2

This variant examined what would happen if functionality was almost (but not quite) removed and social interaction was strongly reduced but not eliminated during training [39]. Based on studies with children [29,40], the point was specifically to examine the role of joint focusing of attention between bird and trainer on the object whose label was to be learned. Here, a single trainer faced away from a bird, who was within reach of the targeted object—for example, a key—that was suspended by a string, in view of the bird. The trainer talked about the object, emphasizing its label, "Look, a shiny *key*!" "Do you want *key*?", etc. These sentences carefully framed the label, allowing for repeated label use while minimizing possible habituation [2,41]. The trainer had no visual or physical contact with the parrot or object but responded to the bird's vocal behavior. Specifically, if a bird attempted to utter the targeted label, it would receive vocal praise ("That's right, it's a *key*!"), but never the object, although it could see the object to which the label referred. Two birds, Alo and Kyaaro, received such training on two labels in this variant, which was compared with their training on two other labels via the standard M/R protocol. Labels for this and all subsequent experiments were chosen according to the same criteria used for the audiotape study. We attempted to provide equivalent exposure to the several labels being trained (see [39]). Both birds learned the labels trained with the standard M/R protocol (and learned fairly quickly) but failed to learn the labels trained with the variant. The comparison showed that the training provided in this variant, not the birds' basic abilities, affected their acquisition.

5.4. Basic Videotape Input

In "basic video" sessions, Kyaaro and Alo received input that demonstrated reference but avoided social interaction and minimized functionality in different ways than did M/R variant 2; again, labels trained in this manner were contrasted with other labels trained with the standard M/R procedure. Here, we again made videotapes of M/R sessions with Alex and two humans, then exposed the juveniles to those tapes [35], but now with the video portion intact. A zoom lens enabled us to include life-size images of Alex and the targeted objects in addition to the somewhat smaller images of the entire training scenario (the object, Alex, and two humans). Tapes also retained normal patterns of breaks for nonvocal exchanges (e.g., when trainers preened Alex) and trainers' departures by using, respectively, scenes of such nonvocal interactions or a blank screen. The audio portion of the video was again analyzed to ensure that it was not degraded compared to that of Alex "live". During sessions, each juvenile parrot was placed on a perch in a separate room so that the videos were viewed in isolation; thus, no direct social interaction with any humans occurred, although the birds saw the humans interacting with Alex and jointly attending to the object to be labeled. By watching a tape of a human or Alex produce a particular sound and either receive an object or be scolded, the juveniles saw but did not directly experience the effect of a vocalization. Videos, therefore, demonstrated reference but lacked all physical social interaction and clear functionality for the subject, who was now simply watching the video and could not receive any reward. (NB: our use of a cathode ray tube (CRT) screen—standard at the time—might have made learning from video impossible because of flicker fusion effects (e.g., [42]), distance, size of image, or lack of UV input [43]. Subsequent studies described below [44,45] take some of these issues into account). Again, birds learned the labels that were trained in M/R sessions, but not those trained via this video technique.

5.5. Video Variant 1

Because of research that (at least at the time I was performing these experiments) suggested that normal children learn more about labels from video instruction if they watch with an interactive adult who emphasizes important points, asks the children questions about what they are viewing, responds to the children's questions, and even corrects the children's comments (e.g., [46], but see [47] for children with language impairments), we decided to repeat the video study, this time providing the parrots with a trainer who would provide at least some interaction during viewing [48]. In our study, however, the trainer merely ensured—as described below—that the birds attended to the input and provided no physical referential reward if the bird identified something onscreen. That is, a bird that attempted the label received vocal praise but not the object. Trainers would provide social approbation for viewing and point to the screen, making comments ("Look what Alex has!"), but would not repeat new labels, ask questions, or relate content to other sessions [11]. Thus, the birds were not provided any real assistance in extracting information from the video. The amount of social interaction was limited and amount of functional meaning was the same as in basic video sessions. As before, labels trained in this manner were contrasted with labels trained via the M/R protocol; Kyaaro and Alo failed to learn labels introduced in this manner, although here they actually did make a few attempts at producing the label. Lack of the object as a reward, however, extinguished this behavior [48], and the majority of their vocalizations were not label attempts but rather utterances used to attract the attention of the trainer (e.g., "hello") or labels for other objects that they had previously acquired during M/R sessions. As before, they learned the labels taught via the M/R protocol.

5.6. Video Variant 2

So far, all these experiments had eliminated most of the functionality from input—that is, the concrete reason for learning new labels as requests for desired items. To see what would happen in the absence of social interaction but with the possibility of functionality, in another experiment using videotaped input, we now added a system that could reward a parrot for attempting the label [44]. The system was controlled by a trainer in another room who monitored the parrot's utterances through headphones. If the bird attempted the label, the trainer activated a pulley, lowering the object from a container to a spot within the bird's reach. We audiotaped sessions to test for interobserver reliability of the parrots' utterances during training. Of course, the bird had to make an attempt in order to be rewarded, and neither Kyaaro nor Alo did so. Again, no new labels were acquired, possibly because they had previously learned that attempts were not useful. From the birds' point of view this experiment did not differ from the basic video, and by ignoring even their general squawks in that study, we may have extinguished their attempts at vocal behavior in the absence of a human. In retrospect, we could have modeled one session where a trainer received a reward for making an attempt, so that the birds would have understood the possibility of receiving an object in this situation via the pulley system. Thus, a variant of this experiment remains a future possibility.

5.7. Video Variants 3 and 4

Here we decided to examine the possibility that additional, though not full, social interaction—that is, a single interactive human—plus functionality and reference might, when added to video, engender learning [44], as it had to some extent with children [46,47].

Here the human co-viewer was more interactive, uttering targeted labels (but only in response to the actions on the video), querying Kyaaro and now Griffin about the video objects, and providing the object if the birds attempted the label, but live modeling was absent. Then, testing the assumption that the parrots might have failed to learn from video simply because they lost interest after seeing the same video repeatedly, we repeated the same amount of interaction and functionality for Griffin, but now with a live feed from sessions with Alex [44], so that the video input varied from session to session. Again, birds in these sessions experienced training on one set of labels in the video conditions and training with different labels in standard M/R sessions. Although Kyaaro did produce a few attempts at a targeted label during his recorded video sessions, he failed to do so either frequently or clearly enough for any acquisition to be noted; thus, he effectively learned nothing from his video sessions but, again, learned from M/R sessions [44]. Griffin behaved similarly in the recorded video sessions, although he made more attempts than did Kyaaro and interacted with the object more often; nevertheless, his vocalizations at the end of training were not distinctive enough to be tested. During live-feed video sessions, he made more attempts on one word ("chalk") than he had during recorded sessions, but fewer for another ("bear"). The study was contaminated, however, because Alex (probably because of the additional exposure during live feeds) began to use "chalk" more frequently around trainers and in Griffin's presence. We therefore did test Griffin on these labels as well as those acquired via M/R training; he failed to use the appropriate labels to identify the former objects but succeeded with the latter.

5.8. LCD Video Training

As noted above, our use of a CRT screen for all of our previous video work might have made learning difficult, if not impossible, because the screen did not reproduce UV (i.e., objects did not look the same as in real-life to parrots, who have UV vision [43]) and because parrots' flicker-fusion rate would make the video seem like a flashing series of still photos. We were not overly concerned about the UV issue, because an earlier study [49] found that parrots, like humans, respond to videos of their toys despite possible color distortion. To test whether CRT screen flicker could have prevented learning, we ran a study with Griffin and Arthur using a liquid crystal display (LCD) monitor, which operates at refresh rate high enough (~200 Hz) to exceed the parrots' flicker rate [50]. We replicated the protocol of video variant 1, because we were not attempting to determine optimal conditions for learning with video, but rather were determining if flicker-fusion was the critical issue. Both Arthur and Griffin produced numerous labels previously acquired during M/R training during the LCD video presentations, but almost no attempts at the actual labels being targeted [45]. Thus, although flicker-fusion might contribute to the lack of learning during video presentation, its elimination did not enable birds to learn from video under conditions of limited social interaction.

5.9. Single Trainer Input

In yet another attempt to determine what was critical for referential label acquisition in Grey parrots, an experiment with Griffin eliminated the model. Here, we contrasted standard M/R sessions with those in which a trainer labeled and handled objects to which she and Griffin jointly attended [51]. In Griffin's presence, the human uttered phrases involving a targeted label (e.g., "Look at *x*!" or "I have an *x*!"), concurrently manipulating the corresponding item. Griffin thus saw behavior somewhat related to that in M/R sessions, where a trainer would say, "Yes, *you* have an *x*!" when rewarding the model/rival. The solo trainer demonstrated the association between label and item but did not model functional label use or errors. As in M/R sessions, Griffin was also queried about the object ("What's here?"). Any of his attempts at the label would be noted and rewarded with the item; thus, only if he attempted a label would he experience its functionality. As in M/R training, if he erred on a label, a trainer would tell him to "Try again," "Say better," etc. We ensured that he was queried at the same rate and in the same way as in M/R training, so these solos differed only in lack of human-human modeling and were not a series of drills. Interestingly, Griffin did not utter any targeted label during or immediately after 50 sessions but did so after two or three subsequent M/R sessions involving the same items. Because birds usually need about 20 M/R sessions to fully and clearly produce a novel label (slightly faster if that label has significant phonetic overlap with an existent label [11]), Griffin had likely acquired labels during the solo-trainer sessions, but did not understand how to *use* them until he observed their usage modeled [51]. Such behavior differed from that in standard M/R sessions, where he learned new labels and their use concurrently, which he again did with the labels taught using the M/R protocol in this experiment.

5.10. M/R with a Minimally Interactive Conspecific

Here we wanted to see what might happen if a parrot experienced full reference and functionality, modeling, corrective feedback, but no role reversal, and if lack of role reversal might be mitigated by the use of a conspecific [51]. Thus, Griffin was trained with Alex and one human in an approximation of the standard M/R system. Alex could act as a model, and because he occasionally erred, could be corrected; however, at this stage he could not query Griffin. We note that he sometimes had done so informally, by interrupting Griffin's M/R sessions (e.g., if we were asking Griffin about the color of an item and Griffin had mumbled his response, Alex might insert "Say better!"), but he did not actually pose specific questions at that time. (Note that Alex did understand the use of questions, and asked *trainers* for labels for novel items, colors, shapes, etc.; maybe he understood who was an appropriate source of information, and that Griffin was not? Only several years later were we able to train Alex to query Griffin). Unlike most of the other modifications to the M/R procedure, this one had some success. Griffin learned two labels trained in these sessions. Acquisition was, however, notably and considerably slower than in the standard M/R procedure, requiring about twice as many sessions for acquisition to occur. Many possible reasons exist for this slower rate of acquisition aside from the lack of role reversal; these are discussed in detail in [51]. Another possibility, however, was that Griffin saw the interactions as a form of duetting and bonding between the human and Alex—who was definitely the dominant parrot in the lab—and maybe Griffin, as the subordinate, feared to interrupt.

6. Implications of the Experiments on M/R Training Variants

The numerous experiments described above clearly could not examine every combination of various possible levels of input with respect to reference, functionality, and social interaction. For example, we did not test what would happen if, in the standard M/R condition, models never erred and thus the birds never saw corrective feedback. (Given studies at the time with pigeons and starlings on other types of tasks, showing how observing models' errors improved performance [52–54], we suggest that learning would have occurred after training without error demonstrations and feedback, but at a slower pace). However, our experiments overall do at least show the importance of specific types of input (Table 1): that two live models who demonstrate reference, functionality, joint attention, and extensive social interaction are necessary but not entirely sufficient for referential label acquisition [11,28]. Why necessary but not sufficient? We deduce this training is necessary because such learning did not occur through the use of (a) audio tapes in isolation; (b) a single trainer who acted referentially but did not demonstrate functionality; (c) a single trainer who also did not interact with the bird; (d) videotapes featuring two trainers (and a parrot) who exhibited the same behavior as live M/R models, including role reversal and corrective feedback of errors; (e) two live models with reversal with no functionality and extremely limited reference; or (f) a solo trainer plus the same videotaped material [35,39,44,48]. Todt's training [8], which lacked explicit referentiality, role reversal, and error demonstration, engendered facile learning, but without generalization to individuals or situations beyond the training situation. Our data also show that in the absence of two live models, role reversal and error demonstration are also insufficient [35]. We deduced that live two-trainer modeling, including functionality, referentiality, and demonstration of corrective feedback of a model's errors but without role reversal, engendered learning but at a far slower pace [51]. Thus, the many elements of input are each necessary, but only the presence of *all* together provides sufficient means for full and rapid acquisition to occur; lack of any one of them may result in various levels of failure.

Table 1. Components and results of various types of tutoring regimes.

Regime	Reference	Functionality	Social Interaction	Parrot	Learning?
M/R	Yes	Yes	Full model	All	Yes
Audiotape	No	No	No	Alo, Kyaaro	No
M/R variant 1	No	No	Full model	Alex	Mimicry
M/R variant 2	Yes	Minimal	Minimal	Alo, Kyaaro	No
Basic video	Yes	Partial	No	Alo, Kyaaro	No
Video variant 1	Yes	Partial	Minimal	Alo, Kyaaro	Minimal
Video variant 2	Yes	Potential	No	Alo, Kyaaro	No
Video variants 3, 4	Yes	Potential	Some	Kyaaro, Griffin	No
LCD video	Yes	Partial	No	Kyaaro, Griffin	No
Single trainer	Yes	Potential	Yes, no model	Griffin	Hidden
Conspecific	Yes	Yes	Yes, no reversal	Griffin	Slow

7. Why Do We Need Interspecies Communication?

I have argued previously (summarized in [4,5,55,56]) that training on symbolic representation, and the form of interspecies communication it engenders, affects the ease with which animals learn and can be tested on certain concepts. Furthermore, as did Premack before me [3], I suggest that a system of two-way communication may enable a researcher to teach a concept that an animal subject may not easily acquire by other means, and, most importantly, that acquisition of symbolic representation can even affect how subjects manipulate information.

Premack [3] presented evidence for chimpanzees, and I for Grey parrots [4,56], showing that symbolic reference enables (but, of course, cannot guarantee) abstract thought. Thus, once a nonhuman understands that a symbol can be used to represent an object, an action, an attribute, etc., the subject can then mentally manipulate that symbol. Such mental manipulation releases thought processes to, for example, enable formation of concepts, insight, and transfer to novel situations. A subject can plan, then mentally—rather than physically—test the consequences of choices. I have previously described, in detail, the several studies on Grey parrots in which I believe symbolic representation and interspecies communication have facilitated learning (e.g., [4,5,56]) and, thus, will only summarize this material here.

7.1. Studies Facilitated by Interspecies Communication

In some studies, symbolic processing plays an important role not specifically because it enables more complex processing, but because it facilitates training and/or eliminates confounds that can detract from the claims of success in other species. In each of these studies, the parrot could be taught a label for a concept, rather than being required to derive the concept over thousands of trial-and-error tests, and then could be exposed to transfer trials to examine how well the concept was understood.

For example, in studies of optical illusions, humans are simply asked what they see. However, such is rarely the case for nonhumans who are tested to determine how their perception of the same stimuli compares to that of humans. Nonhumans instead are generally subjected to intensive training procedures to enable them to discriminate the initial stimulus, followed by tests on their recognition of similar patterns; results often are contingent upon statistical averaging over hundreds of trials of pecking/touching behavior to a very limited set of choices. The data thus are often highly variable and dependent upon details of the experimental design (reviewed in [57]). Alex, in contrast, could simply be shown pictures of the Müller–Lyer illusion (Brentano version), in which the two horizontal lines were of different colors, and, as he had previously acquired the concept of relative size [58], be asked, "What color bigger/smaller?" [57]. Likewise, once another Grey parrot, Griffin, had learned labels for shapes, and thus that a vocal label could represent an object, he could understand how two symbols (e.g., one vocal and one visual), which separately represent the same object, can then represent each other (a form of equivalence [59]) and thus how, for example, a three-dimensional entity can be represented by a two-dimensional drawing, so that he could appropriately identify pictures of occluded objects and Kanizsa figures (i.e., other optical illusions) [60]. (Possibly, one might also therefore argue that this study was critically dependent upon symbolic representation; see below).

It is also likely that Griffin's understanding of symbolic representation assisted him in other experiments. That is, he might have used symbols as place-markers to assist in tasks requiring memory [61,62] and evaluation of probability [63]. Even limited understanding of symbolic representation may have helped other Grey parrots in tasks such as Piagetian liquid conservation [64] and inference by exclusion [65,66], where memory for specific quantities or objects, respectively, is also of considerable importance.

7.2. Studies Critically Dependent upon Symbolic Representation

As noted above, Premack [3] has argued that acquisition of symbolic representation not only enables nonhumans to be tested more easily, but also may actually enable them to perform tasks that are failed by nonhumans lacking such representation. Some of the tasks I describe below I had previously listed in the former category (e.g., see [56]), but now I believe they may fall into the latter. Here, I believe that symbolic representation has been a crucial tool for studying the extent of Grey parrot cognitive capacity.

For example, most nonhumans have considerable difficulty demonstrating an understanding of relational concepts, such as 'bigger–smaller', unlike tasks involving concrete concepts such as color, where something that is "red" is always "red", in relational tasks the sample that is, for example, the smaller object on one trial (e.g., a tennis ball versus a basketball) can be the bigger object on another (e.g., the tennis ball versus a golf ball). Most subjects are trained to derive one concept over large numbers of trials (i.e., by being rewarded for choosing only the bigger item) and then the other through large numbers of reversals (i.e., now being rewarded for choosing only the smaller)—in this training paradigm, however, both concepts may actually never be acquired, in that a subject without symbolic reference may simply learn "choose X" versus "avoid X" (see [67,68] for a discussion). In contrast, my Grey parrot, Alex, who learned to respond to "What color bigger/smaller?" for three sets of items, was then able to transfer his responses, without additional training, to a large number of sets involving sizes outside the training paradigm and to totally novel objects with respect to shape, color, and material [58]; he also spontaneously transferred his understanding of the concept to questions for which the responses involved material rather than color, and spontaneously transferred use of the label "none" (acquired in a different study, [69]) to respond appropriately when the two objects were of the same size [58].

The ability to comprehend concepts of same-different is the specific one that Premack [3] used to buttress his argument for the effect of symbolic representation on nonhuman cognitive processing. Details of that argument have been the subject of an entirely separate paper [4], but the point is that nonhumans lacking symbolic representation do not fully comprehend same–different, whereas those with symbolic representation do. The issue is that same–different is not merely recognition of identity versus non-identity (i.e., whether two stimuli are completely equal or completely unequal in every possible aspect) or a difference in entropy (i.e., in overall randomness) between stimuli sets (e.g., [70]). Rather, it is a task that, according to Premack's stringent criteria [3], requires a feature analysis of the objects being compared, recognition that objects can simultaneously exhibit attributes that involve both similarity and difference, and the ability to understand which attributes are being targeted based on questions of either similarity or difference. As noted in [6,7], an appropriate response requires a subject to (a) attend to multiple aspects of two different objects; (b) determine, from a verbal question, whether the response is to be based on sameness or difference; (c) determine, from the exemplars, exactly what is same or different (i.e., what are their colors/shapes/materials?); and then (d) produce, verbally, the label for the hierarchical category of the appropriate attribute (i.e., "color", "shape", "matter"). The task is therefore a clear instance in which symbolic reference is critical for success. Not only did the Grey parrot, Alex, succeed [4,71] but he transferred his knowledge to a completely novel instantiation of the same-different concept (i.e., something beyond his familiar comparisons of color/shape/material) without additional training. After he learned to use the symbols "same" and "different" to represent the relations amongst specific categories for object pairs, he spontaneously understood how the same–different relationship applied to an abstract representation of a novel situation—as when, queried for the first time "What color bigger?" for two equally-sized items, he asked "What's same?" and then responded "none" (see above, [58]). Such fluid response ability requires symbolic, referential, interspecies communication.

Another topic in which symbolic representation plays a major role is number concepts. Lacking symbols for specific quantities, individuals may still demonstrate some sense of number. Using what is termed the Approximate Number System (ANS), almost every non-human tested plus pre-verbal children can distinguish exact quantities for sets up to about three [72]. The ANS, however, cannot identify quantities precisely for amounts much above that number; to do so, individuals need actual symbols [73]. Such understanding of numerical symbols was once thought to be limited to humans, but several nonhumans—two apes and the Grey parrot, Alex—learned to use Arabic numerals to distinguish quantities up to eight or nine (reviewed in [34]). These subjects could produce the correct numerical label in the presence of specific sets, could identify the correct set in the presence of the corresponding Arabic numeral, understood how to add small quantities [33,74], could learn [75] or infer [32] the ordinality of the numerals after demonstrating an understanding of their cardinality, and all had a zero-like concept (reviewed in [34]). The parrot could also infer the cardinality of a novel number after learning its ordinal value in a number line [34]. Such abilities can be approached by the ANS, but are not fully possible without exact symbolic reference [73].

8. Discussion and Conclusions

Only a very small number of individual nonhumans have acquired the form of symbolic representation described here—a few nonhuman primates, a few marine mammals, a few Grey parrots (summarized in [76])—and such studies are, for the most part, no longer in progress [76]. Some dogs have acquired some labels as well (note [77]). The primary point of this paper was not to review the overall field of interspecies communication, but rather to discuss the processes involved in the acquisition of such communication, specifically from the standpoint of a set of studies on Grey parrots. I suggest that the reason that so few experiments in the field have been successful probably lies less in the capacities of the various nonhumans that were trained, but rather more with the capacities of the humans performing the training. Such training must be done with extreme attention to detail so that all the elements of input that are both necessary and sufficient are present, otherwise learning will either not occur or be incomplete: the full extent of any nonhuman capacity for symbolic representation will not be achieved.

It is important to recognize that none of the nonhuman individuals in these studies on symbolic representation acquired all the complexities that are inherent in human language (even with what one hopes was fully appropriate training); much is yet to be learned about the reasons why. Nevertheless, the acquired level of interspecies communication in this unique subset did enable an examination of their cognitive capacities in ways that would have been much more difficult, if not impossible, by other means. Again, I have not tried to review the entire field of animal cognition, but have focused on my specific area of expertise, the abilities of Grey parrots. Given that these birds are evolutionarily separated from mammals by over 320 million years of evolution [78], and yet have demonstrated abilities in many ways comparable to those of 5–8 year old children (e.g., [4,5,32–34,36–39,62–66], one wonders what other individuals of various species might also demonstrate advanced cognition if tested in an appropriate manner?

Funding: The research described herein was supported by various grants over the course of over 40 years from the National Science Foundation (USA), the Harry Frank Guggenheim Foundation, and donors to *The Alex Foundation*.

Institutional Review Board Statement: This paper reviews studies performed during the course of over 40 years at a large number of different universities. Most recently, research occurred at Harvard University under their IACUC Protocol ID: 13-06-164-2.

Informed Consent Statement: Not Applicable.

Conflicts of Interest: The author declares no conflict of interest. The funders had no role in the design of the study; in the collection, analyses, or interpretation of data; in the writing of the manuscript, or in the decision to publish the results.

References

1. Deacon, T.W. *The Symbolic Species: The Co-Evolution of Language and the Brain*; W.W. Norton & Co: New York, NY, USA, 1997.
2. Pepperberg, I.M. Functional vocalizations an African Grey parrot (*Psittacus erithacus*). *Z. Tierpsychol.* **1981**, *55*, 139–160. [CrossRef]
3. Premack, D. The codes of man and beast. *Behav. Brain Sci.* **1983**, *6*, 125–167. [CrossRef]
4. Pepperberg, I.M. How do a pink plastic flamingo and a pink plastic elephant differ? Evidence for abstract representations of the relations of same-different in a Grey parrot. *Curr. Opin. Behav. Sci.* **2021**, *37*, 146–152. [CrossRef]
5. Pepperberg, I.M. Nonhuman and nonhuman-human communication: Some issues and questions. *Frontiers*. in press.
6. Marks, J.M. *What It Means to be 98% Chimpanzee: Apes, People, and Their Genes*; University of California Press: Berkeley, CA, USA, 2005.
7. Ridgeway, S.H. The central nervous system of the bottlenosed dolphin. In *The Bottlenosed Dolphin*; Leatherwood, S., Reeves, R.R., Eds.; Academic Press: San Diego, CA, USA, 1990; pp. 69–97.
8. Todt, D. Social learning of vocal patterns and modes of their applications in Grey parrots. *Z. Tierpsychol.* **1975**, *39*, 178–188. [CrossRef]
9. Boosey, E.J. The African Grey parrot. *Avic. Mag.* **1947**, *53*, 39–40.
10. Lenneberg, E.H. Biological aspects of language. In *Communication, Language, and Meaning*; Miller, G.A., Ed.; Basic Books: New York, NY, USA, 1973; pp. 49–60.
11. Pepperberg, I.M. *The Alex Studies*; Harvard University Press: Cambridge, MA, USA, 1999.
12. Mowrer, O.H. *Learning Theory and Personality Dynamics*; Ronald Press: New York, NY, USA, 1950.
13. Grosslight, J.H.; Zaynor, W.C. Verbal behavior and the mynah bird. In *Research in Verbal Behavior and Some Neurophysiological Implications*; Salzinger, K., Salzinger, S., Eds.; Academic Press: New York, NY, USA, 1967; pp. 5–9.
14. Bandura, A. Analysis of modeling processes. In *Psychological Modeling*; Bandura, A., Ed.; Aldine-Atherton: Chicago, IL, USA, 1971; pp. 1–62.
15. Krashen, S.D. *Principles and Practices of Second Language Acquisition*; Pergamon Press: Oxford, UK, 1982.
16. Bandura, A. The role of imitation. In *Social Learning and Personality Development*; Bandura, A., Walters, R.H., Eds.; Holt, Rinehart, & Winston: New York, NY, USA, 1963; pp. 47–108.
17. Bandura, A. Psychotherapy based upon modeling principles. In *Handbook of Psychotherapy and Behavior Change: An Empirical Analysis*; Bergin, A.D., Garfield, S.L., Eds.; John Wiley & Sons: New York, NY, USA, 1971; pp. 653–708.
18. Piaget, J. Language and thought from the genetic point of view. *Acta Psychol.* **1954**, *10*, 51–60. [CrossRef]
19. Kuczaj, S.A. On the nature of syntactic development. In *Language Development: Syntax and Semantics*; Kuczaj, S.A., Ed.; Erlbaum: Hillsdale, NJ, USA, 1982; pp. 37–71.
20. Leung, A.; Tunkel, A.; Yurovsky, D. Parents fine-tune their speech to children's vocabulary knowledge. *Psychol. Sci.* **2021**. [CrossRef] [PubMed]
21. Wang, J.; Yang, Y.; Macias, C.; Bonawitz, E. Children with more uncertainty in their intuitive theories seek domain-relevant information. *Psychol. Sci.* **2021**, *32*, 1147–1156. [CrossRef] [PubMed]
22. Brown, I. Role of referent concreteness in the acquisition of passive sentence comprehension through abstract modeling. *J. Expt. Child. Psychol.* **1976**, *22*, 185–199. [CrossRef]
23. Lock, A. *The Guided Reinvention of Language*; Academic Press: London, UK, 1980.
24. Walker, J.M.T. Authoritative classroom management: How control and nurturance work together. *Theor. Pract.* **2009**, *48*, 122–129. [CrossRef]
25. Pepperberg, I.M. Social modeling theory: A possible framework for understanding avian vocal learning. *Auk* **1985**, *102*, 854–864.

26. Smith, W.J. Animal communication and the study of cognition. In *Cognitive Ethology: The Minds of Other Animals*; Ristau, C.A., Ed.; Erlbaum: Hillsdale, NJ, USA, 1991; pp. 209–230.
27. Pepperberg, I.M. A review of the effects of social interaction on vocal learning in African Grey parrots (*Psittacus erithacus*). *Neth. J. Zool.* **1993**, *43*, 104–124. [CrossRef]
28. Pepperberg, I.M. Vocal learning in Grey parrots: A brief review of perception, production, and cross-species comparisons. *Brain Lang.* **2010**, *115*, 81–91. [CrossRef]
29. Tomasello, M.; Todd, J. Joint attention and early lexical acquisition style. *First Lang.* **1983**, *4*, 197–212. [CrossRef]
30. Pepperberg, I.M.; Sherman, D.V. Proposed use of two-part interactive modeling as a means to increase functional skills in children with a variety of disabilities. *Teach. Learn. Med.* **2000**, *12*, 213–220. [CrossRef] [PubMed]
31. Pepperberg, I.M. An interactive modeling technique for acquisition of communication skills: Separation of >labeling= and >requesting= in a psittacine subject. *Appl. Psycholing.* **1988**, *9*, 59–76. [CrossRef]
32. Pepperberg, I.M.; Gordon, J.D. Numerical comprehension by a Grey parrot (*Psittacus erithacus*), including a zero-like concept. *J. Comp. Psychol.* **2005**, *119*, 197–209. [CrossRef]
33. Pepperberg, I.M. Grey parrot (*Psittacus erithacus*) numerical abilities: Addition and further experiments on a zero-like concept. *J. Comp. Psychol.* **2006**, *120*, 1–11. [CrossRef]
34. Pepperberg, I.M.; Carey, S. Grey Parrot number acquisition: The inference of cardinal value from ordinal position on the numeral list. *Cognition* **2012**, *125*, 219–232. [CrossRef]
35. Pepperberg, I.M. Vocal learning in Grey parrots (*Psittacus erithacus*): Effects of social interaction, reference, and context. *Auk* **1994**, *111*, 300–313. [CrossRef]
36. Fuson, K.C. *Children's Counting and Concepts of Number*; Springer: New York, NY, USA, 1988.
37. Pepperberg, I.M. Cognition in an African Grey parrot (*Psittacus erithacus*): Further evidence for comprehension of categories and labels. *J. Comp. Psychol.* **1990**, *104*, 41–52. [CrossRef]
38. Pepperberg, I.M. Proficient performance of a conjunctive, recursive task by an African Grey parrot (*Psittacus erithacus*). *J. Comp. Psychol.* **1992**, *106*, 295–305. [CrossRef]
39. Pepperberg, I.M.; McLaughlin, M.A. Effect of avian-human joint attention on allospecific vocal learning by Grey parrots (*Psittacus erithacus*). *J. Comp. Psychol.* **1996**, *110*, 286–297. [CrossRef] [PubMed]
40. Baldwin, D.A. Infants' contributions to the achievement of joint reference. *Child. Dev.* **1991**, *62*, 875–890. [CrossRef] [PubMed]
41. Endress, A.D.; Nespor, M.; Mehler, J. Perceptual and memory constraints on language acquisition. *Trends Cogn. Sci.* **2009**, *13*, 348–353. [CrossRef] [PubMed]
42. Ikebuchi, M.; Okanoya, K. Male zebra finches and Bengalese finches emit directed songs to the video images of conspecific females projected onto a TFT display. *Zool. Sci.* **1999**, *16*, 63–70. [CrossRef]
43. Bowmaker, J.K.; Heath, L.A.; Wilkie, S.E.; Das, D.; Hunt, D.M. Middle-wave cone and rod visual pigments in birds: Spectral sensitivity and opsin structure. *Investig. Ophth. Vis. Sci.* **1996**, *37*, S804.
44. Pepperberg, I.M.; Gardiner, L.I.; Luttrell, L.J. Limited contextual vocal learning in the Grey parrot (*Psittacus erithacus*): Effect of co-viewers on videotaped instruction. *J. Comp. Psychol.* **1999**, *113*, 158–172. [CrossRef]
45. Pepperberg, I.M.; Wilkes, S.R. Lack of referential vocal learning from LCD video by Grey parrots (*Psittacus erithacus*). *Interact. Stud.* **2004**, *5*, 75–97. [CrossRef]
46. Lemish, D.; Rice, M.L. Television as a talking picture book: A prop for language acquisition. *J. Child. Lang.* **1986**, *13*, 251–274. [CrossRef]
47. Rice, M.L. Children with specific language impairment: Toward a model of teachability. In *Biological and Behavioral Determinants of Language Development*; Krasnegor, N.A., Rumbaugh, D.M., Schiefelbusch, R.L., Studdert-Kennedy, M., Eds.; Erlbaum: Hillsdale, NJ, USA, 1991; pp. 447–480.
48. Pepperberg, I.M.; Naughton, J.R.; Banta, P.A. Allospecific vocal learning by Grey parrots (*Psittacus erithacus*): A failure of videotaped instruction under certain conditions. *Behav. Proc.* **1998**, *42*, 139–158. [CrossRef]
49. Rutledge, D.; Pepperberg, I.M. *Video Studies of Same/Different*; Unpublished raw data; Northwestern University: Evanston, IL, USA, 1988.
50. Boström, J.E.; Haller, N.K.; Dimitrova, M.; Ödeen, A.; Kelber, A. The flicker fusion frequency of budgerigars (*Melopsittacus undulatus*) revisited. *J. Comp. Physiol. A* **2017**, *203*, 15–22. [CrossRef] [PubMed]
51. Pepperberg, I.M.; Sandefer, R.M.; Noel, D.; Ellsworth, C.P. Vocal learning in the Grey parrot (*Psittacus erithacus*): Effect of species identity and number of trainers. *J. Comp. Psychol.* **2000**, *114*, 371–380. [CrossRef] [PubMed]
52. Vanayan, M.H.; Robertson, A.; Biederman, G.B. Observational learning in pigeons: The effects of model proficiency on observer performance. *J. Gen. Psychol.* **1985**, *112*, 349–357. [CrossRef]
53. Biederman, G.B.; Vanayan, M. Observational learning in pigeons: The function of quality of observed performance in simultaneous discrimination. *Learn. Motiv.* **1988**, *19*, 31–43. [CrossRef]
54. Templeton, J.J. Learning from others' mistakes: A paradox revisited. *Anim. Behav.* **1998**, *55*, 79–85. [CrossRef] [PubMed]
55. Pepperberg, I.M. Symbolic communication in nonhumans. In *APA Handbook of Comparative Psychology*; Call, J., Pepperberg, I.M., Snowdon, C.T., Zentall, T.R., Eds.; American Psychological Association Press: Washington, DC, USA, 2017; pp. 663–680.
56. Pepperberg, I.M. The comparative psychology of intelligence: Some thirty years later. *Front. Psychol.* **2020**, *11*, 973. [CrossRef]

57. Pepperberg, I.M.; Vicinay, J.; Cavanagh, P. Processing of the Müller-Lyer illusion by a Grey parrot (*Psittacus erithacus*). *Perception* **2008**, *37*, 765–781. [CrossRef]
58. Pepperberg, I.M.; Brezinsky, M.V. Acquisition of a relative class concept by an African Grey parrot (*Psittacus erithacus*): Discriminations based on relative size. *J. Comp. Psychol.* **1991**, *105*, 286–294. [CrossRef]
59. Pepperberg, I.M. Ordinality and inferential abilities of a Grey parrot (*Psittacus erithacus*). *J. Comp. Psychol.* **2006**, *120*, 205–216. [CrossRef]
60. Pepperberg, I.M.; Nakayama, K. Robust representation of shape by a Grey parrot (*Psittacus erithacus*). *Cognition* **2016**, *153*, 146–160. [CrossRef] [PubMed]
61. Pepperberg, I.M.; Pailian, H. Evolution of mechanisms underlying visual working memory manipulation: When 'bird-brain' is a compliment. In Proceedings of the Paper Presented at the Vision Science Society Annual Conference, St. Petersburg, FL, USA, 24 May 2017.
62. Pailian, H.; Carey, S.; Halberta, J.; Pepperberg, I.M. Age and species comparisons of visual mental manipulation ability as evidence for its development and evolution. *Sci. Rep.* **2020**, *10*, 7689. [CrossRef]
63. Clements, K.; Gray, S.L.; Gross, B.; Pepperberg, I.M. Initial evidence for probabilistic learning by a Grey parrot (*Psittacus erithacus*). *J. Comp. Psychol.* **2018**, *132*, 166–177. [CrossRef]
64. Cornero, F.M.; Hartsfield, L.A.; Pepperberg, I.M. Piagetian liquid overconservation in Grey parrots (*Psittacus erithacus*). *J. Comp. Psychol.* **2020**, *143*, 197–210. [CrossRef] [PubMed]
65. Pepperberg, I.M.; Koepke, A.; Livingston, P.; Girard, M.; Hartsfield, L.A. Reasoning by inference: Further studies on exclusion in Grey parrots (*Psittacus erithacus*). *J. Comp. Psychol.* **2013**, *127*, 272–281. [CrossRef]
66. Pepperberg, I.M.; Gray, S.L.; Cornero, F.M.; Mody, S.; Carey, S. Logical reasoning by a Grey parrot (*Psittacus erithacus*)? *A case study of the disjunctive syllogism. Behaviour* **2019**, *156*, 409–445.
67. Hochmann, J.-R.; Carey, S.; Mehler, J. Infants learn a rule predicated on the relation same but fail to simultaneously learn a rule predicated on the relation different. *Cognition* **2018**, *177*, 49–57. [CrossRef]
68. Hochmann, J.-R.; Mody, S.; Carey, S. Infants' representations of same and different in match- and non-match-to-sample tasks. *Cogn. Psychol.* **2016**, *86*, 87–811. [CrossRef] [PubMed]
69. Pepperberg, I.M. Acquisition of the concept of absence by an African Grey parrot: Learning with respect to questions of same/different. *J. Exp. Anal. Behav.* **1988**, *50*, 553–564. [CrossRef] [PubMed]
70. Young, M.E.; Wasserman, E.A. Entropy and variability discrimination. *J. Exp. Psychol. Learn. Mem. Cogn.* **2001**, *27*, 278–293. [CrossRef]
71. Pepperberg, I.M. Acquisition of the same/different concept by an African Grey parrot (*Psittacus erithacus*): Learning with respect to categories of color, shape, and material. *Anim. Learn. Behav.* **1987**, *15*, 423–432. [CrossRef]
72. Odic, D.; Starr, A. An introduction to the Approximate Number System. *Child. Dev. Persp.* **2018**, *12*, 223–229. [CrossRef]
73. Halaberda, J.; Feigenson, L. Developmental change in the acuity of the 'number sense': The approximate number system in 3-, 4-, 5-, 6-year olds and adults. *Dev. Psychol.* **2008**, *44*, 1457–1465. [CrossRef]
74. Boysen, S.T.; Berntson, G.G. Numerical competence in a chimpanzee (*Pan troglodytes*). *J. Comp. Psychol.* **1989**, *103*, 23–31. [CrossRef] [PubMed]
75. Boysen, S.T.; Berntson, G.G.; Shreyer, T.A.; Quigley, K.S. Processing of ordinality and transitivity by chimpanzees (*Pan troglodytes*). *J. Comp. Psychol.* **1993**, *107*, 208–215. [CrossRef]
76. Pepperberg, I.M. Animal language studies: What happened? *Psychol. Bull. Rev.* **2017**, *24*, 181–185. [CrossRef] [PubMed]
77. Griebel, U.; Oller, D.K. Vocabulary learning in a Yorkshire terrier: Slow mapping of spoken words. *PLoS ONE* **2012**, *7*, e30182. [CrossRef] [PubMed]
78. Hedges, S.B.; Parker, P.H.; Sibley, C.G.; Kumar, S. Continental breakup and the ordinal diversification of birds and mammals. *Nature* **1996**, *381*, 226–229. [CrossRef] [PubMed]

animals

MDPI

Essay

Invasive Research on Non-Human Primates—Time to Turn the Page

Maria Padrell [1], Miquel Llorente [1] and Federica Amici [2,3,*]

[1] Departament de Psicologia, Facultat d'Educació i Psicologia, Universitat de Girona, 17004 Girona, Spain; mariapadrell@gmail.com (M.P.); miguel.llorente@udg.edu (M.L.)
[2] Department of Human Behavior, Ecology and Culture, Max Planck Institute for Evolutionary Anthropology, Deutscher Platz 6, D-04103 Leipzig, Germany
[3] Faculty of Life Science, Institute of Biology, University of Leipzig, Talstrasse 33, D-04103 Leipzig, Germany
* Correspondence: amici@eva.mpg.de

Simple Summary: Despite increasing ethical concerns, primates are still often used in invasive research (i.e., laboratory research that causes body manipulations causing them pain or distress and not aimed at directly improving their well-being). Here, we will review previous studies showing that primates have complex behaviour and cognition, and that they suffer long-term consequences after being used in invasive research. We will discuss the ethical problems that invasive research on primates posit, the legal protection that they are, to date, granted in different countries, and summarize the past and current attempts to ban this kind of research on primates. We will conclude why, in our opinion, invasive research on primates should be banned, and non-invasive methods should be considered the only possible approach to the study of primates.

Abstract: Invasive research on primates (i.e., laboratory research that implies body manipulations causing pain or distress that is not aimed to directly improve the individuals' well-being) has a long history. Although some invasive studies have allowed answering research questions that we could not have addressed with other methods (or at least not as quickly), the use of primates in invasive research also raises ethical concerns. In this review, we will discuss (i) recent advances in the study of primates that show evidence of complex behaviour and cognition, (ii) welfare issues that might arise when using primates in invasive research, (iii) the main ethical issues that have been raised about invasive research on primates, (iv) the legal protection that primates are granted in several countries, with a special focus on the principle of the 3Rs, and (v) previous and current attempts to ban the use of primates in invasive research. Based on this analysis, we suggest that the importance of a research question cannot justify the costs of invasive research on primates, and that non-invasive methods should be considered the only possible approach in the study of primates.

Keywords: welfare; ethics; legal protection; principle of the 3Rs; invasive research ban

Citation: Padrell, M.; Llorente, M.; Amici, F. Invasive Research on Non-Human Primates—Time to Turn the Page. *Animals* **2021**, *11*, 2999. https://doi.org/10.3390/ani11102999

Academic Editors: Lucia Regolin, Maria Loconsole and Sue W. Margulis

Received: 9 August 2021
Accepted: 16 October 2021
Published: 19 October 2021

1. Introduction

In this study, we will use the term invasive research to refer to all research that (i) is conducted in the lab, (ii) implies stressful and/or painful body manipulations of individuals, and (iii) is not aimed to increase their welfare. Such body manipulations can range from inoculation with infectious agents and surgery to drug testing and blood sampling, but exclude veterinarian applications that directly benefit the animals [1]. According to Directive 2010/63/EU of the European Parliament [2], invasive procedures can have different degrees of severity including (i) non-recovery procedures (i.e., procedures performed under general anaesthesia, at the end of which the animal is killed before recovering consciousness); (ii) mild procedures (i.e., providing animals "short-term mild pain, suffering or distress", and "no significant impairment of the well-being or general condition of the animals"); (iii) moderate procedures (i.e., causing "short-term moderate pain, suffering or distress, or long-lasting mild pain, suffering or distress" or a "moderate impairment of the well-being or general condition of the animals"), and (iv) severe procedures (i.e., causing "severe pain, suffering or distress, or long-lasting moderate pain, suffering or distress" or a "severe impairment of the well-being or general condition of the animals"). In our definition, we purposely exclude instances of research conducted in zoos, research facilities using observational methods, and research implying body manipulations of wild animals (e.g., trapping) as they posit different ethical concerns that have been extensively addressed elsewhere [3,4]. Here, we will therefore mainly focus on using animals in neuroscience, biomedical and translational research, and other related research areas. We further note here that, although some researchers prefer to refer to the killing of animals at the end of invasive procedures with the term "sacrifice", which entails a positive connotation (see e.g., [5]), we will stick to the term "killing" throughout the paper because this term conveys, in our opinion, a more objective perspective and is also the term routinely used in legal and technical documents (e.g., [2]).

Invasive research on non-human primates (hereafter, NHPs) has a long history. Studies on their internal biology date back to Classical times, but the systematic inclusion of NHPs in invasive research only started in the last century [6]. Since then, NHPs have become the object of a variety of studies in disciplines such as medicine, neuroanatomy, genetics, and physiology. Although the relevance and importance of some of these studies has been questioned by some researchers, mostly from research areas other than neuroscience, biomedical, and translational research [7,8], it is undeniable that invasive research has allowed answering research questions that could not have been addressed with other methods, or at least not as effectively [9].

The exact number of NHPs currently used worldwide in invasive research is unknown, but estimates suggest that more than 100,000 NHPs are used every year, mostly in the USA [10]. The Scientific Committee on Health, Environmental and Emerging Risks reported that in 2011, 11 million animals (including approximately 6000 NHPs) were used in scientific procedures in the European Union [11]. In the same report, the European Commission estimated that 8898 procedures were conducted on NHPs in 2014, mostly for the development and safety testing of pharmaceuticals and medical devices, treatment and prevention of infectious diseases, and neuroscience [11]. According to the information reposted by the European Member States, the most frequently used species in invasive research were crab-eating macaques (*Macaca fascicularis*; 80%), marmosets and tamarins (Callitrichidae: 8%), and rhesus monkeys (*Macaca mulatta*; 7%) [11]. More recently, the report on the statistics of the use of animals for scientific purposes in the European Union estimated that 8,235 NHPs were used for the first time for scientific purposes in 2017 [12]. In the United States, the annual report of animal usage informed that 70,797 NHPs were housed in research facilities used in regulated activities; 26,137 of these animals were used in painful procedures and received pain relieving, or similar, drugs, while 802 were subject to painful procedures but did not receive drugs because they would adversely affect results [13].

Although NHPs only constitute a small part of all the animals currently used in scientific procedures, it is evident that their use in invasive research is still widespread [10,11]. Understanding whether this is ethically acceptable is therefore a compelling moral imperative for scientists from different research areas. This is even more important if one considers that around 48% of NHP species (including some that are commonly used in invasive procedures, such as crab-eating macaques and several species of marmosets and tamarins) are classified as Threatened by the IUCN. In the following sections, we will discuss which moral statuses and legal protections should be granted to NHPs in line with the most recent scientific developments. To reflect the diversity of the readers of this Journal, we will do that by referring to the current situation of invasive research on NHPs in several areas of the world without restricting our analysis to the European Union case. First, we will summarize recent advances in the study of NHPs that show evidence of highly complex behaviour and cognition that might question their use in invasive research.

Second, we will discuss welfare issues that might arise when using NHPs in invasive research. Third, we will review the main ethical issues that have been raised on invasive research on NHPs. Fourth, we will examine the legal protections that are, to date, granted to NHPs in several countries, specifically focusing on the gap between theory and practice. Fifth, we will provide a brief overview of the past and current attempts to ban the use of NHPs in invasive research. In the conclusions, we will discuss why, in our opinion, non-invasive methods should be considered the only possible current approach to the study of NHPs.

2. Evidence of Complex Behavioural and Cognitive Life in NHPs

According to some authors, the moral status and the legal protection granted to a species should mainly depend on its behavioural and cognitive complexity [14,15]. Although the great majority of animals experience pain and suffering and are therefore negatively affected by invasive procedures, the additional traits of complex behaviour, emotion, and cognition may increase such suffering beyond ethically acceptable levels. In line with this view, Directive 2010/63/EU of the European Parliament states that the welfare of animals used in invasive research must be improved "by raising the minimum standards for their protection in line with the latest scientific developments" ([2], p. 1). Therefore, understanding the behaviour, emotion, and cognition of NHPs is an essential endeavour to ensure them adequate legal protection by reducing the "pain, suffering, distress and lasting harm" that they experience during invasive procedures [2]. Moreover, new scientific knowledge on behavioural, emotional, and cognitive complexity may support a ban on invasive research on NHPs by, for instance, showing that such complexity does not allow a reduction of pain and suffering below ethically acceptable levels, with full replacement being the only viable option.

To date, NHPs are known to interact in complex ways with their physical environment. Frugivorous monkeys and great apes, for instance, use spatial memory to locate fruit trees in the forest [16,17]. Chimpanzees (*Pan troglodytes*) may retain information on larger fruit trees across seasons [18], recall past events after several years [19], and show exceptional memory skills [20]. Similarly, several monkey species that are widely used in invasive procedures, such as rhesus monkeys and marmosets (*Callithrix jacchus*), form long-term memories of past events that include identity, spatial, and temporal components that suggest the existence of episodic-like memory [21,22]. Several monkey and ape species are also sophisticated tool users; macaques, for instance, can use multiple tools in complex sequences, showing cognitive flexibility and foresight [23]. Chimpanzees are also proficient tool manufacturers, having successfully solved meta-tool tasks that require a flexible and hierarchical organization of behaviour [24,25] and recalled the techniques required to manufacture tools after several years [26]. Similarly, there is evidence that chimpanzees can plan ahead by, for instance, producing and gathering stones that they require for later use [27]. In monkeys, planning abilities may be not as complex, although several species are known to plan one or two steps ahead in sequencing tasks (e.g., rhesus monkeys [28,29] and capuchin monkeys, *Sapajus apella* [30]) and might even anticipate future events (e.g., squirrel monkeys, *Saimiri sciureus* [31]). NHPs also have a complex understanding of the objects surrounding them and partially appreciate causal relationships, although there are likely differences in these abilities across species [32].

NHPs also have very complex social lives. Most diurnal NHP species live in large groups that include many females and one to several males [33] who can form long-term social bonds with each other, which are crucial to increasing their fitness (e.g., savannah baboon, *Papio cynocephalus*, and Assamese macaques, *Macaca assamensis* [34–38]). Several NHPs also recognize the social relationships among other group members: monkeys, for instance, know if others have strong social bonds and triadic awareness, understanding the characteristics of social relationships among conspecifics [39,40]. Monkeys can also simultaneously classify group members according to their rank and kinship (e.g., baboons, *Papio hamadryas*, and Japanese macaques, *Macaca fuscata* [41,42]), relying on multiple information sources (e.g., white-faced capuchin monkeys, *Cebus capucinus* [43]). In some NHP species, individuals cooperate with each other by rearing infants together (e.g., Callitrichidae [44]), hunting in group (chimpanzees [45]), sharing food (monkeys and apes [46]), or forming coalitions with specific group members (several monkey and ape species [47,48]). Moreover, monkeys and apes communicate with each other in complex ways, using visual, olfactory, tactile, and auditory signals [49]. For example, some species use specific alarm calls depending on the predator approaching (monkeys and apes [50]), vocalize intentionally (chimpanzees [51]), and even take other group members' knowledge into account when informing them of the presence of danger (chimpanzees [52]). NHPs are also known to anticipate others' behaviour and strategically withhold information from group members. Several species, for instance, can account for their conspecifics' position and visual perspective when competing over food (several monkey and ape species [53–57]). Some species can also anticipate the behaviour of others and understand that the perception/beliefs of other individuals may differ from their own (great apes [58,59]; Japanese macaques [60]). Moreover, several NHP species (including great apes, macaques, and capuchin monkeys) can learn from each other and can transmit this knowledge across generations, giving rise to animal cultures (see [61,62] for reviews).

NHPs also show complex emotional lives. As compared to other mammals, monkeys and apes have relatively slow life histories with long life spans and late ages at weaning and sexual maturity [63]. Therefore, they spend a long time with their mothers, who are crucial for building affective attachment and promote their healthy socio–emotional development. For monkeys and apes, mothers affect the development of their offspring's exploration and play style, affiliative and aggressive patterns, and social preferences and future parenting style as well as provide them with opportunities for social learning and ensure their emotional welfare [64,65]. Moreover, both monkeys and apes may show empathic concern and consolation [66]. Chimpanzees and Tonkean macaques (*Macaca tonkeana*), for instance, show affiliative behaviour toward group members that have been the victim of others' aggression: they preferentially groom conspecifics that have been attacked, especially when they are kin or closely bonded and, in doing so, they decrease the victims' distress and improve their well-being [67,68]. Some authors have also claimed that NHPs engage in cognitively complex forms of prosociality, spontaneously helping others to reach their goals (chimpanzees [69–71]; capuchin monkeys [72]). According to some researchers, NHPs may also show inequity aversion (capuchin monkeys [73]), fairness-related behaviours (several monkey and ape species [74]), and responses to injury and death that closely resemble those shown by humans (e.g., great apes and Sichuan snub-nosed monkeys, *Rhinopithecus roxellana* [75–77]).

The Primate order includes more than 500 species from around 80 genera [78,79], and it is therefore clear that there are important differences across NHP species in terms of the complexities of their cognitive and emotional lives. However, although great apes usually appear to have the most sophisticated abilities, experimental evidence shows that a variety of NHP species "have rich subjective lives filled with intention and emotion" [80] (p. 15). Therefore, these studies provide a view of NHPs as a taxon with highly complex behavioural, cognitive, and socio-emotional lives, long-term memories and a sense of future, diversified social relationships, sensitivity to others' emotions, empathic concern, and the ability to adopt others' points of view [66,81,82].

3. Welfare Considerations

Some of the welfare concerns that have been raised for NHPs are the same as those associated with the use of other vertebrates in invasive research, including housing in poor environments that do not meet the species' behavioural necessities, lack of control over the aversive conditions, exposure to painful and stressful procedures, and death [83–85]. As for other species, NHPs used in invasive research may also experience additional sources of stress, depending on the protocols used and the severity of the invasive procedures (see e.g., [2]) including transport, social isolation, food and water deprivation, withdrawal from drugs, and repeated surgical procedures [86].

The use of NHPs in invasive research also raises further welfare concerns. Having complex cognitive and emotional skills implies that NHPs may be especially sensitive to the negative consequences of invasive procedures. This might depend on two main reasons. First, NHPs have long-term memories of past events, and some species are also known to plan ahead [19,27]. This implies that they can recall previous stressful experiences for many years, and perhaps imagine their future, amplifying the stress and/or pain they experience. Second, NHPs have very complex social lives. This implies that they have specific needs and requirements that make it very challenging to provide them with decent welfare conditions in a laboratory setting. For example, they typically have long life spans and can thus spend years as experimental subjects [80]; they have complex social relationships that are extremely difficult to reproduce in a laboratory [87]; and they need appropriate environments that promote complex species-typical behaviours and provide sufficient cognitive stimulation [88–90].

Unfortunately, living conditions in experimental settings hardly satisfy NHPs' most basic needs. Monkeys up to 15 kg, for instance, can be legally housed in cages as small as 0.5 m^3 in the USA, 0.63 m^3 in China, and 3.6 m^3 in Europe, with metal grid floor being allowed outside Europe [91]. In most countries, NHPs used in invasive procedures are given some form of enrichment activity, such as "objects to manipulate, a variety of foods, [...] behavioural training or time in exercise rooms", and, in some cases, "access to music, television to watch, or touch-screen computers" ([91], p. 4). Although some researchers contend that these forms of enrichment increase NHPs' welfare (e.g., [91]), such artificial activities provide no ecologically valid experience and no sufficient stimulation from a primatological perspective. Moreover, although "all NHPs are housed with visual and auditory contact with conspecifics" ([91], p. 4), and thus not necessarily in groups, this is certainly not enough to satisfy their complex social needs. Furthermore, depending on the experimental procedures and on the legal requirements of the countries in which NHPs are housed (see below), NHPs may also experience suffering from early maternal separation [92–95]. In Europe, for instance, macaques are usually weaned around 8 months [91], a developmental stage in which infants are still largely dependent on their mothers under natural conditions [96], whereas transgenic monkeys are usually separated from their mothers from birth (see [91]) with a strong negative impact on their behaviour and physiology [97].

Several studies have shown how study subjects that do not undergo death during invasive experiments do suffer long-term psychological effects as a consequence of their traumatic experiences in laboratories [98]. These studies have been mostly conducted on NHPs that have been moved to sanctuaries after spending several years as laboratory subjects and are therefore mostly limited to research on chimpanzees [99]. Traumatic experiences due to invasive research, for instance, are linked to the occurrence of abnormal behaviours similar to posttraumatic stress disorder and depression [100,101]. Traumatized chimpanzees are 19 times more likely to be diagnosed with depression and 88 times more likely to show posttraumatic stress disorder than their wild conspecifics [100]. These problems persist for years, even after animals have been released from laboratories and moved to sanctuaries [98,100]. Although environmental enrichment may reduce the occurrence of some of these negative consequences (e.g., stereotypies and social impairment), it appears to have no effect on more severe forms (e.g., self-harm) [102]. Therefore, invasive research may clearly have a strong impact on NHP welfare, producing negative long-term effects on their physical and physiological well-being.

4. Ethical Issues Linked to the Use of NHPs in Invasive Research

The use of NHPs in invasive research is mostly considered necessary to increase our knowledge and improve our health and quality of life [103–110]. By allowing discoveries that are important for humans (e.g., reducing pain or increasing their life expectancy and well-being), invasive research provides humans with benefits that would overcome the costs inflicted to animals in terms of pain, suffering, or death, thus justifying their use (see [14,111] for a discussion).

From an ethical perspective, the main issue is whether we have the right to inflict pain and death to NHPs if this provides humans with important benefits. Even if NHPs were perfect models to study humans, as most researchers contend [103–110], is causing stress, pain, and/or death to NHPs ethically justified under the premise that human benefits outweigh NHP suffering [112]? For some authors, such a utilitarian approach posits several problems. Its main limit is that it implies an assessment of all the costs and benefits linked to invasive research [113]. First, it is often hard to reliably predict the benefits that a research project really provides, and, indeed, the majority of invasive studies on animals fail to translate to humans [94,111,114]. Second, it is extremely difficult to objectively compare the costs and benefits that involve different species [115–119]. As noted by Arnason [115], benefits to humans often appear to compensate the costs to NHPs only if NHP interests are highly disregarded. Third, quantifying the costs of invasive research to NHPs is no easy endeavour. In theory, all procedures reducing the welfare of NHPs imply some costs, from distress and the impossibility to express the species–specific behavioural repertoire to malnutrition, discomfort, pain, injury, isolation, diseases, and negative valence emotions (e.g., fear, anxiety, and anger), depending on the exact procedures used and their severity (see [120]; see previous sections). These sources of stress can have a strong negative impact on NHP health and welfare, as discussed above [94,121].

Even if the costs and benefits of invasive research could be objectively measured, the utilitarian approach may have intrinsic limits because it assumes the potential ethical validity of invasive research as long as the benefits to humans outweigh the costs to NHP (see [94]). Such an approach is not used in human research and has been rightly strongly condemned, both ethically and legally, when having been used in specific historical contexts [122]. Killing or harming human participants (without their consent and providing them with no direct benefits) is ethically and legally unacceptable, no matter how important the benefits for the rest of humanity might be. In other words, human research is based on the principle of deontology, which implies that all humans have an inherent value and are thus considered more than just a means to an end. According to some researchers, there is no reason why the principle of deontology should not be applied also to other species, as any living organism has an inherent value and deserves ethical consideration [123]. Given that species other than humans have a complex perceptual, emotional, and cognitive life (see above), they should also be granted the right not to be harmed or killed for the benefit of others [94]. This approach implies reconsidering the moral statuses of other species in research, i.e., the extent to which NHPs matter for their own sake (see below). Moreover, according to some researchers, additional principles of human research ethics should be applied to NHPs [115], including Beauchamp and Childress's [124] four principles on autonomy, non-maleficence, beneficence and justice [94], the principles of consent and autonomy [125–127], and dissent by the study subjects [128]. For instance, it has been proposed that NHPs should be granted the same legal protection as human subjects who cannot provide informed consent through the assignment of legal guardians and higher safeguards [127].

5. Legal Protection Granted to NHPs around the World

Despite recent advances in animal research ethics, the protection that is legally granted to NHPs in research is still largely based on utilitarian principles (i.e., the use of NHPs is allowed if the expected benefits to humans overcome the harm caused to the study subjects). In particular, most national legislations regulating invasive research are currently based on the principle of the 3Rs, which was first formulated by Russell and Burch in 1959 [129]. The principle of the 3Rs pursues (i) the replacement of animals in invasive research through alternative methods requiring no living animals, (ii) the reduction in the number of study subjects involved in research projects, and (iii) the refinement of all the experimental procedures in order to increase animal welfare and reduce their suffering [129]. Historically, the principle of the 3Rs has played a crucial role in drawing attention to the problems of using animals in research and in suggesting possible ways to address them and still underpins the legislation on the protection of research animals in several countries. Below, we discuss four examples (i.e., Europe, USA, Japan, and China) in which the principle of the 3Rs has been differently implemented depending on the pre-existing legal and cultural background. Indeed, there are, to date, no common international regulations for invasive research on NHPs as they exist for research on humans (see [91] for a discussion).

In Europe, the use of NHPs in research is regulated by Directive 2010/63/EU of the European Parliament, which aims to eliminate major disparities in the use of research animals across European countries and ensure a minimum level of protection for study subjects [2]. The Directive in particular rules the replacement and reduction of animals and the refinement of their conditions whenever they are used in "procedures", i.e., invasive or non-invasive practices for experimental, scientific, or educational purposes that may cause them suffering, pain, distress, or long-term harm [2]. According to Article 4 of the Directive, alternative methods (e.g., cell lines, modelling, or simulations) should be used whenever possible instead of live animals [2]. If there is no alternative method available, researchers should ensure the highest standards of animal welfare and care and reduce the number of animals to a minimum (without compromising the objectives of the project). During procedures, one should reduce pain, suffering, distress, or long-term harm as much as possible, carefully handling the animals, providing high-quality species-specific housing, husbandry, and enrichment activities as well as anaesthesia and pain relief as needed. According to the Directive, the use of NHPs in procedures raises further ethical concerns as it is especially worrying for the general public and requires special care to comply with the behavioural and social needs that NHPs face in captive settings [2]. Hence, procedures on NHPs can be carried out only if there are no alternative methods to reach the same objectives, they are for basic research projects aimed to preserve the species, or if it might have potentially high benefits for humans (e.g., the avoidance, prevention, or treatment of diseases or abnormalities and the development, manufacture, or testing of the effectiveness and safety of drugs, food, and other substances or products) [2].

Necessarily, EU Directive 2010/63/EU is a compromise among different interests. On the one hand, for instance, it states that animals have an intrinsic value which must be respected and that they should always be treated as sentient creatures; it states that procedures causing long-term "severe pain, suffering or distress" should not be allowed; and it identifies the full replacement of research animals as its final goal [2]. On the other hand, the EU Directive considers the use of live animals in invasive research to still be necessary [2] and allows NHPs to be routinely harmed and killed during procedures when the results of the project *might* provide enough benefits to humans and no other alternatives are available. More than ten years after its promulgation, indeed, almost 10 million animals are still used every year in scientific procedures in Europe with around 30,000 belonging to "species of particular public concern", i.e., dogs, cats, and NHPs [12].

In the USA, there is a long history of invasive research with little attention to the welfare of animals [130]. This situation started to change in 1985 when several amendments to the Animal Welfare Act (1966) [131], which is the only federal USA law regulating the use of research animals, required the United States Department of Agriculture (USDA) to implement measures ensuring the psychological well-being of NHPs in research [130]. With these amendments, for the first time, research protocols on NHPs must be reviewed by special committees that also include non-scientist members, and researchers must formally check for the possibility of replacing NHPs, reducing the number of animals used, and refining their living conditions by, for instance, ensuring a captive environment that promotes their psychological well-being [131]. However, these amendments still leave research facilities with a high degree of freedom, as they can, for example, develop their own plans to ensure the welfare of study animals. If a research facility receives funding from the Public Health Service (PHS), however, it also must comply with the PHS policy on human care and use of laboratory animals. This policy implies the institution of special committees to review research proposals on animals and to inspect animal facilities as well as requires researchers to comply with the Guide for the Care and Use of Laboratory Animals [132]. According to the Guide, researchers must minimize the distress and pain inflicted to research animals, sedate or anesthetize animals if the procedures cause relevant pain or distress, and provide living conditions that are appropriate for the species and that guarantee health and well-being to the animals [132]. However, the Guide also introduces several exceptions when these exceptions are justified for scientific purposes (e.g., sedation or anaesthesia may be not used during painful or distressful procedures if necessary for reaching the project's goals) [132]. In May 2019, a spending bill by a committee in the House of Representatives raised serious concerns about animal welfare in research laboratories in the USA and questioned (i) the possibility to really ensure animal well-being and (ii) the validity of animals as models in biomedical research [133]. The bill also urged the National Institutes of Health (NIH) to take concrete steps to reduce and replace NHPs used in invasive research and to regularly report on the progress to Congress [133,134]. In contrast to other NHPs, chimpanzees are offered stronger legal protection in the USA. A study commissioned by the NIH considered the use of chimpanzees for biomedical research inappropriate, especially in consideration of the ethical issues raised by their phylogenetic closeness to humans [135]; so, since 2015, the NIH has completely banned invasive research on chimpanzees [136].

Except for the Animal Welfare Act (1966), which sets relatively low standards of protection for research animals [131], all the restrictions to invasive research in the USA only apply to projects funded by the PHS. Moreover, even when researchers must comply with the Guide for the Care and Use of Laboratory Animals [132], several exceptions are granted depending on the aims of the study. For instance, if researchers aim to specifically test for the effects of social deprivation on NHPs, the obligation of ensuring the social well-being of the study subjects does not have to be fulfilled [130]. In our opinion, if laws seriously ensured welfare standards for NHPs in invasive research, studies on social deprivation should simply not be allowed because they do not comply with these standards. This introduces a certain degree of freedom in the application of the rules that should ensure the welfare of research animals. Indeed, even if the Guide suggests that NHPs should be housed in social groups [132], in 2007, only 38% of the monkeys in indoor labs shared their cage with at least one other individual [130,137]. Moreover, researchers have often noted how abnormal behaviour that signals severe stress in NHPs (e.g., self-harm behaviour) has not decreased since the introduction of the Animal Welfare Act [138], suggesting that the measures taken so far are far from sufficient to really ensure the "psychological well-being" of NHPs. Despite these problems, in 2000, the NIH still recommended an *increase* in the number of NHPs used in invasive research as they would be "crucial" for biomedical and behavioural research and, indeed, according to the USDA, the number of NHPs used for research reached 75,825 in 2017, an increase of 22% as compared to in 2015 [133].

In Japan, the use of animals in research is regulated by the Law for the Humane Treatment and Management of Animals (Law No. 105, 1973; see [139]). According to this law, researchers should avoid purposelessly killing or inflicting harm and injuries to animals, and they should treat them "properly" [139]. Following an amendment in 2005, the Law further requires researchers to implement the principle of the 3Rs, reducing and replacing research animals when possible and refining their living conditions [139]. For research animals, the Law further refers to the Standards Relating to the Care and Management of Experimental Animals [140]. The Standards consider the use of animals in invasive research necessary but also foster the implementation of the 3Rs [139,141]. In 2006, the Science Council of Japan (2006) published the Guidelines for Proper Conduct of Animals Experiments, which provides some very general advices on the aspects that researchers should consider when conducting research on animals [142]. The Guidelines, for instance, attribute the responsibility of properly conducting animal experiments to the directors of the research institutions and suggest that researchers should take into account animal welfare and reduce the pain and distress caused to the study subjects as allowed by their research aims [142].

Crucially, all these restrictions are not legally binding, but they are "promoted appropriately with the understanding of the people" ([142], p. 1). Research institutions, for instance, should voluntarily establish internal regulations on animal research that reflect the general principles expressed in the Law, the Standards, and the Guidelines, but there is no obligation to doing that, and there is extensive freedom in how these standards are effectively implemented in different laboratories. Similarly, there is no obligation to report the number of NHPs used for research purposes [143]. Therefore, animal research in Japan appears to generally prioritize the researchers' right to academic freedom over the animals' rights to health and well-being. For some authors, Japanese legislation on invasive research is surprisingly permissive and, although this may depend on the religious and cultural peculiarities of the country [141,142], it is also true that part of Japanese society actually demands more stringent rules and transparency in the use of research animals by, for instance, claiming access to experimental protocols on NHPs and other animals (see [143]).

In China, the Ministry of Science and Technology (MOST) is responsible for regulating the use of NHPs and other animals in invasive research, in line with the Statute on the Administration of Laboratory Animal Use approved by the State Council in 1988 [144,145]. In 2006, the MOST also issued the Guideline on the Humane Treatment of Laboratory Animals [146], which requires animal research facilities to orient their practices to the principles of the 3Rs and animal welfare and to institute specific committees to review and control the ethics of research projects requiring the use of animals [144,145]. The Guideline, for instance, requires researchers to inflict as little pain as possible during their experiments and to consider the species' welfare when housing animals [146]. However, the Guideline only provides very general recommendations and, being mere regulations of the MOST (and not laws), failure to respect them implies lower sanctions as compared to what happens in other countries [147] (see [144]). Through time, these Statutes and Guidelines have been implemented and complemented through a variety of regional, provincial, and municipal laws, guidelines, and policies [144,145], leading to a complex interplay of different layers that leave a certain flexibility to animal research facilities [147,148]. The Provincial Department of Science and Technology, for instance, is responsible for the use of research animals at the provincial level, but local Administration Offices of Laboratory Animal Use enforce the regulations and provide licenses to both the researchers and research institutions that breed or use animals for research [145].

The Chinese case is especially relevant for the international implications it has. Biomedical research is crucial for the Chinese economy and it is continuously expanding [145,149,150]. In 2011, China had around 40 authorized NHP breeding centres employed more than 100,000 people in animal research labs, used around 25,000 NHPs, and exported 25,000 more to other institutions in Europe and the USA (see [145]). Given that Western countries usually have a much stricter legislative system on invasive research and higher standards for animal welfare, several researchers and companies in Western countries have outsourced animal testing in China, or they have moved (or threatened to move) to China to conduct their studies [147,151]. In this way, researchers face lower legal standards and can conduct studies they would not be allowed to run in Europe or the USA [149,152]. Recently, for instance, the neuroscientist Nikos Logothetis left his position as a director of the Max Planck Institute for Biological Cybernetics in Germany (where he conducted invasive research on NHPs to mainly study the neural mechanisms of perception) to co-direct a new International Center for Primate Brain Research in Shanghai, which will house around 6,000 NHPs, including transgenic monkeys [150]. In 2014, videos of lame and bleeding monkeys secretly recorded in his lab led to an official investigation; although Logotethis and his colleagues were relieved from all accusations and charges at the end, public pressure and growing skepticism on invasive research led the neuroscientist and his colleagues to move their research groups to China [150]. Similarly, after the 2019 USA spending bill raised concerns about animal welfare in research laboratories (see above), the president of the National Association for Biomedical Research, Matthew Bailey, expressed his concern about the interference of Congress and stated that, in this way, "research will be more likely to move to other countries" [134].

These cases have fuelled the debate on the consequences of increasing the protection of research animals. For some people, increasing the legal protection of NHPs and other animals might lead scientists to simply leave for countries with lower standards of animal welfare [11,153,154]. This reasoning echoes economic discussions on the need to lower the protection of workers in Western countries to make them more competitive on the global market and avoid the delocalization of industries and other economic activities. However, a race to the bottom is no answer to the risks implicit in having different welfare standards across the world. Increasing the protection of NHPs is an ethical question that goes beyond merely practical considerations, and that concerns the scientific community and the general public alike (see [153]). Moving (or menacing to move) to another country to conduct invasive research and thus circumvent national legislation has ethical implications that cannot be ignored [147] and that should lead to practical consequences (see below).

6. Brief Overview of Attempts to Stop Invasive Research on NHPs

To date, pressure to increase the legal protection of NHPs and to drastically reduce or ban their use in invasive research mainly comes from three different sources: (i) campaigns or petitions by animal organizations with support from public opinion; (ii) parliamentary questions and spending bills by political groups within national institutions; and (iii) scientific publications by authors who are often experts in ethics, philosophy, or alternative methods.

Campaigns by animal organizations are mainly aimed to inform the public about the state of the art of NHP research and raise concern about the current practices. Following a long-running campaign by Cruelty Free International, for instance, most airlines have stopped transporting NHPs for use in invasive research [155]. Moreover, animal organizations often rely on groups of lawyers that can initiate legal actions to protect specific animals. The Animal Legal Defense Fund, for instance, recently filed two lawsuits against the USDA for its failure to protect the psychological well-being of primates used in biomedical research (New England Anti-Vivisection Society vs. Goldentyer [156]) and for inconsistent inspections of NHP research facilities (Animal Legal Defense Fund and Rise for Animals vs. USDA [157]). Similarly, in 2015 the "Stop Vivisection" movement, with a document signed by 1.17 million citizens, demanded that the European Commission abrogate Directive 2010/63/EU on the protection of animals used in invasive research and ban their use in Europe. The European Commission reiterated that, "for the time being, animal experimentation remains important for protecting human and animal health, and for maintaining an intact environment", but it also stated that Europe is "working towards the ultimate goal of full replacement of animals" ([158], p. 10). Although these campaigns often fail to reach the ultimate goal of banning invasive research, they engage public opinion and maintain high pressure on institutions to raise welfare standards for research animals.

Parliamentary questions and spending bills by political groups within national institutions have a similar function. Often, they are proposed by institutional minorities and have little chance to really achieve their goals. The spending bill approved by a committee in the USA House of Representatives in 2019, for instance, strongly urged research institutions in the USA to more effectively reduce and replace NHPs in invasive research and to report on their progress to Congress. However, the final version of the bill negotiated by the Senate and the House of Representatives was much duller, recognizing the importance of NHPs in invasive research and moderating the requests advanced [134]. Still, these initiatives contributed to raising awareness in the public and making researchers feel that they are accountable for their practices. In 2019, the Environmental Protection Agency of the USA announced the goal of a reduction of invasive studies on mammals by 30% by 2025 and its complete ban by 2035 [134]. Similarly, the National Institutes of Health (NIH) in the USA has organized two workshops on NHP research in the last four years, including one on the science and ethics of biomedical research [134]. Although 2035 is a far-off date, and invasive studies will still be allowed pending administrators' approval despite the lack of real advances following the NIH workshops (see [134]), these events force researchers to at least justify their use of NHPs and explore alternatives.

Although most researchers consider invasive research as still being necessary [103–110], there are several scientific groups and research centres specifically working on alternatives to animal testing. These include the Center for Alternatives to Animal Testing at the Johns Hopkins University in the USA and the University of Konstanz in Germany, which support the creation, validation, and use of alternative approaches to animal research, education, and product-safety testing and publish the peer-reviewed journal ALTEX, which is focused on alternatives to animal experiments. Other research centres include the European Union Reference Laboratory for alternatives to animal testing in Ispra, Italy, which coordinates and implements validation studies of alternative methods, and the non-profit organization Americans for Medical Advancement, which opposes the view that animals are valid models on which to conduct research and fosters the use of alternative scientific procedures. Moreover, there is also a long tradition of studies authored by experts in ethics or philosophy clearly positioned against the use of NHPs in invasive research [94,115,159,160]. These studies often call for the need to overcome the principle of the 3Rs and recognize a higher moral status for NHPs. Some researchers, for instance, demand that NHPs be granted fundamental rights, such as the right to life and bodily and mental integrity, and that these rights be defended before a court through legal representatives [159].

The positions of primatologists, anthropologists, comparative psychologists, and other experts on NHP behaviour and cognition (hereafter referred to as "NHP experts") with regards to the use of NHPs in invasive research is often not clear. In some cases, NHP experts have been consulted during the preparation of reports or opinions on the use of NHPs in invasive research, but their role has usually been very marginal [130,131]. The Weatherall's report from 2006 [8] and the review by Bateson and colleagues from 2011 [7], for instance, are some of the most famous reports on the use of NHPs, but there are virtually no NHP experts among their authors. The Weatherall's report was written by a working group consisting of experts "drawn from outside the active non-human primate research community", in the own authors' words ([8], p. 15). In the Bateson review, eight out of the nine panel members were neuroscientists, physicists, or professors in medical areas with no experience in NHP behaviour, and the chairman of the panel was Patrick Bateson, an eminent behavioural biologist with vast knowledge in several research areas but little experience with important ethological, behavioural, and cognitive aspects of NHPs [7] (also see [161]).

To our knowledge, there is also no international working group led by NHP experts aimed at banning the use of NHPs in invasive research. The only notable exception is perhaps the Great Ape Project, an international movement of ethicists, anthropologists, and primatologists who call for the extension to great apes of fundamental human rights (i.e., right to life, freedom, and non-torture [162]). Despite its limit to focus only on great apes, which can be largely justified by the limited knowledge we had on other NHPs at the time the movement started, the Great Ape Project crucially moves beyond the principle of the 3Rs, fostering the recognition of a higher moral status for great apes. Other projects have had a similar focus on chimpanzees and have resulted in a series of positive changes in terms of legal protection granted to this species, at least in some countries. In Japan, for instance, two organizations, SAGA and GAIN, have obtained the complete ban of invasive procedures on chimpanzees, thanks to the extensive support of the public and the active involvement of primatologists [163]. These organizations foster the post-mortem use of chimpanzees and the use of other non-invasive procedures, and they are currently keeping a database of all the apes in Japan that are used in invasive procedures in order to gradually extend them the protection now granted to chimpanzees [163]. Moreover, these organizations have helped establish a sanctuary for laboratory chimpanzees dismissed from biomedical research, in which chimpanzees are provided with life-long care in social groups and more natural settings [164–166]. Similarly, chimpanzees in the USA are better protected than other NHP species; although there is no federal law prohibiting the ownership of chimpanzees, their use in invasive procedures has been completely banned since 2015, and they are individually registered in a database to better monitor their welfare [136,167,168].

7. Conclusions and Future Perspectives—Beyond the 3Rs

To date, abundant experimental evidence has shown the complexity of the social, emotional, and cognitive lives of monkeys and apes and the long-term effects that invasive research can have on their physical and physiological well-being [169–171]. These scientific advances clearly show that invasive research implies huge costs for NHPs, and therefore raise crucial ethical issues on the use of NHPs in invasive research. Still, the legal protection of NHPs is generally inspired by the principle of the 3Rs, although there is a lot of variation across countries in the practical implementation of this principle. Currently, the pressure to improve NHP legal protection is mainly exerted by animal organizations supported by public opinion, political groups within national institutions, or researchers working on ethics, philosophy, or alternative methods. These often call for a switch from a utilitarian approach (and the principle of the 3Rs) to a deontological approach that questions the general validity of NHP research from an ethical perspective [14,84].

Despite claims by several institutions around the world on the long-term goal of replacing NHPs in invasive research (e.g., [2]), the situation for NHPs has not changed much. Actually, NHPs are still abundantly used worldwide in invasive research, and the refinement measures that are generally used cannot ensure, in our opinion, the welfare of animals with such complex socio-emotional and cognitive lives. Therefore, despite the benefits that invasive research on NHPs may have for humans (including their use as models to test preventive and therapeutic strategies during pandemics), we argue that their use in invasive procedures should be immediately banned. The European Union establishes the full replacement of research animals as its final goal [2], but it also identifies the reasons why the fulfilment of this goal might still require time, including "the risk aversive nature of society" (which would be resistant to new alternative methods), the lack of funds to develop and validate alternative research methods, and "factors related to scientific practice and career progression where dynamics such as competition, the reputation and track record of researchers […] and entrenchment discourage switching from NHPs to alternative […] models" ([11], p. 21). These issues, however, can be easily addressed. As scientists, we can drive the change by ensuring that resources are preferentially devoted to projects using alternative methods, asking researchers to publish more precise details when invasive procedures are used, urging journal editors to ask the opinions of primatologists concerning welfare aspects during the review process, and/or otherwise encouraging researchers to conduct non-invasive studies. By massively diverting research funds to non-invasive research, for instance, it would be possible to foster new research approaches that might emerge to be as beneficial for humans as is invasive research. As NHP experts, we can use experimental evidence to contribute concerted legal opinions and reports worldwide and develop a concrete institutional plan with clear timelines to successfully ban invasive research on NHPs.

Overall, we are convinced that ethical and scientific reasons call for the immediate ban of NHPs in invasive research. Such an approach calls for an overcoming of the principle of the 3Rs; after more than six decades from its original formulation, we think it is time to move forward and re-discuss our current goals based on the competencies and knowledge we have acquired after all these years. Reducing the number of NHPs and refining their living conditions are measures that cannot adequately guarantee their welfare and should thus be considered ethically inappropriate. NHPs need to be granted life and health in the first place, and these rights cannot be waived for research purposes no matter how beneficial it might be for humans. This is even more important when considering that most NHP species are currently threatened with extinction [172]. We are responsible for preserving biodiversity and protecting natural habitats, whereas they play a key role as ecosystem engineers [173]. Based on the advances in the study of NHP behaviour and cognition, and on ethical considerations, we believe that the time has come to definitively stop non-invasive research on NHPs, and that non-invasive behavioural and experimental methods should be considered the only possible approach to the study of NHPs.

Author Contributions: M.P., M.L. and F.A. read literature, conceptualized, and wrote the paper. All authors have read and agreed to the published version of the manuscript.

Funding: Generalitat de Catalunya (2017 SGR-1040 to ML), and the Universitat Rovira i Virgili (2019PFR-URV91 to ML). ML is a Serra Húnter Fellow.

Institutional Review Board Statement: Not applicable (review).

Data Availability Statement: Not applicable (review).

Acknowledgments: We are very grateful to Filippo Aureli for his extensive and generous feedback on previous versions of this MS, to Tetsuro Matsuzawa for his brilliant research and outstanding ethical engagement, and to one other anonymous reviewer for his/her valuable comments. We would like to thank the guest editors of this Special Issue for their extensive support throughout the revision process, and the academic and assistant editors.

Conflicts of Interest: The authors declare no conflict of interest.

References

1. Conlee, K.M.; Hoffeld, E.H.; Stephens, M.L. A demographic analysis of primate research in the United States. *Altern. Lab. Anim.* **2004**, *32*, 315–322. [CrossRef]
2. European Union. *Directive 2010/63/EU of the European Parliament and of the Council of 22 September 2010 on the Protection of Animals Used for Scientific Purposes*; European Union: Brussels, Belgium, 2010. Available online: https://eur-lex.europa.eu/legal-content/EN/ALL/?uri=CELEX:32010L0063 (accessed on 10 March 2021).
3. Fedigan, L.M. Ethical issues faced by field primatologists: Asking the relevant questions. *Am. J. Primatol.* **2010**, *72*, 754–771. [CrossRef]
4. Malone, N.M.; Fuentes, A.; White, F.J. Subjects of knowledge and control in field primatology. *Am. J. Primatol.* **2010**, *72*, 779–784. [CrossRef]
5. Lynch, M.E. Sacrifice and the transformation of the animal body into a scientific object: Laboratory culture and ritual practice in the neurosciences. *Soc. Stud. Sci.* **1988**, *18*, 265–289. [CrossRef]
6. Honess, P. A brief history of primate research: Global health improvements and ethical challenges. *Arch. Med. Biomed. Res.* **2015**, *2*, 151–157. [CrossRef]
7. Bateson, P.; Johansen-Berg, H.; Jones, D.K.; Keverne, E.B.; Matthews, P.M.; Milner, A.D.; Prescott, M.; Raga, I.; Shattock, R.; Strauss, J., III; et al. *Review of Research Using Non-Human Primates*; Biotechnology and Biological Sciences Research Council, Medical Research Council and Wellcome Trust: 2011. Available online: https://mrc.ukri.org/documents/pdf/bateson-review-of-non-human-primates/ (accessed on 18 October 2021).
8. Weatherall, D. *The Use of Nonhuman Primates in Research*; Academy of Medical Sciences: 2006. Available online: https://mrc.ukri.org/documents/pdf/the-use-of-non-human-primates-in-research/ (accessed on 18 October 2021).
9. Botting, J.H.; Morrison, A.R. Animal research is vital to medicine. *Sci. Am.* **1997**, *276*, 83–85. [CrossRef] [PubMed]
10. SCHER. The Need for Non-Human Primates in Biomedical Research, Production and Testing of Products and Devices. Brussels. 2009. Available online: http://ec.europa.eu/health/ph_risk/committees/04_scher/docs/scher_o_110.pdf (accessed on 10 March 2021).
11. Vermeire, T.; Epstein, M.; Hoet, P.; Krätke, R.; Testai, E.; Badin, R.A.; Flecknell, P.; Hudson-Shore, M.; Jones, D.; Langermans, J.A.; et al. *Final Opinion on the Need for Non-Human Primates in Biomedical Research, Production and Testing of Products and Devices, Update 2017*; Scientific Committee on Health, Environmental and Emerging Risks (SCHEER), European Commission: Brussels, Belgium, 2017. Available online: https://ec.europa.eu/health/sites/health/files/scientific_committees/scheer/docs/scheer_o_004.pdf (accessed on 10 March 2021).
12. European Commission. *2019 Report on the Statistics on the Use of Animals for Scientific Purposes in the Member States of the European Union in 2015–2017*; European Commission: Brussels, Belgium, 2020. Available online: https://op.europa.eu/en/publication-detail/-/publication/04a890d4-47ff-11ea-b81b-01aa75ed71a1 (accessed on 10 March 2021).
13. USDA. *Annual Report Animal Usage by Fiscal Year*; U.S. Department of Agriculture, Animal and Plant Health Inspection Service: Washington, DC, USA, 2018.
14. Fenton, A. On the need to redress an inadequacy in animal welfare science: Toward an internally coherent framework. *Biol. Philos.* **2012**, *27*, 73–93. [CrossRef]

15. Jaworska, A.; Tannenbaum, J. The grounds of moral status. In *The Stanford Encyclopedia of Philosophy*; Zalta, E.N., Ed.; Department of Philosophy, Stanford University: Stanford, CA, USA, 2017. Available online: https://plato.stanford.edu/entries/grounds-moral-status/ (accessed on 10 March 2021).
16. Janson, C.H.; Byrne, R.W. What wild primates know about resources—Opening up the black box. *Anim. Cogn.* **2007**, *10*, 357–367. [CrossRef]
17. Zuberbühler, K.; Janmaat, K.R.L. Foraging cognition in non-human primates. In *Primate Neuroethology*; Platt, L.M., Ghazanfar, A.A., Eds.; Oxford University Press: Oxford, UK, 2010; pp. 64–83.
18. Janmaat, K.R.L.; Ban, S.D.; Boesch, C. Chimpanzees use long-term spatial memory to monitor large fruit trees and remember feeding experiences across seasons. *Anim. Behav.* **2013**, *86*, 1183–1205. [CrossRef]
19. Lewis, A.; Berntsen, D.; Call, J. Long-term memory of past events in great apes. *Curr. Dir. Psychol. Sci.* **2019**, *28*, 117–123. [CrossRef]
20. Inoue, S.; Matsuzawa, T. Working memory of numerals in chimpanzees. *Curr Biol.* **2007**, *17*, 1004–1005. [CrossRef]
21. Hoffman, M.L.; Beran, M.J.; Washburn, D.A. Memory for "what", "where", and "when" information in rhesus monkeys (*Macaca mulatta*). *J. Exp. Psychol.* **2009**, *35*, 143–152. [CrossRef]
22. De Castro, V.; Girard, P. Location and temporal memory of objects declines in aged marmosets (*Callithrix jacchus*). *Sci. Rep.* **2021**, *11*, 9138. [CrossRef] [PubMed]
23. Hihara, S.; Obayashi, S.; Tanaka, M.; Iriki, A. Rapid learning of sequential tool use by macaque monkeys. *Physiol. Behav.* **2003**, *78*, 427–434. [CrossRef]
24. Boesch, C.; Boesch-Achermann, H. *The Chimpanzees of the Taï Forest: Behavioural Ecology and Evolution*; Oxford University Press: Oxford, UK, 2000.
25. Sanz, C.M.; Call, J.; Boesch, C. (Eds.) *Tool Use in Animals: Cognition and Ecology*; Cambridge University Press: Cambridge, UK, 2013.
26. Vale, G.L.; Flynn, E.G.; Pender, L.; Price, E.; Whiten, A.; Lambeth, S.P.; Schapiro, S.J.; Kendal, R.L. Robust retention and transfer of tool construction techniques in chimpanzees (*Pan troglodytes*). *J. Comp. Psychol.* **2016**, *130*, 24–35. [CrossRef] [PubMed]
27. Osvath, M. Spontaneous planning for future stone throwing by a male chimpanzee. *Curr. Biol.* **2009**, *19*, 190–191. [CrossRef] [PubMed]
28. Beran, M.J.; Pate, J.L.; Washburn, D.A.; Rumbaugh, D.M. Sequential responding and planning in chimpanzees (*Pan troglodytes*) and rhesus macaques (*Macaca mulatta*). *J. Exp. Psychol.* **2004**, *30*, 203–212. [CrossRef] [PubMed]
29. Scarf, D.; Danly, E.; Morgan, G.; Colombo, M.; Terrace, H.S. Sequential planning in rhesus monkeys (*Macaca mulatta*). *Anim. Cogn.* **2011**, *14*, 317–324. [CrossRef] [PubMed]
30. Beran, M.J.; Parrish, A.E. Sequential responding and planning in capuchin monkeys (*Cebus apella*). *Anim. Cogn.* **2012**, *15*, 1085–1094. [CrossRef]
31. Naqshbandi, M.; Roberts, W.A. Anticipation of future events in squirrel monkeys (*Saimiri sciureus*) and rats (*Rattus norvegicus*): Tests of the Bischof-Kohler hypothesis. *J. Comp. Psychol.* **2006**, *120*, 345–357. [CrossRef] [PubMed]
32. Cacchione, T.; Rakoczy, H. Comparative metaphysics: Thinking about objects in space and time. In *APA Handbook of Comparative Psychology: Perception, Learning, and Cognition*; Call, J., Burghardt, J.M., Pepperberg, I.M., Snowdon, C.T., Zentall, T., Eds.; American Psychological Association: Washington, DC, USA, 2017; pp. 579–599.
33. Mitani, J.C.; Call, J.; Kappeler, P.M.; Palombit, R.A.; Silk, J.B. (Eds.) *The Evolution of Primate Societies*; University of Chicago Press: Chicago, IL, USA, 2012.
34. Archie, E.A.; Tung, J.; Clark, M.; Altmann, J.; Alberts, S.C. Social affiliation matters: Both same-sex and opposite-sex relationships predict survival in wild female baboons. *Proc. R. Soc. B* **2014**, *281*, 20141261. [CrossRef] [PubMed]
35. Schülke, O.; Bhagavatula, J.; Vigilant, L.; Ostner, J. Social bonds enhance reproductive success in male macaques. *Curr. Biol.* **2010**, *20*, 2207–2210. [CrossRef] [PubMed]
36. Silk, J.B.; Alberts, S.C.; Altmann, J. Social bonds of female baboons enhance infant survival. *Science* **2003**, *302*, 1231–1234. [CrossRef] [PubMed]
37. Silk, J.B.; Beehner, J.C.; Bergman, T.J.; Crockford, C.; Engh, A.L.; Moscovice, L.R.; Wittig, R.M.; Seyfarth, R.M.; Cheney, D.L. The benefits of social capital: Close social bonds among female baboons enhance offspring survival. *Proc. R. Soc. B* **2009**, *276*, 3099–3104. [CrossRef] [PubMed]
38. Silk, J.B.; Beehner, J.C.; Bergman, T.J.; Crockford, C.; Engh, A.L.; Moscovice, L.R.; Wittig, R.M.; Seyfarth, R.M.; Cheney, D.L. Strong and consistent social bonds enhance the longevity of female baboons. *Curr. Biol.* **2010**, *20*, 1359–1361. [CrossRef]
39. Cheney, D.L.; Seyfarth, R.M. The representation of social relations by monkeys. *Cognition* **1990**, *37*, 167–196. [CrossRef]
40. Perry, S.; Barrett, H.C.; Manson, J.H. White-faced capuchin monkeys show triadic awareness in their choice of allies. *Anim. Behav.* **2004**, *67*, 165–170. [CrossRef]
41. Bergman, T.J.; Beehner, J.C.; Cheney, D.L.; Seyfarth, R.M. Hierarchical classification by rank and kinship in baboons. *Science* **2003**, *302*, 1234–1236. [CrossRef] [PubMed]
42. Schino, G.; Tiddi, B.; Di Sorrentino, E.P. Simultaneous classification by rank and kinship in Japanese macaques. *Anim. Behav.* **2006**, *71*, 1069–1074. [CrossRef]
43. Kajokaite, K.; Whalen, A.; Panchanathan, K.; Perry, S. White-faced capuchin monkeys use both rank and relationship quality to recruit allies. *Anim. Behav.* **2019**, *154*, 161–169. [CrossRef]

44. Burkart, J.M.; Hrdy, S.B.; Van Schaik, C.P. Cooperative breeding and human cognitive evolution. *Evol. Anthropol.* **2009**, *18*, 175–186. [CrossRef]
45. Boesch, C. Cooperative hunting roles among Tai chimpanzees. *Hum. Nat.* **2002**, *13*, 27–46. [CrossRef]
46. Jaeggi, A.V.; Van Schaik, C.P. The evolution of food sharing in primates. *Behav. Ecol. Sociobiol.* **2011**, *65*, 2125. [CrossRef]
47. Kapsalis, E. Matrilineal kinship and primate behavior. In *Kinship and Behavior in Primates*; Chapais, B., Berman, C.M., Eds.; Oxford University Press: Oxford, UK, 2004; pp. 153–175.
48. Silk, J.B. Kin selection in primate groups. *Int. J. Primatol.* **2002**, *23*, 849–875. [CrossRef]
49. Liebal, K.; Oña, L. Different approaches to meaning in primate gestural and vocal communication. *Front. Psychol.* **2018**, *9*, 478. [CrossRef] [PubMed]
50. Cheney, D.L.; Seyfarth, R.M. Flexible usage in primate social calls. *Proc. Natl. Acad. Sci. USA* **2018**, *115*, 1974–1979. [CrossRef] [PubMed]
51. Schel, A.M.; Machanda, Z.; Townsend, S.W.; Zuberbühler, K.; Slocombe, K.E. Chimpanzee food calls are directed at specific individuals. *Anim. Behav.* **2013**, *86*, 955–965. [CrossRef]
52. Crockford, C.; Wittig, R.M.; Mundry, R.; Zuberbühler, K. Wild chimpanzees inform ignorant group members of danger. *Curr. Biol.* **2012**, *22*, 142–146. [CrossRef]
53. Hare, B.; Call, J.; Tomasello, M. Do chimpanzees know what conspecifics know? *Anim. Behav.* **2001**, *61*, 139–151. [CrossRef] [PubMed]
54. Hare, B.; Call, J.; Agnetta, B.; Tomasello, M. Chimpanzees know what conspecifics do and do not see. *Anim. Behav.* **2000**, *59*, 771–785. [CrossRef]
55. Canteloup, C.; Poitrasson, I.; Anderson, J.R.; Poulin, N.; Meunier, H. Factors influencing deceptive behaviours in Tonkean macaques (*Macaca tonkeana*). *Behaviour* **2017**, *154*, 765–784. [CrossRef]
56. Byrne, R.W.; Whiten, A. *Machiavellian Intelligence: Social Expertise and the Evolution of Intellect in Monkeys, Apes and Humans*; Oxford University Press: Oxford, UK, 1988.
57. Amici, F.; Call, J.; Aureli, F. Variation in withholding information in three monkey species. *Proc. R. Soc. B* **2009**, *276*, 3311–3318. [CrossRef]
58. Kano, F.; Krupenye, C.; Hirata, S.; Tomonaga, M.; Call, J. Great apes use self-experience to anticipate an agent's action in a false-belief test. *Proc. Natl. Acad. Sci. USA* **2019**, *116*, 20904–20909. [CrossRef]
59. Krupenye, C.; Kano, F.; Hirata, S.; Call, J.; Tomasello, M. Great apes anticipate that other individuals will act according to false beliefs. *Science* **2016**, *354*, 110–114. [CrossRef] [PubMed]
60. Hayashi, T.; Akikawa, R.; Kawasaki, K.; Egawa, J.; Minamimoto, T.; Kobayashi, K.; Kato, S.; Hori, Y.; Nagai, Y.; Iijima, A.; et al. Macaques exhibit implicit gaze bias anticipating others' false-belief-driven actions via medial prefrontal cortex. *Cell Rep.* **2020**, *30*, 4433–4444. [CrossRef] [PubMed]
61. Boesch, C. From material to symbolic cultures: Culture in primates. In *The Oxford Handbook of Culture and Psychology*; Valsiner, J., Ed.; Oxford University Press: Oxford, UK, 2012.
62. Cacchione, T.; Amici, F. Insights from comparative research on cultural learning. *Prog. Brain Res.* **2020**, *254*, 247–270. [PubMed]
63. Harvey, P.; Martin, R.D.; Clutton-Brock, T.H. Life histories in comparative perspective. In *Primate Societies*; Smuts, B.B., Cheney, D.L., Seyfarth, R.M., Wrangham, R.W., Struhsaker, T.T., Eds.; The University of Chicago Press: Chicago, IL, USA, 1987.
64. Maestipieri, D. Maternal influences on primate social development. *Behav. Ecol. Sociobiol.* **2018**, *72*, 130. [CrossRef]
65. Cassidy, J.; Shaver, P.R. *Handbook of Attachment: Theory, Research, and Clinical Applications*; Guilford Publications: New York, NY, USA, 2018.
66. De Waal, F.B.M.; Preston, S.D. Mammalian empathy: Behavioural manifestations and neural basis. *Nat. Rev. Neurosci.* **2017**, *18*, 498–509. [CrossRef] [PubMed]
67. Fraser, O.N.; Stahl, D.; Aureli, F. Stress reduction through consolation in chimpanzees. *Proc. Natl. Acad. Sci. USA* **2008**, *105*, 8557–8562. [CrossRef] [PubMed]
68. Palagi, E.; Dall'Olio, S.; Demuru, E.; Stanyon, R. Exploring the evolutionary foundations of empathy: Consolation in monkeys. *Evol. Hum. Behav.* **2014**, *35*, 341–349. [CrossRef]
69. Melis, A.P.; Warneken, F.; Jensen, K.; Schneider, A.C.; Call, J.; Tomasello, M. Chimpanzees help conspecifics obtain food and non-food items. *Proc. Biol. Sci.* **2011**, *278*, 1405–1413. [CrossRef] [PubMed]
70. Warneken, F.; Hare, B.; Melis, A.P.; Hanus, D.; Tomasello, M. Spontaneous altruism by chimpanzees and young children. *PLoS Biol.* **2007**, *5*, e184. [CrossRef] [PubMed]
71. Warneken, F.; Tomasello, M. Altruistic helping in human infants and young chimpanzees. *Science* **2006**, *311*, 1301–1303. [CrossRef] [PubMed]
72. Barnes, J.L.; Hill, T.; Langer, M.; Martinez, M.; Santos, L.R. Helping behaviour and regard for others in capuchin monkeys (*Cebus apella*). *Biol. Lett.* **2008**, *4*, 638–640. [CrossRef]
73. Brosnan, S.; de Waal, F.B.M. Monkeys reject unequal pay. *Nature* **2003**, *425*, 297–299. [CrossRef]
74. Brosnan, S. Justice- and fairness-related behaviors in nonhuman primates. *Proc. Natl. Acad. Sci. USA* **2013**, *110*, 10416–10423. [CrossRef]
75. Anderson, J.R.; Gillies, J.R.; Lock, L.C. *Pan* thanatology. *Curr. Biol.* **2010**, *20*, 349–351. [CrossRef]

76. Biro, D.; Humle, T.; Koops, K.; Sousa, C.; Hayashi, M.; Matsuzawa, T. Chimpanzee mothers at Bossou, Guinea carry the mummified remains of their dead infants. *Curr. Biol.* **2012**, *20*, 351–352. [CrossRef]
77. Yang, B.; Anderson, J.R.; Li, B.G. Tending a dying adult in a wild multi-level primate society. *Curr. Biol.* **2016**, *26*, R403–R404. [CrossRef]
78. International Union for Conservation of Nature, IUCN Red List of Threatened Species, Version 2021–2022. Available online: www.iucnredlist.org (accessed on 22 September 2021).
79. Integrated Taxonomic Information System (ITIS), Order Primates. Available online: www.itis.gov (accessed on 22 September 2021).
80. Prescott, M.J. Ethics of primate use. *Adv. Sci. Res.* **2010**, *5*, 11–22. [CrossRef]
81. De Waal, F.B.M. With a little help from a friend. *PLoS Biol.* **2007**, *5*, e190. [CrossRef] [PubMed]
82. De Waal, F.B.M. The antiquity of empathy. *Science* **2012**, *336*, 874–876. [CrossRef] [PubMed]
83. Bassett, L.; Buchanan-Smith, H.M. Effects of predictability on the welfare of captive animals. *Appl. Anim. Behav. Sci.* **2007**, *102*, 223–245. [CrossRef]
84. Broom, D.M. Animal welfare: Concepts and measurement. *J. Anim. Sci.* **1991**, *69*, 4167–4175. [CrossRef] [PubMed]
85. Olsson, A.S.; Robinson, P.; Pritchett, K.; Sandøe, P. Animal research ethics. In *Handbook of Laboratory Animal Science: Essential Principles and Practices*; Hau, J., van Hoosier, G.L., Eds.; CRC Press: Boca Raton, FL, USA, 2003; pp. 13–30.
86. Conlee, K.M.; Rowan, A.N. The case for phasing out experiments on primates. *Hastings Cent. Rep.* **2012**, *42*, S31–S34. [CrossRef]
87. Hannibal, D.; Bliss-Moreau, E.; Vandeleest, J.; McCowan, B.; Capitanio, J. Laboratory rhesus macaque social housing and social changes: Implications for research. *Am. J. Primatol.* **2016**, *79*, 1–14. [CrossRef] [PubMed]
88. Clark, F. Cognitive enrichment and welfare: Current approaches and future directions. *Anim. Behav. Cogn.* **2017**, *4*, 52–71. [CrossRef]
89. Clark, F. Great ape cognition and captive care: Can cognitive challenges enhance well-being? *Appl. Anim. Behav. Sci.* **2011**, *135*, 1–12. [CrossRef]
90. Meehan, C.L.; Mench, J.A. The challenge of challenge: Can problem solving opportunities enhance animal welfare? *Appl. Anim. Behav. Sci.* **2007**, *102*, 246–261. [CrossRef]
91. Mitchell, A.S.; Hartig, R.; Basso, M.A.; Jarrett, W.; Kastner, S.; Poirier, C. International primate neuroscience research regulation, public engagement and transparency opportunities. *NeuroImage* **2021**, *229*, 117700. [CrossRef]
92. Zhang, X.L.; Pang, W.; Hu, X.T.; Li, J.L.; Yao, Y.G.; Zheng, Y.T. Experimental primates and non-human primate (NHP) models of human diseases in China: Current status and progress. *Zool. Res.* **2014**, *35*, 447–464.
93. Ferdowsian, H.; Merskin, D. Parallels in sources of trauma, pain, distress, and suffering in humans and nonhuman animals. *J. Trauma Dissociation* **2012**, *13*, 448–468. [CrossRef]
94. Carvalho, C.; Gaspar, A.; Knight, A.; Vicente, L. Ethical and scientific pitfalls concerning laboratory research with non-human primates, and possible solutions. *Animals* **2018**, *29*, 12. [CrossRef]
95. Latham, N.; Mason, G. Maternal deprivation and the development of stereotypic behavior. *Appl. Anim. Behav. Sci.* **2008**, *110*, 84–108. [CrossRef]
96. Amici, F.; Kulik, L.; Langos, D.; Widdig, A. Growing into adulthood—How social bonds develop in immature rhesus macaques (*Macaca mulatta*). *Behav. Ecol. Sociobiol.* **2019**, *73*, 18. [CrossRef]
97. Kalin, N.H.; Carnes, M. Biological correlates of attachment bond disruption in humans and nonhuman primates. *Prog. Neuropsychopharmacol. Biol. Psychiatry* **1984**, *8*, 459–469. [CrossRef]
98. Lopresti-Goodman, S.M.; Bezner, J.; Ritter, C. Psychological distress in chimpanzees rescued from laboratories. *J. Trauma Dissociation* **2015**, *16*, 349–366. [CrossRef] [PubMed]
99. Ross, S.R.; Leinwand, J.G. A review of research in primate sanctuaries. *Biol. Lett.* **2020**, *16*, 20200033. [CrossRef]
100. Ferdowsian, H.R.; Durham, D.L.; Kimwele, C.; Kranendonk, G.; Otali, E.; Akugizibwe, T.; Mulcahy, J.B.; Ajarova, L.; Johnson, C.M. Signs of mood and anxiety disorders in chimpanzees. *PLoS ONE* **2011**, *6*, e19855. [CrossRef]
101. Ferdowsian, H.R.; Durham, D.L.; Johnson, C.M.; Brüne, M.; Kimwele, C.; Kranendonk, G.; Otali, E.; Akugizibwe, T.; Mulcahy, J.B.; Ajarova, L. Signs of generalized anxiety and compulsive disorders in chimpanzees. *J. Vet. Behav.* **2012**, *7*, 353–361. [CrossRef]
102. McCowan, B.; Rommeck, I. Bioacoustic monitoring of aggression in group-housed rhesus macaques. *J. Appl. Anim. Welf. Sci.* **2006**, *9*, 261–268. [CrossRef]
103. Anestidou, L.; Johnson, A.F. (Eds.) *Care, Use, and Welfare of Marmosets as Animal Models for Gene Editing-Based Biomedical Research: Proceedings of a Workshop*; National Academies Press: Washington, DC, USA, 2019.
104. Bernardi, S.; Salzman, C.D. The contribution of nonhuman primate research to the understanding of emotion and cognition and its clinical relevance. *Proc. Natl. Acad. Sci. USA* **2019**, *116*, 26305–26312. [CrossRef] [PubMed]
105. Capitanio, J.P.; Emborg, M.E. Contributions of non-human primates to neuroscience research. *Lancet* **2008**, *371*, 1126–1135. [CrossRef]
106. Friedman, H.; Haigwood, N.; Ator, N.; Newsome, W.; Allan, J.S.; Golos, T.G.; Kordower, J.H.; Shade, R.E.; Goldberg, M.E.; Bailey, M.R.; et al. The critical role of nonhuman primates in medical research. *Pathog. Immun.* **2017**, *2*, 352–365. [CrossRef] [PubMed]
107. Mitchell, A.S.; Thiele, A.; Petkov, C.I.; Roberts, A.; Robbins, T.W.; Schultz, W.; Lemon, R. Continued need for non-human primate neuroscience research. *Curr. Biol.* **2018**, *28*, 1186–1187. [CrossRef]

108. Okano, H.; Kishi, N. Investigation of brain science and neurological/psychiatric disorders using genetically modified non-human primates. *Curr. Opin. Neurobiol.* **2018**, *50*, 1–6. [CrossRef]
109. Roberts, A.C.; Clarke, H.F. Why we need nonhuman primates to study the role of ventromedial prefrontal cortex in the regulation of threat- and reward-elicited responses. *Proc. Natl. Acad. Sci. USA* **2019**, *116*, 26297–26304. [CrossRef]
110. Yeung, K.R.; Chiu, C.L.; Pears, S.; Heffernan, S.J.; Makris, A.; Hennessy, A.; Lind, J.M. A cross-sectional study of ageing and cardiovascular function over the baboon lifespan. *PLoS ONE* **2016**, *11*, e0159576.
111. Regan, T. Empty cages: Animal rights and vivisection. In *Contemporary Debates in Applied Ethics*; Cohen, A.I., Wellman, C.H., Eds.; Blackwell Publishing: Oxford, UK, 2005; pp. 77–90.
112. Ferdowsian, H.; Fuentes, A. Harms and deprivation of benefits for nonhuman primates in research. *Theor. Med. Bioeth.* **2014**, *35*, 143–156. [CrossRef]
113. Quigley, M. Non-human primates: The appropriate subjects of biomedical research? *J. Med. Ethics* **2007**, *33*, 655–658. [CrossRef]
114. Hackam, D.G.; Redelmeier, D.A. Translation of research evidence from animals to humans. *JAMA* **2006**, *296*, 1731–1732. [CrossRef]
115. Arnason, G. The emergence and development of animal research ethics: A review with a focus on nonhuman primates. *Sci. Eng. Ethics* **2020**, *26*, 2277–2293. [CrossRef] [PubMed]
116. Arnason, G.; Clausen, J. On balance: Weighing harms and benefits in fundamental neurological research using nonhuman primates. *Med. Healthc. Phil.* **2016**, *19*, 229–237. [CrossRef] [PubMed]
117. Nordgren, A. *For Our Children: The Ethics of Animal Experimentation in the Age of Genetic Engineering*; Rodopi: Amsterdam, The Netherlands, 2010.
118. Gagneux, P.; Moore, J.J.; Varki, A. The ethics of research on great apes. *Nature* **2005**, *437*, 27–29. [CrossRef] [PubMed]
119. Beauchamp, T.L.; Morton, D.B. The upper limits of pain and suffering in animal research: A moral assessment of the European Union's legislative framework. *Camb. Q. Healthc. Ethics* **2015**, *24*, 431–447. [CrossRef] [PubMed]
120. Brambell, F.W. *Report of the Technical Committee to Enquire into the Welfare of Animals Kept under Intensive Livestock Husbandry Systems*; Her Majesty's Stationery Office: London, UK, 1965.
121. Pryce, C.R.; Rüedi-Bettschen, D.; Dettling, A.C.; Feldon, J. Early life stress: Long-term physiological impact in rodents and primates. *News Physiol. Sci.* **2002**, *17*, 150–155. [CrossRef] [PubMed]
122. Pellegrino, E. The Nazi doctors and Nuremberg: Some moral lessons revisited. *Ann. Intern. Med.* **1997**, *127*, 307–308. [CrossRef] [PubMed]
123. Regan, T. *The Case for Animal Rights*; University of California Press: Berkeley, CA, USA, 1983.
124. Beauchamp, T.L.; Childress, J.F. *Principles of Biomedical Ethics*; Oxford University Press: Oxford, UK, 2013.
125. Fenton, A. Can a chimp say "no"? *Camb. Q. Healthc. Ethics* **2014**, *23*, 130–139. [CrossRef]
126. Beauchamp, T.L.; Wobber, V. Autonomy in chimpanzees. *Theor. Med. Bioeth.* **2014**, *35*, 117–132. [CrossRef] [PubMed]
127. Wendler, D. Should protections for research with humans who cannot consent apply to research with nonhuman primates? *Theor. Med. Bioeth.* **2014**, *35*, 157–173. [CrossRef]
128. Kantin, H.; Wendler, D. Is there a role for assent or dissent in animal research? *Camb. Q. Healthc. Ethics* **2015**, *24*, 459–472. [CrossRef] [PubMed]
129. Russell, W.M.S.; Burch, R. *The Principles of Humane Experimental Technique*; Methuen: London, UK, 1959.
130. Gluck, J.P. Moving beyond the welfare standard of psychological well-being for nonhuman primates: The case of chimpanzees. *Theor. Med. Bioeth.* **2014**, *35*, 105–116. [CrossRef]
131. *Animal Welfare Act (1966). Public Law 89-544, Enacted by the 89th United States Congress*; USDA: Washington, DC, USA, 1966.
132. National Research Council. *Guide for the Care and Use of Laboratory Animals*; National Academies Press: Washington, DC, USA, 2011.
133. Reardon, S. US lawmakers propose plan to reduce primate research at National Institutes of Health. *Nature* **2019**. [CrossRef] [PubMed]
134. 2020 U.S. spending bill restricts some animal research, pushes for lab animal retirement. 2019. Available online: https://www.sciencemag.org/news/2019/12/2020-us-spending-bill-restricts-some-animal-research-pushes-lab-animal-retirement (accessed on 10 March 2021).
135. Institute of Medicine. *Chimpanzees in Biomedical and Behavioral Research: Assessing the Necessity*; The National Academies Press: Washington, DC, USA, 2011.
136. Kaiser, J. An end to U.S. chimp research. *Science* **2015**, *350*, 1013. [CrossRef] [PubMed]
137. Baker, K.C.; Weed, J.L.; Crockett, C.C.; Bloomsmith, M.A. Survey of environmental enrichment programs for laboratory primates. *Am. J. Primatol.* **2007**, *69*, 377–394. [CrossRef]
138. Balcombe, J.; Ferdowsian, H.R.; Durham, D. Self-harm in laboratory-housed primates: Where is the evidence that the animal welfare act amendment has worked? *J. Appl. Anim. Welf. Sci.* **2011**, *14*, 361–370. [CrossRef]
139. Kurosawa, T.M. Japanese regulation of laboratory animal care with 3Rs. *AATEX* **2007**, *14*, 317–321.
140. Prime Minister's Office. Standards relating to the care and management of experimental animals (Notice No. 6, 1980). *Exp. Anim.* **1982**, *31*, 228–231.
141. Kagiyama, N.; Nomura, T. Japanese regulations on animal experiments. In *Development of Science-Based Guidelines for Laboratory Animal Care*; ILAR, National Research Council, Ed.; National Academy Press: Washington, DC, USA, 2004; pp. 50–56.

142. Science Council of Japan. Guidelines for Proper Conduct of Animals Experiments. 2006. Available online: http://www.scj.go.jp/en/animal/ (accessed on 10 March 2021).
143. Matsuda, Y. Recent trends in the number of laboratory animals used in Japan. *Altern. Lab. Anim.* **2004**, *32*, 299–301. [CrossRef]
144. Kong, Q.; Qin, C. Laboratory animal science in China: Current status and potential for the adoption of three R alternatives. *Altern. Lab. Anim.* **2010**, *38*, 53–69. [CrossRef]
145. Bayne, K.; Wang, C. Oversight of animal research in China. In *Laboratory Animals*; Guillen, G., Ed.; Academic Press: Cambridge, MA, USA, 2014; pp. 243–266.
146. Ministry of Science and Technology. Guideline on Humane Treatment of Laboratory Animals. 2006. Available online: www.most.gov.cn/zfwj/zfwj2006/200512/t20051214_54389.htm (accessed on 18 October 2021).
147. Cao, D. Ethical questions for research ethics: Animal research in China. *J. Anim. Ethics* **2018**, *8*, 138.
148. Cao, D. *Animals in China: Law and Society*; The Palgrave Macmillan: London, UK, 2015.
149. Cyranoski, D. Monkey kingdom. *Nature* **2016**, *532*, 300–302. [CrossRef]
150. Vogel, G. Monkey facility in China lures neuroscientist. *Science* **2020**, *367*, 496–497. [CrossRef]
151. Xia, C.; Gautam, A. Biopharma CRO industry in China: Landscape and opportunities. *Drug Discov. Today* **2015**, *20*, 794–798. [CrossRef]
152. European Commission. National Competent Authorities for the implementation of Directive 2010/63/EU on the protection of animals used for scientific purposes. In *A Working Document on the Development of a Common Education and Training Framework to Fulfil the Requirements under the Directive*; European Commission: Brussels, Belgium, 2018. Available online: http://ec.europa.eu/environment/chemicals/lab_animals/pdf/guidance/education_training/en.pdf (accessed on 10 March 2021).
153. Bayne, K.; Turner, P.V. Animal welfare standards and international collaborations. *ILAR J.* **2019**, *60*, 86–94. [CrossRef]
154. Chatfield, K.; Morton, D. The Use of Non-human Primates in Research. In *Ethics Dumping*; Schroeder, D., Cook, J., Hirsch, F., Fenet, S., Muthuswamy, V., Eds.; Springer Briefs in Research and Innovation Governance; Springer: Cham, Switzerland, 2018.
155. Cruelty Free International. Available online: https://www.crueltyfreeinternational.org/what-we-do/cargo-cruelty/stop-airlines-transporting-monkeys-animal-experiments (accessed on 10 March 2021).
156. Animal Legal Defense Fund. New England Anti-Vivisection Society vs. Goldentyer. 2020. Available online: https://aldf.org/case (accessed on 10 March 2021).
157. Animal Legal Defense Fund. Animal Legal Defense Fund and Rise for Animals vs. USDA. 2020. Available online: https://aldf.org/case (accessed on 10 March 2021).
158. European Commission. *Communication from the Commission on the European citizens' Initiative "Stop Vivisection"*; European Commission: Brussels, Belgium, 2015. Available online: https://ec.europa.eu/environment/chemicals/lab_animals/pdf/vivisection/en.pdf (accessed on 10 March 2021).
159. Walker, R.L.; Eggel, M. From mice to Monkeys? Beyond orthodox approaches to the ethics of animal model choice. *Animals* **2020**, *10*, 77. [CrossRef] [PubMed]
160. Fasel, R.; Blattner, C.; Mannino, A.; Baumann, T. Grundrechte für Primaten. *Sentience Politics* **2016**, *1*, 1–18.
161. Greek, R.; Hansen, L.A.; Menache, A. An analysis of the Bateson review of research using nonhuman primates. *Med. Bioeth.* **2011**, *1*, 3–22. [CrossRef]
162. Cavalieri, P.; Singer, P. (Eds.) *The Great Ape Project: Equality Beyond Humanity*; Fourth Estate: London, UK, 1993.
163. Matsuzawa, T. SAGA and GAIN for great apes. *Primates* **2016**, *57*, 1–2. [CrossRef]
164. Morimura, N.; Idani, G.; Matsuzawa, T. The first chimpanzee sanctuary in Japan: An attempt to care for the "surplus" of biomedical research. *Am. J. Primatol.* **2011**, *73*, 226–232. [CrossRef] [PubMed]
165. Matsuzawa, T. WISH cages: Constructing multiple habitats for captive chimpanzees. *Primates* **2020**, *61*, 139–148. [CrossRef] [PubMed]
166. Hayashi, M.; Sakuraba, Y.; Watanabe, S.; Kaneko, A.; Matsuzawa, T. Behavioral recovery from tetraparesis in a captive chimpanzee. *Primates* **2013**, *54*, 237–243. [CrossRef] [PubMed]
167. Reardon, S. US government gives research chimps endangered-species protection. *Nature* **2015**. [CrossRef]
168. ChimpCARE. Available online: https://chimpcare.org/ (accessed on 22 September 2021).
169. Ahktar, S. Animal welfare and animal pain: Can pain sometimes be worse for them than for us. In *The Oxford Handbook of Animal Ethics*; Beauchamp, T.L., Frey, R.G., Eds.; Oxford Handbooks Online: Oxford, UK, 2011.
170. Bermond, B. A neuropsychological and evolutionary approach to animal consciousness and animal suffering. *Anim. Welf.* **2001**, *10*, 47–62.
171. Lea, S.E.G. Anticipation and memory as criteria for special welfare consideration. *Anim. Welf.* **2001**, *10*, 195–208.
172. Estrada, A.; Garber, P.A.; Rylands, A.B.; Roos, C.; Fernandez-Duque, E.; Di Fiore, A.; Nekaris, K.A.-I.; Nijman, V.; Heymann, E.W.; Lambert, J.E.; et al. Impending extinction crisis of the world's primates: Why primates matter. *Sci. Adv.* **2017**, *3*, e1600946. [CrossRef] [PubMed]
173. Chapman, C.A.; Bonnell, T.R.; Gogarten, J.F.; Lambert, J.E.; Omeja, P.A.; Twinomugisha, D.; Wasserman, M.; Rothman, J.M. Are Primates Ecosystem Engineers? *Int. J. Primatol.* **2012**, *34*, 1–14. [CrossRef]

animals

MDPI

Article

I Wanna Draw Like You: Inter- and Intra-Individual Differences in Orang-Utan Drawings

Marie Pelé [1,*], Gwendoline Thomas [2], Alaïs Liénard [2], Nagi Eguchi [3], Masaki Shimada [3] and Cédric Sueur [4,5]

1 Anthropo-Lab, ETHICS EA7446, Lille Catholic University, 59000 Lille, France
2 UFR LLSHS, Université Sorbonne Paris Nord, 95100 Paris, France; thomas.gwenndoo@gmail.com (G.T.); alais.lienard@edu.univ-paris13.fr (A.L.)
3 Department of Animal Sciences, Teikyo University of Science, Uenohara 409-0193, Japan; nagi.1999.1118@i.softbank.jp (N.E.); masakishimada@japan.email.ne.jp (M.S.)
4 Université de Strasbourg, CNRS, IPHC UMR 7178, F-67000 Strasbourg, France; cedric.sueur@iphc.cnrs.fr
5 Institut Universitaire de France, 75231 Paris, France
* Correspondence: marie.pele@univ-catholille.fr

Simple Summary: Drawing has increasingly been proposed as an enrichment activity for captive primates in zoological parks and research institutes. The monkeys and apes are free to use the materials at their disposal and are not constrained or conditioned to show this behaviour. This provides a good opportunity to collect drawings by non-human primates and allows for comparative studies between hominids. This study is based on 749 drawings recovered from five orang-utans (*Pongo pygmaeus*) at Tama Zoological Park in Japan, where caretakers regularly facilitated drawing activities for the apes. Analyses showed that individuals differ in their drawing style, especially in the colours used, the space they filled, and the shapes they drew. One individual, Molly, did more complex drawings than other individuals and drew differently according to the seasons and her age. This study is the first to reveal such individual differences and can give some clues about the emergence of drawings in human beings.

Abstract: This study analyses 749 drawings by five female Bornean orang-utans (*Pongo pygmaeus*) at Tama Zoological Park in Japan. We searched for differences between individuals but also tried to identify possible temporal changes among the drawings of one individual, Molly, who drew almost 1300 drawings from 2006 to 2011. An analysis of the drawings was carried out after collecting quantitative and qualitative variables. Our findings reveal evidence of differences in the drawing style of the five individuals as well as creative changes in Molly's drawing style throughout her lifetime. Individuals differed in terms of the colours used, the space they filled, and the shapes (fan patterns, circles, or loops) they drew. Molly drew less and less as she grew older, and we found a significant difference between drawings produced in winter, when orang-utans were kept inside and had less activity, and those produced during other seasons. Our results suggest that the drawing behaviour of these five orang-utans is not random and that differences among individuals might reflect differences of styles, states of mind, and motivation to draw.

Keywords: primate cognition; scribbles; evolutionary anthropology; art; aesthetics

check for updates

Citation: Pelé, M.; Thomas, G.; Liénard, A.; Eguchi, N.; Shimada, M.; Sueur, C. I Wanna Draw Like You: Inter- and Intra-Individual Differences in Orang-Utan Drawings. *Animals* **2021**, *11*, 3202. https://doi.org/10.3390/ani11113202

Academic Editors: Lucia Regolin and Charles Snowdon

Received: 14 September 2021
Accepted: 8 November 2021
Published: 9 November 2021

1. Introduction

Drawing behaviour has been studied in non-human primate species such as chimpanzees (*Pan troglodytes*), gorillas (*Gorilla gorilla*), orang-utans (*Pongo pygmaeus*), capuchin monkeys (*Cebus apella*), and rhesus macaques (*Macaca mulatta*) (for a review, see [1]). Drawing has increasingly been proposed as an enrichment activity for captive primates in zoological parks and research institutes. The monkeys and apes are free to use the materials at their disposal and are not constrained or conditioned to show this behaviour. This provides a good opportunity to collect drawings by non-human primates and allows for comparative studies between hominids [2].

This study is based on 1433 drawings recovered from five orang-utans (*Pongo pygmaeus*) at Tama Zoological Park in Japan, where caretakers regularly facilitated drawing activities for the apes. Orang-utans (*Pongo* sp.) are phylogenetically close to humans, with 97% common genetic heritage [3]. This makes them ideal candidates to help us understand the origins of drawing behaviour. They can develop highly sophisticated cognitive abilities [4] and complex tool use [5–7] that are both at least comparable to those observed in chimpanzees. They are also well known for their curiosity [8]. The wrists of orang-utans are more flexible than those of chimpanzees, making it easier for them to draw as they can bend their hands backwards [9]. In our drawings database, the daily number of drawings varied significantly between individuals, despite the same opportunities to draw, showing that they had different levels of motivation for this activity. This observation leads us to our first question: Do orang-utans show differences in their drawing/marking behaviour? More specifically, it is interesting to note that one female, called Molly, drew almost 1300 drawings in her last five years of life. She had the opportunity to draw regularly from her 54th birthday onwards (in 2006, at Tama Zoo). Moreover, the colours and page filling in Molly's drawings might have been affected by aspects of her daily life, such as the identity of her caretaker and events in her environment [10]. This leads us to our second question: Was there any temporal evolution (due to environmental changes and/or Molly's own development) in her drawing behaviour?

Previous studies on captive orang-utans and other primates showed that they will continue to draw even in the absence of rewards [10–13]. These findings are consistent with the Gestalt principle found in young children, which links the scribbling activity to a discovery of motor play activity [14]. Like in humans, spontaneous drawings indicate an intrinsic interest in exploratory and manipulative play for captive non-human primates [10,14]. Moreover, when tracks have already been drawn on the paper, further scribblings are added, suggesting that visible tracks have some kind of reinforcing value [12]. Schiller [15] worked with an 18-year-old female chimpanzee called Alpha, who showed a keen interest in drawing. In order to study figure formation and position, different stimuli (squares or scribbles) were drawn on paper by researchers. In terms of figure production, the results showed that Alpha mainly used two types of strokes: short dashes and almost parallel broad zigzag strokes (also referred to as the fan pattern). The findings for the placement of figures revealed that when a single figure was positioned off-centre, Alpha drew in the largest open space, producing what Schiller calls a 'sort of balance between her markings and the presented figure' (p. 104). A similar experiment by Morris [16] with a one-year-old male chimpanzee named Congo showed similar results: three-quarters of the 40 free drawings on blank sheets of paper showed marks on spaces that had previously been empty, and half of the drawings featured marks that were concentrated at the centre of the paper. Although these results are only descriptive (like those found for Alpha), they do seem to support Schiller's findings concerning the strong tendency to mark a central figure and to position marks in the blank space opposite an offset figure, as well as the inclination to simply enjoy scribbling. Smith [17] was the first to use quantitative methods in order to analyse chimpanzee drawings. All his results were consistent with the previously cited findings. Later, Boysen et al. [11] continued the stimulus-drawing test in chimpanzees by presenting 18 different figures to three chimpanzees. Like Smith [17], Boysen attained a point where the presence of any stimulus figure on the page elicited more centralised markings than in cases where the sheet of paper was blank. The notion of the centre of a page therefore seems to be an important point to take into consideration when studying the emergence of drawing capacities.

Kinematic aspects could be seen as precursors of a graphic representation [18], yet drawings by non-human primates include different types of marks, such as the straight lines, curves, loops, or hook-like strokes observed in drawings by chimpanzees [12]. Although chimpanzees only develop the skill to use a mark-making instrument at the age of

20–23 months, these marks can be observed before the age of 13–23 months through the use of touchscreens. In a comparative study, a set of 396 pictures was collected, made up of 40 drawings by chimpanzees, 153 by gorillas, 146 by orang-utans, and 57 non-figurative drawings by children up to and including the age of four [19]. Zeller [19] noticed that the main features of the orang-utans' patterns were diagonals, arcs, and curvilinear designs, while those of the gorillas contained mainly arcs and open curving lines, and a very high proportion of dots. In contrast, the chimpanzees' drawings were mainly characterised by the use of straight and jointed lines. Unlike Tanaka, Zeller noted that orang-utans were the only apes that can use a closed loop or circular pattern [19]. This is the most difficult pattern to produce because it requires high levels of motor control. Another important point concerns the use of colours. Findings in children have shown that their choice of colours reflected their emotional state at that time [20], and it seems that this may be the case in individuals of other species, as with the case of the orang-utan called Molly [10]. These results support the hypothesis that there is a choice in the use of colour, the type of strokes, and the use of space, and that drawings from great apes do not result from totally random scribbles [2,19]. Despite their non-figurative nature, drawings produced by apes could therefore provide a great deal of meaningful information.

In this study, we analyse 790 drawings (selected from a total of 1433 drawings) drawn by five orang-utans (Figure 1) and compare them at an interindividual level. Indeed, a study of five chimpanzees by Morris [16] suggests that interindividual differences exist in drawing and that personality may have an impact on the way marks are distributed. We analyse drawings at an individual level in order to assess temporal changes in this behaviour. We expect to find preferences pertaining to the use of colours, as well as a trend of centring drawings, the use of curved strokes and circles inside the drawings, and possibly a number of differences in marking behaviour between individuals. We also expect possible temporal changes in the drawings of one individual, Molly, who drew almost 1300 drawings from 2006 to 2011. To our knowledge, this is the first study analysing such a large number of drawings by orang-utans, and indeed by non-human primates in general.

Figure 1. Examples of two drawings for each of the five orang-utans.

2. Materials and Methods

2.1. Subjects and Collection of Drawings

From 2006 to 2016, 1433 drawings by five female orang-utans at Tama Zoological Park in Tokyo were collected by the caretakers (Table 1). More information about individuals can be found in the Great Apes Information Network (GAIN [21]). We analysed 749 drawings (26/26 drawings by Gypsy, 16/16 by Julie, 32/32 by Yuki, 60/60 by Kiki, and 656/1299 by Molly). White, high-quality paperboard (272 × 242 mm) and 16 different coloured crayons were given to the orang-utans as permanently available enrichment. Several paperboards were simultaneously provided to orang-utans to avoid competition, but no conflict was observed given the small number of drawings for four of the five individuals. The sixteen colours were always available each drawing day, with crayons being replaced when necessary. The drawing activity was not part of a research protocol implemented at the zoo but was for enrichment [22]. The drawings were recovered for analysis after the drawing activities, meaning that we had no control over the methodology. However, zookeepers were present during all the drawing session in order to watch the orang-utans' behaviour and to ensure the identity of the drawer for each drawing; they did not encourage them to draw in anyway. Individuals did not provide the same number of drawings each year, suggesting different motivations between apes and over time. These different numbers of drawings seem to show that the animals did not simply copy each other as, most of the time, the other individuals did not copy Molly when she was drawing. Moreover, no sharing was observed, showing how apes lose interest after finishing their drawings [16]: they just dropped their drawings or even tore them up. These damaged drawings were not considered for analyses. When orang-utans finished a drawing, they could take another paperboard given at their disposal. However, individuals, including Molly, never created more than two drawings per day. We pseudo-randomly selected 656 drawings by Molly to obtain approximately equal numbers for each three-month period from 2006 to 2011 (about 30 drawings for each of the 19 periods; Molly died in 2011), meaning that we randomly selected a drawing in the first period of three months, then the second period, etc., until the 19th one; then we selected a second drawing for the first period again, then the second period, etc., until we had about 30 drawings in each period. The orang-utans were housed in social groups in outdoor and indoor enclosures with environmental enrichment. Food (pellets) and water were supplied ad libitum. Fruits and vegetables were distributed twice daily.

Table 1. Information on individuals.

Name	Date of Birth	Birthplace	Number of Drawings	Percentage of Drawings	Stay at Tama Zoo	Drawing Period
Molly	1 January 1952	Wild	1299	91%	2005–2011	2006–2011
Gypsy	1 January 1955	Wild	26	2%	1958–2017	2007–2014
Julie	6 May 1965	Captivity	16	1%	2005–Today	2009–2014
Yuki	1 January 1970	Wild	32	2%	2008–2015	2008–2010
Kiki	21 October 2000	Wild	60	4%	2007–Today	2010–2016

2.2. Ethics Note

The Tama Zoological Park Ethics Board approved this non-invasive behavioural study, which complied with the Code of Ethics of the Japanese Association of Zoos and Aquariums.

2.3. Data Collection

To ensure that our observations were as accurate as possible, we used the Gimp 2.10.22 software to apply a 10 × 10 grid [14] to every drawing. Torn sheets (41 drawings) were not analysed due to the risk of missing values. Two observers (GT and AL) made the measurements with a correlation of $0.964 \leq r \leq 1$ for all variables (Figure S1 in Supplementary Materials). A third observer (MP) then confirmed the qualitative variable measurements. Disagreements were resolved by the intervention of the third observer.

2.3.1. Quantitative Variables

We first looked for non-exclusive quantitative variables (Figure S2) as defined in previous studies [14]: (1) the coverage rate, defined as the number of cells containing one or more strokes out of the total number of cells; (2) the overlap rate, defined as the number of cells containing strokes of at least two different colours that overlap, divided by the coverage rate and multiplied by 100; (3) the solid colour rate, defined as the number of drawing grid cells that were covered at rates of 50% or more, divided by 100; and (4) the distance to the centre, defined as the absolute distance between the centre of an ellipse surrounding the design and the centre of the grid. Three other indices were measured via the Gimp 2.10.22 software, namely, (5) the number of colours used, (6) the mean deviation of the colour spectrum, and (7) the standard deviation of the colour spectrum.

2.3.2. Qualitative Variables

We also extracted data from every drawing (8) for the predominant colour used by the individual and the drawings shapes, as defined in previous studies [12,23]: (9) fan patterns, (10) circles, (11) triangles, and (12) loops (Figure S3). These indices are not exclusive, as several can be found in the same drawing. A fan pattern is a stroke making at least three round trips of angles $\leq 45°$. A loop is a curved stroke forming a single distinct angle where it intersects itself. A circle is a curved stroke intersecting itself without distinct angles. A triangle is a flat loop with distinct angles.

2.4. Statistical Analysis

We conducted two types of analyses: a comparative study between individuals and a longitudinal study of data for Molly. For both, we first checked for high correlations (>80%) between our drawing variables using the chart correlation function of the R package PerformanceAnalytics [24]. Our correlation chart revealed no strong correlation between our quantitative variables (Figure S4). A principal component analysis (PCA) was then carried out with the R package FactoMineR [25] in order to reduce our 12 variables and group them into various dimensions that were then further interpreted from a biological perspective. A PCA with Varimax rotation was also carried out [26] but did not explain more variance than classical PCA did. For the comparative study between individuals, the coordinates of each drawing in each dimension were used to compare individuals two-by-two during comparisons of means for each dimension with the functions kruskal_test() and pairwise.wilcox.test() from the R packages coin [27] and stats [28].

For the longitudinal study of data for Molly, we carried out another PCA. Here, we wanted to study the effect of both 3-month periods ($N = 19$ periods) corresponding to seasons ($N = 4$ seasons) on Molly's drawing behaviour. We then applied a multifactorial linear model (LM) for each dimension of our PCA using the function lm() from the R package car [29]. The potential collinearity between our two predictor/predictive variables was tested by the calculation of the variance inflation factor VIF from the R package car. These diagnostics revealed a VIF of <1.1 for both predictors, indicating that there was no notable problem of collinearity. p-values for LM were calculated via Monte Carlo sampling with 10,000 permutations, using the function PermTest() of the R package pgirmess [30]. Permutation tests for LM were well adjusted for the moderate sample size [31] and did not require a normal distribution of model residuals [32].

Pairwise post hoc comparisons for significant LMs were carried out with the function pairwisePermutationTest() from the R package rcompanion [33]. α levels (0.05) were Benjamini–Hochberg corrected for pairwise comparisons and adjusted for permutation tests. Finally, we analysed the main colour used by orang-utans with the function chisq.test() of the R package stats. We only reported differences where $p < 0.05$. All statistical analyses were done with R, version 4.0.3 [28].

3. Results

3.1. Comparative Analyses between Individuals

On average, the coverage rate of the drawings by the five orang-utans was 50 ± 30%, the overlap rate was 20 ± 30%, and the solid colour rate was 10 ± 20%. For the average number of colours used and shapes drawn, the five orang-utans used 3.0 ± 1.8 colours and drew 1.8 ± 2.0 fan patterns, 0.0 ± 0.2 circles, 0.2 ± 0.6 triangles, and 0.7 ± 1.2 loops per drawing. The colour spectrum had a mean of 0.8 ± 0.1 and a standard deviation of 0.1 ± 0.1. Finally, strokes were applied 33.7 ± 25.9 mm from the centre, on average. Variations of each metric per individual are presented in Figure S5.

The three dimensions retained in the PCA had an eigenvalue above 1 and described 63.5% of the explained variance of the dataset. Each metric showed a higher loading in one dimension than in the two others (Table 2, Figure 2). Thus, the first dimension of the PCA (eigenvalue = 3.86, variance = 35.2%) was mainly explained by the filling variables: the recovery rate, the overlap rate, the solid colour rate, the number of fan patterns, the number of colours, and the distance to the centre. Colour variables applied to the second dimension (eigenvalue = 1.92, variance = 17.5%), i.e., the standard deviation and the mean of the colour spectrum. Finally, geometrical shapes were associated with the third dimension (eigenvalue = 1.19, variance = 10.8%), i.e., the number of triangles, loops, and circles.

Table 2. Loadings of the metrics on the three PCA dimensions of our dataset (five orang-utans). Bold values indicate the dimension in which each variable is retained.

	Dim.1	Dim.2	Dim.3
Coverage rate	**0.84**	0.34	−0.02
Overlap rate	**0.86**	0.17	−0.15
Number of colours	**0.72**	0.32	−0.26
Fan patterns	**0.76**	−0.12	−0.05
Circles	0.10	0.14	**0.50**
Triangles	0.08	0.10	**0.73**
Loops	0.33	0.31	**0.50**
Colour mean	−0.39	**0.79**	−0.14
Std. deviation of colour mean	0.38	**−0.82**	0.15
Distance to centre	**−0.61**	−0.31	−0.02
Solid colour rate	**0.74**	−0.37	−0.07

At least one individual was different from the other ones in the first dimension (Kruskal–Wallis: $\chi^2_4 = 50.31$, $p < 0.001$). Pairwise comparisons revealed that Molly had higher values than Gypsy and Kiki. Kiki also differed from Julie and Yuki (see pairwise statistics in Table S1, Figure 3a). At least one individual was different from the other ones in the second dimension ($\chi^2_4 = 150.48$ $p < 0.001$). Pairwise comparisons revealed that Molly had significantly higher values than all other individuals. Julie had lower values than Kiki and Yuki (Table S1, Figure 3b). At least one individual was different from the other ones in the third dimension ($\chi^2_4 = 44.51$, $p < 0.001$), with differences between Gypsy and Molly (Table S1, Figure 3c). Statistical analyses for each of the 12 metrics are given in the Supplementary Materials Section (Tables S2–S4, Figure S5). The metrics of Dimension 1 were all different for at least one individual ($\chi^2_4 > 11.53$, $p < 0.02$), with data for Molly being globally different from the data of the other individuals (more recovery, more overlap, more colours, closer to the centre) and Kiki showing different results to those of Yuki. Indeed, Kiki presented the lowest values for recovering, overlapping, and colours whilst showing the second highest values after Molly. Although the metrics of the second dimension also revealed interindividual differences ($\chi^2_4 > 40.08$, $p < 0.001$), the pairwise comparisons yielded more restricted results with only Molly showing very low values and Kiki showing very high values of standard deviation of the colorimetric profile. This indicates that Molly filled the sheet and showed less contrast, whilst Kiki's drawing showed high contrast due to the few but strong marks on the paper (see Figure 1). Finally, the only difference between individuals in Dimension 3 concerned the number of loops (χ^4_{433} = 25.07, $p < 0.001$), with Molly drawing more loops than Kiki.

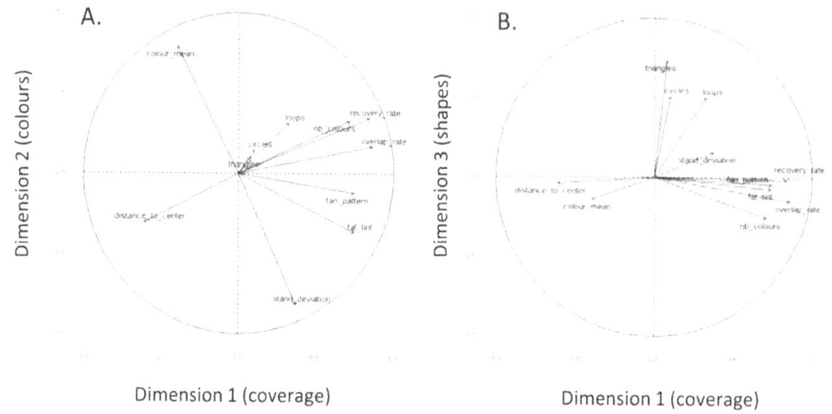

Figure 2. Distribution of variables (each arrow is a variable) for (**A**) Dimension 1 and Dimension 2 and for (**B**) Dimension 1 and Dimension 3. The arrow length indicates the correlation with the dimensions. Dimensions' values are in arbitrary units.

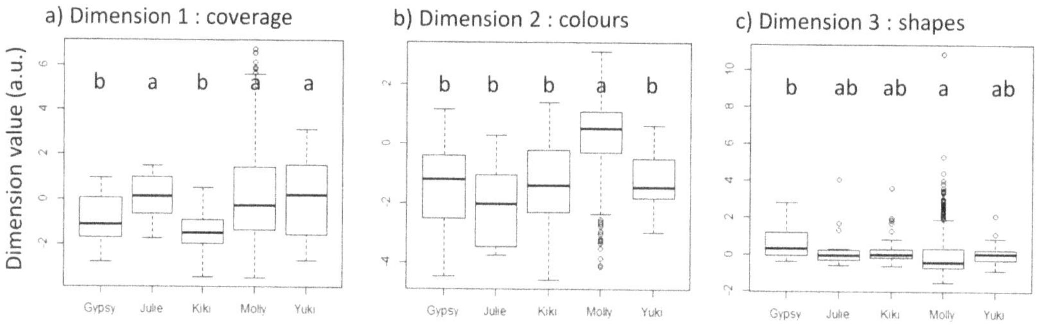

Figure 3. Box plots showing the differences between individuals in each of the three drawing dimensions. Individuals having different letters show statistical differences. a.u. means arbitrary unit.

The main colour found in each drawing revealed a non-random and significant difference in the colours used between individuals (Figure S6). For Kiki (χ^2_{11} = 62,647, $p < 0.001$) and Molly (χ^2_{11} = 289.07, $p < 0.001$), the main colour in their drawings was green (representing 27.45%, and 22.6% of the drawings, respectively). For Gypsy (χ^2_{11} = 41.72, $p < 0.001$), Julie (χ^2_{11} = 29, p = 0.002), and Yuki (χ^2_{11} = 53.2, $p < 0.001$), the main colour in their drawings was red (representing 40%, 33.3%, and 36.67% of the drawings, respectively).

3.2. *Longitudinal Changes in Molly's Drawing*

The number of drawings by Molly over the six years is shown in Figure S7. Three dimensions were retained in the PCA. Their sum represented 64.3% of the explained variance of the dataset (dimension 1: eigenvalue = 4.26, variance = 38.7%; dimension 2: eigenvalue = 1.60, variance = 14.5%; dimension 3: eigenvalue = 1.21, variance = 11%). Metrics are found in the same three dimensions for Molly as for the four other individuals (Table S5). The effect of the 3-month periods and seasons on the three dimensions of the PCA were calculated. There was a significant effect of both seasons (linear model with permutation test: $p < 0.0001$) and periods ($p < 0.0001$) on the first dimension. The filling behaviour of drawings decreased with the periods (linear regression: *t*-value = −4.7, $p < 0.001$, $R^2 = 0.03$, F = 22.73, Figure 4a). Pairwise comparisons revealed that winter had significantly lower values than all other seasons ($p < 0.005$, Figure 4b). There was no significant effect of the seasons or the periods on the second ($p = 0.162$ and $p = 0.869$, respectively) or on the third dimension ($p = 0561$ and $p = 0.769$, respectively). More precisely, Molly's drawings differed according to the seasons and time in terms of the coverage rate ($p = 0.002$ and $p < 0.0001$, respectively), the overlap rate ($p = 0.002$ and $p < 0.0001$), the solid colour rate ($p <0.0001$ and $p < 0.0001$), the distance to the centre ($p = 0.04$ and $p = 0.0002$), and the fan patterns ($p = 0.045$ and $p < 0.0001$), with the winter globally always being different from other seasons. However, the number of colours used did not differ according to time ($p = 0.09$) or the seasons ($p = 0.273$). Although Dimension 2 did not show differences for the time and the period, the mean colorimetric profile showed differences according to the season ($p = 0.009$) with higher values in winter ($p < 0.03$) than other seasons, because Molly filled the paper sheet less in winter, thus making these drawings brighter. Dimension 3 did not show any differences according to time and period, but the number of loops decreased with time ($p = 0.02$, z-score = −2.29) and was higher in summer ($p < 0.045$) compared to other seasons. Other metrics did not show any difference according to the time or seasons.

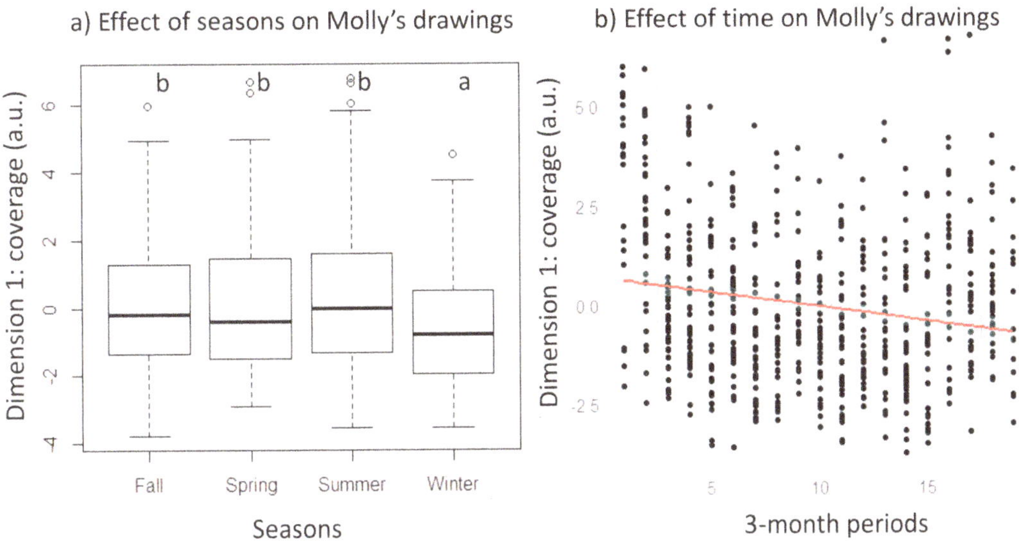

Figure 4. Effect of (**a**) seasons and (**b**) time on Dimension 1 (coverage) of Molly's drawings. For (**b**), linear regression with standard error lines is plotted using ggplot2 R package (method = 'lm' [34]). a.u means arbitrary units.

The analysis of the main colour qualitative variable revealed a non-random and a significant difference in the main colour used by Molly in the four seasons (Figure S8): winter ($X^2_9 = 100.16$, $p < 0.001$), spring ($X^2_9 = 90{,}912$, $p < 0.001$), summer ($X^2_9 = 87{,}385$, $p < 0.001$), and autumn ($X^2_9 = 103.27$, $p < 0.001$). Molly preferred green in summer and winter and swapped to pink for spring and autumn. Purple was used more in spring compared to other seasons.

4. Discussion

We studied drawings by orang-utans in order to consider their cognitive capacities phylogenetically within the framework of other hominoids like humans or chimpanzees. In particular, we looked for differences in the drawing behaviour of five female orang-utans, one of which was in her last years of life. Whilst this sample size is limited, this is the first time that such a quantitative study has been done in orang-utans and more generally in primates showing differences between individuals and observed temporal changes in individuals' drawings. Indeed, many studies on cognition can be done with a small sample size when the study's aim is to show what individuals are capable of and how they differ in their capacities [2,4,35,36]. We used a new way to quantitatively analyse drawing with PCA. Principal component analysis is important and innovative for this kind of behavioural analysis because it allows one to (1) combine many variables having different meanings, (2) increase the potentiality of finding differences between the studied categories (here the individuals and the three-month intervals), (3) and obtain dimension having biological or psychological meaning that could not be found without such an analysis. We used PCA to understand drawing development in humans and found three dimensions, which are diversity, efficiency, and sequentiality, allowing representativeness to be understood in drawings [37,38]. Indeed, we found differences between the scribblings of children and adults in these dimensions because adults have representativeness and intention when they draw such figures, whilst toddlers have not. Therefore, the perspectives of such analyses in drawing are huge and could lead to important results in understanding the emergence of drawing in hominids. In our current study, we found three dimensions that had biological and cognitive meaning: the filling aspect, the colour or contrast aspect, and the shape aspect. This study proved to be exploratory and innovative because of these three dimensions and because it showed interindividual and intraindividual differences in drawings in a species other than humans. These two results seem to show the implication of cognitive aspects going beyond a simple motivation for putting colours on a paperboard.

For most of the studied time, orang-utans used several colours in their drawings. Zeller [19] found that blue was the most commonly used colour used in the drawings of apes (including orang-utans) and children. The main colour used in our orang-utans' drawings differed between individuals. As orang-utans see colours as we do [39], this difference could reflect either an aesthetic or an emotional preference [40]. In children, colours are linked to emotion [40], including in their drawing [20]. This link with emotion may have an evolutionary origin concerning mate choice and competition [41]. Molly, who usually used green and pink colours, mainly chose the colour red when another orang-utan was giving birth [10]. All the individuals drew patterns that were described in Kellogg's children's scribbles classification [23]. Like children, orang-utans can draw multiple lines (called fan patterns), loops, and circles. Orang-utans have more dexterity than other great apes since they can draw curved lines [19]. Orang-utans have also drawn triangles [42]. Thus, in a way, orang-utans draw better than other non-human apes. The drawing technique used by orang-utans might also be interesting. Previous studies showed that orang-utans often hold the drawing tool between their fingers. However, one female was regularly observed laying the pastel on the sheet before rolling it with her hand [42]. This behaviour could explain some large fan patterns observed in orang-utan drawings. In many studies, primates demonstrated an ability to draw fan patterns. We can cite the chimpanzees Congo and Bella, and the capuchin monkey Pablo [16]. Drawings by chimpanzees, human children, and orang-utans therefore have a lot in common, and these new findings may enrich the phylogeny of drawing behaviour among primates.

Our comparative study revealed differences in drawing behaviour among orang-utans, especially for the filling dimension. Here, it is not simply a question of one individual standing out from other individuals, but several differences observed between the five individuals: the drawings by Molly were the most complex (more amply filled than pictures by other individuals, with the use of more shapes and colours) followed by those drawn by Yuki. Kiki also showed differences to the other individuals with her simple but strongly marked drawings (i.e., one colour used, pressing hard on the crayon). We can attribute these differences to personality, motivation, or even to different cognitive abilities between individuals. Indeed, in humans, great interindividual and interindividual differences are observed in drawings [38]. The same could be observed in non-human primates [1] as suggested in chimpanzees [2]. Studies in human children suggest that these interindividual differences might be due to varying levels of cognitive skill maturation [43] and the different speeds at which children learn to draw [2]. According to Willats [44], there may be an interrelation between a child's stage in their drawing skill development and their increasing comprehension of their living world. Our orang-utans have different life experiences. While Molly, who did more complex drawings, was born in the wild and had lived in two zoological parks and given birth four times, Kiki, who produced fewer complex drawings, was born in captivity and quickly moved to Tama Zoological Park, where she gave birth to one baby. Molly had also lost her sight in one eye between 1993 and 1996 (personal communications from Mr. Kurotori, Tama zookeeper). These different experiences could perhaps explain the differences found in the way they drew. We can also mention the age difference between Molly, who was 54 years old at the beginning of her drawing period, and Kiki, who was the youngest of the study group at just 10 years old. Kiki's minimalist use of colours and space in the paper was evidence of either her drawing style or a lack of motivation or interest in the drawing activity. However, Kiki was not the least experienced of the individuals in terms of drawing, even if she was the youngest: she produced 60 drawings, whereas 44-year-old orang-utan Julie produced just 16 drawings. As in humans, younger apes can possess an intrinsic motivation to draw but can be more interested in the objects (crayons and paperboards) than the drawing [2,12]. However, with age, complexity of drawing can increase. In human beings, Martinet et al. [2] showed that adults add more details in their drawing compared to children. This was confirmed by analysing different drawings metrics [37]. A parallel could be made between humans and non-human apes for drawing as is done for other activities, such as the social ones [45,46].

Data for Molly showed a higher mean and a lower standard deviation of the colour spectrum. This indicates a higher diversity in Molly's drawings, many of which are bright and have lower levels of contrast. Indeed, although Molly did sometimes fill the sheet, particularly with fan patterns—as described for the chimpanzee Congo [16]—many of her drawings were almost empty. Other orang-utans showed smaller patterns, which contrasted with the white background. Moreover, Molly seemed to press less on the crayons than the other individuals, which explains the lower contrast (lowest standard deviation of the colour spectrum) in comparison to data for other individuals. The drawings of our orang-utans (and especially those drawn by Molly) confirm the findings of Smith [17], who was the first to report that chimpanzees tended to draw near the centre of the page. Zeller [19] confirmed this finding for other apes. The interindividual differences we describe in orang-utans' drawing in terms of the way individuals draw suggest the existence of different personalities, as first suggested by Morris [16]. The wide difference in the number of productions per individual also shows different levels of interest in the activity; this finding confirms previous observations in orang-utans and in chimpanzees [11,12,19].

The longitudinal study then demonstrated that Molly's drawings evolved over time and between seasons. She used fewer colours, less space, and drew further from the centre as the years progressed. At the end of her life, Molly had fat deposits above her eyes and had to use her hands to lift it out of the way, meaning that she could use her hands less to draw. She also became blind in her left eye, which diminished her visual field. However, this impairment did not decrease her drawing capacities much and she still showed more complex drawings than other individuals at the end of her life. Regarding the differences observed in winter, the results tend to show that Molly's frame of mind changed. The orang-utans in our study group do not go outside in winter, and the weather in Japan is very cold for them. They might become bored, and their motivation may decrease. In this way, our results show that Molly did not systematically use the same main colour in her drawings for all seasons: while she preferred purple as the main colour in spring (in 23.4% of her drawings), she used it much less in summer (7.25%), autumn (3.27%), and winter (5.45%). As Hanazuka and his colleagues [10] hypothesised for non-human primates, drawings could be a window into the internal state of the orang-utan. Moreover, there are no visitors during winter. Evidence that Molly draws more loops and changes the colour she uses in summer could be a cue indicating a good mood due to the weather and the presence of more visitors. Further research is needed to assess whether the number of loops is just a reflection of Molly's motivation or of a more cognitive aspect such as symbolism [47].

However, this study is limited by the enrichment origin of these drawings, which restricts the possibilities of controlled conditions. Several drawings were covered with stains and were dirty, with some shapes overlapping. It is difficult to be precise and objective when measuring variables such as the shapes present in the drawing or the predominant colour, especially where water had blurred the patterns. In the future, we could therefore use touchscreen tablets instead of paper [2]. The use of a tablet requires a genuine trust between the researcher and the animal established from an early age so as not to put the researcher in danger [48]. Alternatively, this can also be easily done with secure and safe touchscreens installed outside the caging. However, animals do not have to confound with the usual cognitive training and tests, which seem to be difficult [1] as indicated by a previous study aiming to make two chimpanzees draw a line [49]. Drawing on tablets has several advantages, such as the extraction of more important information such as temporal data. This method will reveal differences between individuals in terms of the order of colours used but also the fractality of the patterns [2]. It would also be interesting to study drawings by males to see if these preferences change according to sex, although this may be difficult as males have previously demonstrated less interest in this task [42]. Another question pertains to the part of the body orang-utans use to draw. Indeed, some apes use their hands while others use their mouth [42]. We also need to increase the sample size, studying more orang-utans including males and studying other species such as chimpanzees, gorillas, or even some species that have never been studied, such as gibbons. However, this sample size did not impair our results as our analyses were able to show for the first time interindividual and intraindividual differences in the drawings of a non-human species.

5. Conclusions

Overall, the fact that orang-utans draw freely without constraint shows that they are capable of doing so, as shown in previous studies [10,19]. Our study used a systematic and complete analysis to show that apes exhibit differences in their drawings. Applying this methodology to different ape species could allow for an understanding of behavioural and cognitive differences in drawings linked to other cognitive (e.g., sociality, curiosity) or personality (anxiety, boldness) aspects. To go further in the understanding of this heterogeneity, it would be interesting to study these species (chimpanzees, orang-utans, gorillas, and gibbons, which have never been studied concerning drawing) and individuals further, but also to study mother–offspring pairs in order to identify a possible transmission of the motivation or ability to draw. Cultural transmission of habits has been shown, for example, in chimpanzees [50] and in orang-utans [51]. This type of study would contribute to our understanding of the origins of drawing in humans.

Supplementary Materials: The following are available online at https://www.mdpi.com/article/10.3390/ani11113202/s1. Figures S1–S8 and Tables S1–S5. Figure S1: Inter-observer correlations for quantitative variables in the classical analysis, Figure S2: On the left, schematic (top) and actual (bottom) examples of (a) a covered cell, (b) an overlapped cell and (c) solid colour rate. On the right, calculation of the distance to the centre. C is the centre of the paper sheet, M is the centre of the drawing ellipse, and d is the diameter, Figure S3: examples of (a) a fan pattern, (b) a circle, (c) a loop and (d) a triangle, Figure S4: Correlation chart of our quantitative variables, Figure S5: Boxplots of each drawing metric per individual, Figure S6: Frequency of drawings per main colour per individual, Figure S7: Number of drawings according to the seasons over the years, Figure S8: Frequency of Molly's drawings per main colour per season, Table S1: pairwise comparisons tests (p-values) for each dimension between the five individuals, Table S2: pairwise comparisons tests (p-values) for each metric of Dimension 1, Table S3: pairwise comparisons tests (p-values) for each metric of Dimension 2, Table S4: pairwise comparisons tests (p-values) for each metric of Dimension

3, Table S5: Loadings of the metrics on the three PCA dimensions of our dataset for Molly. Bold values indicate the dimension in which each variable is retained.

Author Contributions: Conceptualization, C.S. and M.P. methodology, C.S., G.T. and M.P.; validation, C.S. and M.P.; formal analysis, G.T., A.L., M.P. and C.S.; investigation, G.T., A.L., M.P. and C.S.; resources, N.E. and M.S.; data curation, N.E. and M.S.; writing—original draft preparation, C.S. and G.T.; writing—review and editing, all authors; visualization, C.S., A.L. and G.T.; supervision, C.S., M.P. and M.S.; project administration, C.S., M.P. and M.S.; funding acquisition, C.S., M.P. and M.S. All authors have read and agreed to the published version of the manuscript.

Funding: This project has received financial support from the CNRS through the MITI interdisciplinary programs, the University of Strasbourg through an IDEX Exploratory Research program, and from a KAKENHI program.

Informed Consent Statement: Not applicable.

Data Availability Statement: The data presented in this study are available on request from the corresponding author. The data are not publicly available due to Tama Zoological Park's request and copyright.

Acknowledgments: We are grateful to the Tama Zoological Park in Japan for the providence of orang-utans drawings.

Conflicts of Interest: The authors declare no conflict of interest. The funders had no role in the design of the study; in the collection, analyses, or interpretation of data; in the writing of the manuscript, or in the decision to publish the results.

References

1. Martinet, L.; Pelé, M. Drawing in Nonhuman Primates: What We Know and What Remains to Be Investigated. *J. Comp. Psychol.* **2020**, *135*, 176–184.
2. Martinet, L.; Sueur, C.; Hirata, S.; Hosselet, J.; Matsuzawa, T.; Pelé, M. New Indices to Characterize Drawing Behavior in Humans (Homo Sapiens) and Chimpanzees (*Pan troglodytes*). *Sci. Rep.* **2021**, *11*, 3860. [CrossRef] [PubMed]
3. Locke, D.P.; Hillier, L.W.; Warren, W.C.; Worley, K.C.; Nazareth, L.V.; Muzny, D.M.; Yang, S.-P.; Wang, Z.; Chinwalla, A.T.; Minx, P.; et al. Comparative and Demographic Analysis of Orang-Utan Genomes. *Nature* **2011**, *469*, 529–533. [CrossRef] [PubMed]
4. Dufour, V.; Pelé, M.; Neumann, M.; Thierry, B.; Call, J. Calculated Reciprocity after All: Computation behind Token Transfers in Orang-Utans. *Biol. Lett.* **2009**, *5*, 172–175. [CrossRef] [PubMed]
5. Lethmate, J. Tool-Using Skills of Orang-Utans. *J. Hum. Evol.* **1982**, *11*, 49–64. [CrossRef]
6. Bardo, A.; Cornette, R.; Borel, A.; Pouydebat, E. Manual Function and Performance in Humans, Gorillas, and Orangutans during the Same Tool Use Task. *Am. J. Phys. Anthropol.* **2017**, *164*, 821–836. [CrossRef] [PubMed]
7. Mendes, N.; Steinbeis, N.; Bueno-Guerra, N.; Call, J.; Singer, T. Preschool Children and Chimpanzees Incur Costs to Watch Punishment of Antisocial Others. *Nat. Hum. Behav.* **2018**, *2*, 45–51. [CrossRef] [PubMed]
8. Damerius, L.A.; Burkart, J.M.; van Noordwijk, M.A.; Haun, D.B.; Kosonen, Z.K.; Galdikas, B.M.; Saraswati, Y.; Kurniawan, D.; van Schaik, C.P. General Cognitive Abilities in Orangutans (Pongo Abelii and *Pongo pygmaeus*). *Intelligence* **2019**, *74*, 3–11. [CrossRef]
9. Mackinnon, J. The Behaviour and Ecology of Wild Orang-Utans (*Pongo pygmaeus*). *Anim. Behav.* **1974**, *22*, 3–74. [CrossRef]
10. Hanazuka, Y.; Kurotori, H.; Shimizu, M.; Midorikawa, A. The Effects of the Environment on the Drawings of an Extraordinarily Productive Orangutan (Pongo Pygmaeus) Artist. *Front. Psychol.* **2019**, *10*, 2050. [CrossRef]
11. Boysen, S.T.; Berntson, G.G.; Prentice, J. Simian Scribbles: A Reappraisal of Drawing in the Chimpanzee (*Pan troglodytes*). *J. Comp. Psychol.* **1987**, *101*, 82–89. [CrossRef] [PubMed]
12. Tanaka, M.; Tomonaga, M.; Matsuzawa, T. Finger Drawing by Infant Chimpanzees (*Pan troglodytes*). *Anim. Cogn.* **2003**, *6*, 245–251. [CrossRef]
13. DeLoache, J.S.; Pickard, M.B.; LoBue, V. How very young children think about animals. In *How Animals Affect Us: Examining the Influences of Human–Animal Interaction on Child Development and Human Health*; American Psychological Association: Washington, DC, USA, 2011; pp. 85–99, ISBN 978-1-4338-0865-4.
14. Casti, A.B. Reaction to Stimulus Figures in Chimpanzee Drawings. Master Thesis, Central Washington University, Ellensburg, WA, USA, 2016.
15. Schiller, P.H. Figural Preferences in the Drawings of a Chimpanzee. *J. Comp. Physiol. Psychol.* **1951**, *44*, 101. [CrossRef]
16. Morris, D. *The Biology of Art: A Study of the Picture-Making Behaviour of the Great Apes and Its Relationship to Human Art*; Taylor & Francis: Abingdon, UK, 1962.
17. Smith, D.A. Systematic Study of Chimpanzee Drawing. *J. Comp. Physiol. Psychol.* **1973**, *82*, 406–414. [CrossRef] [PubMed]
18. Adi-Japha, E.; Levin, I.; Solomon, S. Emergence of Representation in Drawing: The Relation between Kinematic and Referential Aspects. *Cogn. Dev.* **1998**, *13*, 25–51. [CrossRef]
19. Zeller, A. 'What's in a Picture?' A Comparison of Drawings by Apes and Children. *Semiotica* **2007**, *2007*, 181–214. [CrossRef]

20. Crawford, E.; Gross, J.; Patterson, T.; Hayne, H. Does Children's Colour Use Reflect the Emotional Content of Their Drawings? *Infant Child Dev.* **2012**, *21*, 198–215. [CrossRef]
21. Matsuzawa, T. SAGA and GAIN for Great Apes. *Primates* **2016**, *57*, 1–2. [CrossRef] [PubMed]
22. Sueur, C.; Pelé, M. Importance of living environment for the welfare of captive animals: Behaviours and enrichment. In *Animal Welfare: From Science to Law*; La Fondation Droit Animal, Éthique et Sciences: Paris, France, 2019; pp. 175–188. Available online: https://www.fondation-droit-animal.org/proceedings-aw/importance-of-living-environment-for-the-welfare-of-captive-animals/ (accessed on 7 November 2021).
23. Kellogg, R. *Analyzing Children's Art*; National Press Books: Palo Alto, CA, USA, 1969.
24. Peterson, B.G.; Carl, P. PerformanceAnalytics: Econometric Tools for Performance and Risk Analysis. 2020. Available online: https://rdrr.io/cran/PerformanceAnalytics/man/PerformanceAnalytics-package.html (accessed on 7 November 2021).
25. Lê, S.; Josse, J.; Husson, F. FactoMineR: A Package for Multivariate Analysis. *J. Stat. Softw.* **2008**, *25*, 1–18. [CrossRef]
26. Kaiser, H.F. The Varimax Criterion for Analytic Rotation in Factor Analysis. *Psychometrika* **1958**, *23*, 187–200. [CrossRef]
27. Hothorn, T.; Hornik, K.; van de Wiel, M.A.; Zeileis, A. Implementing a Class of Permutation Tests: The Coin Package. *J. Stat. Softw.* **2008**, *28*, 1–23. [CrossRef]
28. R Core Team. *R: A Language and Environment for Statistical Computing*; R Foundation for Statistical Computing: Vienna, Austria, 2020.
29. Fox, J.; Weisberg, S. *An R Companion to Applied Regression*, 3rd ed.; Sage: Thousand Oaks, CA, USA, 2019.
30. Giraudoux, P. Pgirmess: Spatial Analysis and Data Mining for Field Ecologists. *R Package Version* **2018**, *1*, 678.
31. Horiguchi, M.; Uno, H. On Permutation Tests for Comparing Restricted Mean Survival Time with Small Sample from Randomized Trials. *Stat. Med.* **2020**, *39*, 2655–2670. [CrossRef]
32. Good, P. *Permutation, Parametric and Bootstrap Tests of Hypotheses*, 3rd ed.; Taylor & Francis: Abingdon, UK, 2005.
33. Mangiafico, S. Rcompanion: Functions to Support Extension Education Program Evaluation. 2021. Available online: https://www.researchgate.net/publication/307608990_rcompanion_Functions_to_support_extension_education_program_evaluation_R_statistical_package (accessed on 7 November 2021).
34. Wickham, H. Ggplot2. *Wiley Interdiscip. Rev. Comput. Stat.* **2011**, *3*, 180–185. [CrossRef]
35. Pelé, M.; Thierry, B.; Call, J.; Dufour, V. Monkeys Fail to Reciprocate in an Exchange Task. *Anim. Cogn.* **2010**, *13*, 745–751. [CrossRef] [PubMed]
36. Krupenye, C.; Kano, F.; Hirata, S.; Call, J.; Tomasello, M. Great Apes Anticipate That Other Individuals Will Act According to False Beliefs. *Science* **2016**, *354*, 110–114. [CrossRef]
37. Martinet, L.; Sueur, C.; Beltzung, B.; Pelé, M. Understanding the Emergence of Drawing Behaviour with Age: A Multi-Metric Analysis. Available online: https://osf.io/aybz6 (accessed on 7 November 2021).
38. Sueur, C.; Martinet, L.; Beltzung, B.; Pelé, M. Making Drawings Speak through Mathematical Metrics. *arXiv* **2021**, arXiv:2109.02276.
39. Tigges, J. On Color Vision in Gibbon and Orang-Utan. *Folia Primatol.* **1963**, 188–198. [CrossRef]
40. Mikellides, B. Colour psychology: The emotional effects of colour perception. In *Colour Design*; Elsevier: Amsterdam, The Netherlands, 2012; pp. 105–128.
41. Rowland, H.M.; Burriss, R.P. Human Colour in Mate Choice and Competition. *Philos. Trans. R. Soc. B Biol. Sci.* **2017**, *372*, 20160350. [CrossRef] [PubMed]
42. Vancatova, M. Creativity and Innovative Behaviour in Primates on the Example of Picture-Making Activity of Apes. Available online: https://nsu.ru/xmlui/bitstream/handle/nsu/3372/05.pdf;sequence=1 (accessed on 7 November 2021).
43. Saito, A.; Hayashi, M.; Takeshita, H.; Matsuzawa, T. The Origin of Representational Drawing: A Comparison of Human Children and Chimpanzees. *Child Dev.* **2014**, *85*, 2232–2246. [CrossRef]
44. Willats, J. *Making Sense of Children's Drawings*; Psychology Press: Hove, UK, 2006; ISBN 978-1-135-62498-9.
45. Almeling, L.; Hammerschmidt, K.; Sennhenn-Reulen, H.; Freund, A.M.; Fischer, J. Motivational Shifts in Aging Monkeys and the Origins of Social Selectivity. *Curr. Biol.* **2016**, *26*, 1744–1749. [CrossRef] [PubMed]
46. Sueur, C.; Quque, M.; Naud, A.; Bergouignan, A.; Criscuolo, F. Social Capital: An Independent Dimension of Healthy Ageing. 2021. Available online: https://ecoevorxiv.org/bg94n/ (accessed on 7 November 2021).
47. Gardner, H.; Wolf, D. *The Symbolic Products of Early Childhood*; Lawrence Erlbaum Associates, Inc.: Mahwah, NJ, USA, 1987.
48. Matsuzawa, T. The 40th Anniversary of the Ai Project: The Commemorative Gift Is a Silk Scarf Painted by Ai the Chimpanzee. *Primates* **2017**, *58*, 261–265. [CrossRef] [PubMed]
49. Iversen, I.H.; Matsuzawa, T. Visually Guided Drawing in the Chimpanzee (*Pan troglodytes*)1. *Jpn. Psychol. Res.* **1996**, *38*, 126–135. [CrossRef]
50. Whiten, A.; Goodall, J.; McGrew, W.C.; Nishida, T.; Reynolds, V.; Sugiyama, Y.; Tutin, C.E.; Wrangham, R.W.; Boesch, C. Cultures in Chimpanzees. *Nature* **1999**, *399*, 682–685. [CrossRef] [PubMed]
51. Van Schaik, C.P.; Ancrenaz, M.; Borgen, G.; Galdikas, B.; Knott, C.D.; Singleton, I.; Suzuki, A.; Utami, S.S.; Merrill, M. Orangutan Cultures and the Evolution of Material Culture. *Science* **2003**, *299*, 102–105. [CrossRef] [PubMed]

MDPI
St. Alban-Anlage 66
4052 Basel
Switzerland
www.mdpi.com

Animals Editorial Office
E-mail: animals@mdpi.com
www.mdpi.com/journal/animals

www.ingramcontent.com/pod-product-compliance
Lightning Source LLC
Chambersburg PA
CBHW040745250926
51850CB00007BB/1413
9783725807994